The Switch

The Switch

An Off and On History of Digital Humans

Jason Puskar

University of Minnesota Press
Minneapolis
London

Portions of chapter 5 are adapted from "Counting on the Body: Techniques of Embodied Digitality," *New Media & Society* 21, no. 10 (2019): 2242–60. Portions of chapter 15 are adapted from "Pistolgraphs: Liberal Techno-Agency and the Nineteenth-Century Camera Gun," *Nineteenth Century Contexts* 36, no. 5 (2014): 517–34.

Published by the University of Minnesota Press
111 Third Avenue South, Suite 290
Minneapolis, MN 55401–2520
http://www.upress.umn.edu

ISBN 978-1-5179-1539-1 (hc)
ISBN 978-1-5179-1540-7 (pb)

A Cataloging-in-Publication record for this book is available from the Library of Congress.

Printed in the United States of America on acid-free paper

32 31 30 29 28 27 26 25 24 23 10 9 8 7 6 5 4 3 2 1

For Erin O'Donnell

I write by speaking to you.

Contents

Part V. Haptic Liberalism

Acknowledgments

Nothing undermines the myth of individual authorship like interdisciplinarity. More people than I can count helped make this book.

My colleagues at the University of Wisconsin–Milwaukee have supported it from the start. I am especially grateful to UWM's Center for 21st Century Studies, which supported this research with two fellowships and access to its rich intellectual community. My thanks too to the Department of English, where my wide-ranging and curious colleagues made this idiosyncratic project feel at home. For many conversations on these topics I am indebted to Aneesh Aneesh, Ivan Ascher, Joe Austin, John Blum, Erica Bornstein, Liam Callanan, Jane Gallop, Richard Grusin, Thomas Haigh, Jennifer Johung, Gloria Kim, Nan Kim, Andrew Kincaid, Elana Levine, Jennifer Marshall, Ted Martin, Annie McClanahan, Walter Benn Michaels, Stuart Moulthrop, Tasha Oren, Marc Roehrle, Nigel Rothfels, Angela Sorby, Florian Springer, and Merry Wiesner-Hanks. Marija Gajdardziska-Josifovska generously supported this research by helping me reserve time to finish it. My interdisciplinary writing group, the Subduction Zone, invested many hours in early drafts, and countless improvements are due to the guidance of Anne Basting, Jennifer Jordan, Aims McGuinness, and Anne Pycha.

I benefited from a year working on this project as a system fellow at the Institute for Research in the Humanities at the University of Wisconsin–Madison, and from feedback from the entire group, especially Susan Stanford Friedman, Lynn Nyhart, and Andrew Zolides.

So many generous people helped me with interdisciplinary research questions. Daniel Novak led me to the literary metaphors of gun cameras in chapter 15. Al Muchka at the Milwaukee Public Museum shared his expertise on the history of firearms and gave me extraordinary access to the museum's Dietz Business Machines Collection. Johannes Lang graciously shared the font he created that imitates the writing from Benjamin Livermore's Permutation Typograph, which appears in chapter 9. Alessandra Gillen was my inside informant on the history and culture of pinball. John Liffen at the Science Museum in London patiently entertained my questions, pointed me toward Cooke and Wheatstone's permutating keyboard, and provided the image of it in chapter 2. The Cinémathèque Suisse helped me track down film and images of the robot Sabor IV. The Museo Nazionale della Scienza e della Tecnologia Leonardo da Vinci helped me identify sources related to Antonio Michela Zucco's 1862 stenography machine. The Breker Auction Team in Cologne, Germany, generously provided an image of that machine. Ivor Hughes helped me in my search for images of nineteenth-century printing telegraphers. Patrick Manning made the photographs of items in my own collection.

I am grateful to the anonymous reviewers of this manuscript from the University of Minnesota Press, who showed me the way to substantive improvements, and to the talented editors at the press. I am especially grateful to Doug Armato, director of the University of Minnesota Press, who encouraged me to pursue this project from the beginning.

I often benefited from comments by those attending talks based on chapters-in-progress, including attendees at events at the Newberry Library's American Literature Seminar, the International Graduate Center for the Study of Culture at the University of Giessen, the Holtz Center for Science and Technology Studies at UW–Madison, and the English departments of UW–Madison and Marquette University. I also gained valuable insights from attendees at many conferences, especially the annual conferences of the Humanities and Technology Association, Interdisciplinary Nineteenth-Century Studies, and the Society for Science, Literature, and the Arts.

Finally, special thanks to my wife Erin and my sons Owen and Jonas, who have endured a long-term anthropological study of our own switching techniques, and who have tolerated my writing about them. This is the rare topic where children can have just as many insights as the experts. All three have been my most valued collaborators.

Introduction

Awake at the Switch

Every day my morning starts with a switch. When my alarm sounds, I grope for the snooze button to claw back ten more minutes of fitful sleep. This sequence has been known to repeat until I press another button to turn off the alarm for good, at which point I reach over and flip a switch to turn on the lamp next to the bed. I am now just barely awake, but already I have operated three different switches, the first three intentional acts of my day. In switching these machines on and off, I am also summoning myself to consciousness, and so, in a real sense that the rest of this book will try to make clear, switching myself on too. The simple argument of this book is that these small, pedestrian, but incredibly common little actions do all sorts of things with, for, and to people. Buttons and switches push back, not just first thing in the morning but throughout the day, at home and at work, for adults and children, and in almost every corner of the developed world today.

Small, finger-activated binary switches are now so ubiquitous that entire fields of activity look drastically different from how they appeared just a century and half ago. As I sit at a computer typing these words on an array of seventy-eight electrical switches that English speakers fancifully term *keys,* it is hard not to wonder how that basic configuration influences the activity that results. Many theorists of media and technology have studied the relationship between people and computers, but the elementary action of touching the keys has interested few.[1] Perhaps it seems so open and obvious that people assume that it must be the part of the computational system least concealed within the black box. But this is not so. Every switch is a black box of its own, so small and simple that it tends to escape notice, which may be why so few have tried to pry off the lid before. In the pages that follow, we will strip this little machine down to its core, tallying its histories, inventorying its applications, and wondering at these strange little marvels that scarcely existed before the mid-nineteenth century. And more than just looking at switches, we also will touch, feel, pry, measure, count, and read them in all their many forms and across a wide range of technical objects, from servant bells to cameras, video games to musical instruments, calculators to light switches, remote controls to the most terrifying example of all: “The Button,” an all-too-telling metaphor for nuclear annihilation. The fact that the term *button* conjures images of Armageddon is just one index of its insidious power.

But I am getting ahead of myself. The alarm just went off, and I have finally touched my feet to the floor, determined to take special note of how many other switches and buttons I might press during the rest of this day. Reader, it will be more than either of us thinks! One early sign of trouble is that I could conceive of no way to count them without a button-based mechanical counter, the kind ticket takers use to tally patrons filing through the door. It is silver, heavy, and pleasingly nonergonomic, and it makes a satisfyingly complex internal click. With that in my left hand, I go about my day, entering my passcode into my cell phone so I can touch a button-shaped icon to summon up the weather forecast. I have now pressed almost a dozen buttons before leaving the edge of my bed. I wake my children by flipping switches to illuminate their rooms and silence their white-noise machines. Downstairs, the switches grow thicker, harder to count. When I open and close the refrigerator, I inadvertently operate a switch that turns off the light. The microwave door opens at the touch of a button and functions by keypad. The toaster has five buttons, and the kitchen stereo many more. Even the lids of the thermoses I pack in my children’s lunches open with a spring-loaded clasp in the form of a flush, surface-mounted button. At the front door, one ancient light switch in my old house is still the push-button kind, and as I herd the children past, I press it to turn off the light on the porch.

It is 7:35 a.m. and I have now flipped many dozens of switches already, but this is nothing compared to what comes next. The car requires dozens more: radio, defroster, windshield washer, electric windows, to say nothing of the concealed pressure switches not designed for human hands but activated inadvertently when I open the door or sit on the seat. At work this continues: elevators, drinking fountains, photocopiers, telephones, and the computer in my office, which now performs too many separate functions to list. Back home, I press still more while cooking, doing laundry, paying bills, washing dishes, or watching television. By the end of the day, I have been awake for almost seventeen hours, and when I finally fall into bed, I press another string of buttons to set my alarm for morning.

In that span of just twenty-four hours (eight of which I spent sleeping), I activated 30,207 switches: on average, about one every three seconds. The vast majority of those—29,715 to be exact—were on computer keyboards, and I pressed most of those while composing and editing documents in a word processor and writing email messages. But this count is deeply flawed. I gave up counting mouse clicks around two hundred, surely a fraction of the total. I regularly forgot to tally other switches that I operated without noticing at the time: ATMs, credit card readers, my office tea kettle. I was more successful with some of the roughly three hundred others I recorded in the course of the day: eighty-seven on the television remote control, eleven to operate the elevators at work, seven to operate two different drinking fountains, four on a microwave oven, and two each on the dishwasher and thermostat.[2] On the same day, I wore a pedometer to see whether I would flip more switches or take more steps. When I climbed into bed at the end of the day, the pedometer tallied 10,250 steps, well above average, but only a third of the number of switches I used that day. If my own tally is any guide, the act of flipping switches exceeds even my most repetitive bodily rhythms, such as my roughly twenty-one thousand blinks or twenty-five thousand breaths each day. At the heart of this book is the simple claim that, if we do anything thirty-thousand times a day, we had better consider carefully just what it is we are doing, and what *it* might be doing to *us.*

After all, whatever switches are doing, they have been at it for a while now. Switches began proliferating in their modern forms around the middle of the nineteenth century, at the start of a dizzying period of technological change. If that seems long enough for five or six generations to reflect on, in truth not many people have bothered. This is understandable. Switches are so small and seemingly trivial that they hardly attract attention. Part of their job is to go unnoticed. As a result, the advent of the binary switch can seem both slow and fast at the same time: slow because switches took many decades to develop; fast, because almost nobody noticed that happening, until one day people looked around and suddenly realized they could not live without them. To call this book *The Switch* is thus to name both its mechanical object of study and the sense of rapid historical transformation that object has occasioned, as new kinds of action replaced earlier ones. Over a span of time that is long to a single person but short to a historian, millions of people swapped traditional forms of media for countless switches, buttons, and keys. Computer keys replaced pen and ink; shutter buttons banished paint brushes; buzzers and bells drowned out human voices. These transformations have been socially wide and culturally deep, but they did not happen overnight. They just seemed to. In fact, people are still in the process of making this switch to switching, for the change has never been either all-encompassing or complete. In calling this book *The Switch,* then, I do not mean to imply that these momentous and decisive changes really were instantaneous, but only that the panoply of modern switches gathered so quietly over a century and a half that, eventually, they caught us unawares. Unexamined changes always seem more sudden than they really are.

The chapters that follow try to slow down the switching, both the mechanical and the historical kind, in order to get a glimpse of what really lurks within. Doing so can bring hidden facts to light. For instance, a simple light switch seems to be the epitome of instantaneous and binary action, but beneath the threshold of human perception other processes unfold in time, such as the flow of electricity at measurable speed, or the gradual onset of illumination in a filament or diode. Between on and off, more is going on than meets the eye. The same is true of the historical switch to binary inputs that is the subject of this book. Upon patient examination, historical changes that seem simple and sudden turn out to be enduring and complex. Slowing down the switch requires lingering over neglected episodes in the history of technology, and over mundane objects like light switches that rarely earn much notice. Slowing down the switch also compels us to pick up the story long before the advent of modern computing, and in binary functions separate from binary code. All of this is doubly important for people whose lives are so thoroughly mediated by modern switches. Anyone who activates thirty thousand of them a day can easily start to imagine that everything else works in roughly the same way, as if all action were instantaneous, as if all change were binary, as if history really could be a flick of a switch.

Structures of Opposition

People who spend less time at the computer probably use fewer switches, but others, such as those who communicate heavily through text messaging on cell phones, work in the field of information technology, or spend hours playing video games, may use even more. And if the keyboard dominates many lives like mine, it pays to remember that other occupations use other kinds of switches almost as intensively. A roofer with a nail gun would tap its triggering switch hundreds and probably thousands of times a day, as would the auto mechanic with air powered tools, the policeman with a radio, the vandal with a can of spray paint, the cashier at the register, the receptionist with a telephone, and the doctor with increasingly computerized diagnostic machines and medical records. Even childhood play seems to be mediated by an astonishing number of binary switches today, from infant toys to video games.

Yet none of these devices existed before about 1840, when early electrical switches and new kinds of mechanical control interfaces first began to appear. Before that, everyday life was resolutely analog for the vast majority of people who had ever lived on earth. Analog processes are defined by the quality of continuous variability, of gradual change along a smooth continuum of possibilities. A gas valve, for instance, can be turned on or off, but between those two extremes it can be continuously varied between potentially unlimited positions. Most machinery with moving parts worked on a rotational principle before the early nineteenth century, such as clocks, pulleys, valves, and wheels. Many other technical activities, such as writing script with a pen, also have analog properties, as each separate letter flows into the next, and as a continuous line makes up each part of each letter in fluid succession. Most areas of life were analog in this basic sense, consisting of gradually interpenetrating interactions, all of them blending into each other in continuously variable processes. That is not to say that continuously variable processes were not often described in starkly binary terms, such as day and night, but just that such binary distinctions were fundamentally rhetorical, and less often strictly organized by machinery. Nobody mistook day and night for strictly binary alternatives, as if there were no variation within or gradations between them, but still they found it useful to draw sharper lines rhetorically.

Indeed, people defined such mutually exclusive alternatives in relation to each other. As structuralist linguists and anthropologists have argued for almost a century, all concepts take shape only in relation to other concepts. There is no free-standing concept of *day* without a corresponding concept of *night*. Structuralists (and the many disciplines influenced by them) have insisted ever since that concepts themselves have been established by the binary oppositions wrought in language. As Ferdinand de Saussure put it: "In language there are only differences."[3] And not just differences, but sharp distinctions between paired terms, or what Saussure also called *oppositions*. These binary oppositions are the minimum relation of difference, and thus the building blocks of language itself. Anthropologists extended this basic insight about human language to broader symbolic systems in human culture: the raw and the cooked, or purity and danger, not to mention those confining dualist categories of human identity, such as male and female, heterosexual and homosexual, black and white.[4]

Binary opposition is structuralism's own deepest structure, but it was hardly a twentieth-century invention. It has its own conceptual affinities with earlier dualisms such as Hegelian dialectics, but it also developed from still earlier dualisms, including Plato's eternal forms and transitory bodies, Aristotle's substances and accidents, Thomas Aquinas's soul and body, René Descartes's mind and matter, and more. To structuralists, this ancient lineage was just evidence that binary opposition truly is the fundament of all meaning, but more recent philosophers and theorists have had their doubts. After all, if concepts are made of arbitrary differences—if the differences are historical rather than essential—then the concept of binary opposition itself might not be universal and essential either. Poststructuralists like Jacques Derrida acknowledge the practical power of binaries but doubt their natural or essential character, the stability of their relation, or the justice of the values they imply. Others have proposed alternatives to binary opposition: Alfred North Whitehead's fluid process, Gilles Deleuze's rhizomes, Donna Haraway's cyborgs, and many more.[5] I think it's fair to say that most theorists today regard binary opposition as, variously, a necessary evil, a crude distortion of vastly more complex realities, an ideology of power, or even a prison house confining all thought and meaning.

I am generally in sympathy with these positions, but like many skeptics of structural dualisms, even the most vocal, I also recognize that it is hard to think at all without binarizing at every turn. Rather than trying to escape or subvert binary opposition in what follows, I am eager to trace it back to its sources, one of which is the prevalence of binary switching itself. No less than linguistic structures that oppose *light* to *dark* or social practices that oppose *white* to *black,* switches enforce a functional binarism at the level of technique. They strip fluid process of continuous variability and render it as starkly differentiated alternatives. Over time, these alternatives come to seem as natural as the structures of language. When I flip the switch on my bedside lamp it dispels the darkness all at once, and at a level of illumination that varies imperceptibly. The gradual transformation of night into day gives way to the sudden exchange of one state for another, with no discernible gradation in between. Process becomes product seemingly instantaneously, and in ways that transform symbolic oppositions in the medium of language into practical and material oppositions in the medium of machinery. At that stage, binarism appears not as code, not as language, not even as a reductive heuristic, but as the operation of the world, the way things work, including, significantly, the way that people function within it. Especially when we consider how many different areas of life binary switching has touched, this amounts to a truly stunning transformation of action, and with it, a corresponding transformation of those agents who like to imagine themselves as action's exclusive source.

One by one, the fluid processes of earlier periods yielded to the binary logic of on and off, and as a result, people began to experience agency itself as a matter of selection between mutually exclusive alternatives. Through all of this, and over many decades, continuously variable processes have gradually transformed into pairs of alternatives, frequently governed by switches. Historians have attended extensively to the ways in which the industrial revolution reorganized social and individual life through new technologies such as steam, new environments such as urban centers, and new industries such as transportation, but few have attended to the transformation of both actions and agents that started around the same time with binary switching.

Change and Choice

The consequences of those transformations will occupy many of the chapters that follow, but for the moment, it bears noting that these changes are closely tied to modern versions of political subjectivity in liberal societies. To put the matter simply at the outset: switches transform change into choice. That is, they transform the fluid dynamics of process into strictly bounded alternatives that people can then experience as derived from human agency. Taylor Dotson has recently gone a step further by suggesting not just that technologies shape certain kinds of choices, but that they ensure that "the act of choosing—however trivial—remains cognitively near." Technologies "help to reproduce the belief that one is an unencumbered choosing self."[6] We might go even further than that. By transforming change into choice, binary switches bring choices into existence where, previously, only a confusing welter of interacting factors would have existed. So it is not just that binary switches multiply opportunities for people to make choices, but that they help structure possible action in forms that can register as choices in the first place. And in doing so, they also constitute choosers, people defined by the act of making choices.

The English idiom "making a choice" even flatters people that they are the creators of their own choices, as if they single-handedly chiseled each decision from dynamic circumstance. It would be more accurate if people said they were "doing a choice," given that choices are substantially made in advance, prestructured long before a person arrives to select among definite options. Making a choice involves multiple agencies and supervening structures, and only in the last instance is it a matter of the decisive intervention of a singular human being. At that moment, however, the whole ensemble has made more than just a choice. It has made a chooser too.

There were other methods for transforming change into choice before modern switching, such as contracts. In preparing any contract, the parties may negotiate as long as they like, but at a key moment laden with disambiguating rituals, such as the vows at a wedding, both contracting parties must either commit or walk away. These are all-or-nothing moments in which ambivalent hopes, desires, forecasts, intents, and determinations of others all must be subsumed within a single individual choice of whether to say "I do." There is no marrying halfway, even for the halfhearted. Similarly, the rigid binary choice of *yea* or *nay* is central to parliamentary procedures, in which a longer process of discussion and debate concludes with an outcome that admits no ambiguous middle. For that reason, as we shall see, voting machines are an important chapter of this history, as they mechanized democratic procedures designed to transform free and open discussion into unambiguous choices. Langdon Winner argued in a foundational article on technology and politics almost four decades ago: "Technological innovations are similar to legislative acts or political foundings that establish a framework for public order that will endure over many generations." Indeed, Winner argued, technology is legislation by other means, not "politics proper" but the more "tangible arrangements of steel and concrete, wires and transistors, nuts and bolts."[7] The political power of transportation or communications systems is hardly in doubt, after a generation of insightful scholarship from the fields of science and technology studies and actor-network theory. But what Winner calls "the seemingly insignificant features on new machines" have not always been read in expressly political terms, including the millions upon millions of switches that control so much of the modern world. One of the main political functions of switches is to naturalize liberal individualism itself by linking discrete, bounded actions unmistakably to the primitive movements of a single human body.

With a level of complexity that will take many pages to unravel, binary switches reconfigure those who use them through three fundamental and overlapping effects. First, by automating some process as some elaboration of on and off, they define actions as limited, bounded, discrete. They turn action into *an* action. They divide up what otherwise might register as more sprawling processes, interactions, or network effects. As a result, they also differentiate that action from all others that it might become entangled with. Switches are machines that create breaks in the flow of action, which can then register to humans as starting points from which entirely new actions can be born. Second, switches attribute these now bounded and delimited actions to individual people. By first distinguishing individual actions, switches make it easier to assign them to individual people, and thus to help people experience themselves as individuals in the first place. Because there are few switches made to be used by more than one person at the same time, and because the actions they automate are so clearly bounded with a start and a stop, switches make it harder to perceive distributed agency, collaborative action, complex interaction, and even chance and accident. Everything starts and stops with a single person, a flattering arrangement for people who sometimes like to think of themselves as autonomous and free. Finally, by hiding everything but the inputs and outputs of automated systems, switches allow people to experience the resulting effects as their own. At the level of input, the effort required can be so slight, the skill so minimal, that initiating an action seems to require little more than willing to do so. At the level of output, in contrast, the action can be so powerful, complex, instantaneous, and distant that agency extends well beyond the limits of the body. This disproportionate scale of willing to doing magnifies and aggrandizes the people involved. For a century and half, the mediation of action along these lines has defined what millions of people have taken for granted as exclusively human agency.

There will be much more to say about all these effects in subsequent chapters, but perhaps it bears noting that these effects are by no means crude delusions or distorting ideologies. They have sometimes even been powerfully humanizing, at least according to those criteria that liberal societies have used to define the human. Some people excluded from the full range of political rights and responsibilities—including women, children, and the disabled—have benefitted tremendously from switching, for it has allowed many of them to approximate an able-bodied, adult-male standard of autonomy and independence, and hence to make a fuller claim on undiminished humanity. For instance, modern computer keyboards have allowed some severely autistic people to communicate in language for the first time, a powerful claim on the status of the human by people who often have been treated inhumanely. Similarly, nineteenth-century women were professionally disabled by being forced to learn an inferior form of handwriting from childhood, but suddenly in the 1880s typing made their writing identical to that of any man. Something might be lost when switches organize both action and the agents so rigidly, but something can be gained as well: a less deniable claim to full equality with that paragon of liberal political personhood, the able-bodied adult man. Switching is thus more specifically a source of certain kinds of liberal political subjectivity, and of the status that subjectivity enjoys. It produces free and autonomous individuals in the sense that it organizes the conduct of people so that their actions appear to be maximally free and empowered, minimally dependent on others.

It does not pay to generalize too much about these political effects at this stage. In some contexts, binary switching had egalitarian results; in others, it seems to have isolated and enfeebled those it promised to empower. Most of the time, the results were decidedly mixed. The clearest evidence for the political relevance of binary switching may be that even adult men eventually found it hard to do without it. Women and the disabled exploited the capacities of switching to demonstrate their equality, but men also needed binary switches to live up to liberalism's high expectations. In a world of fluid process and boundless interaction, who can say where one's own actions begin and end, or what particular consequence arises from them? Eventually, any able-bodied adult man who preferred to think of himself as a free and autonomous liberal stalwart had to rely on assistive technologies too.

Techniques of the Human

The chapters that follow build on decades of scholarship in media studies and the history of technology, including media archeology, posthuman and cyborg studies, and various offshoots of phenomenology, from pragmatism to process philosophy. This study begins with the collected wisdom of many posthumanist scholars, including Haraway, N. Katherine Hayles, Rosi Braidotti, Jane Bennett, and many more, who have compromised any conception of the human as a natural and organic whole. This project commits to two further tasks. First, along with many other scholars, it continues to investigate how the "human" and its humanist subjectivities have been fashioned from an assemblage of bodies, plants, animals, machines, forms of mediation, self-organizing materials, and more. I have already begun to suggest how intimately switching has mediated the human as one part of these assemblages, primarily as that crucial part that helps conceal the fact of assembly. Moreover, this project attempts to learn more about how technological mediation makes the human seem natural and organic in the first place. That point may seem counterintuitive, but as many of the following chapters will show, switching techniques installed models of agency in bodies so subtly and so deeply that they seemed to have been there all along.

One particularly important source for understanding this reconfiguration of the human is the German school of *Kulturtechniken,* or "cultural techniques."[8] The English term unfortunately conceals a telling ambiguity of the German, which referred to something like agricultural engineering in the nineteenth century.[9] Bernhard Siegert reminds us that the first part of this name, *kultur* (culture) derives from Latin *colere,* to cultivate or till.[10] Human culture is thus closely associated with technologies such as plows and the techniques of using them. But more importantly, the second part of the name, from the German noun *Technik,* can refer both to *technology* and to *technique,* a conflation of material objects and human practices little evident in English, which sequesters those separate meanings in different words.[11] In the pages that follow, the term *technique* should be understood with that ambiguity intact, as an assemblage of people and things, a hybrid formation of technological devices and the human practices associated with them. Neither is preeminent. People make machines and put them to work, but machines make people and put them to work too. A technique is nothing other than this collaboration of technical forms and human practices, neither of which means much on its own.

Wherever there are cultures, there are techniques, so we should not suppose that the comparatively recent interfaces considered here corrupted some purer form of humanity from a pretechnical golden age. The *post* in *posthuman* is not a temporal reference, indicating something that came after humanity's natural, organic phase. The category of the human has always been cultural, generated in significant measure through techniques, a process Siegert refers to as *hominization.* He further suggests that this basic assumption marks a key point of difference between German and American strains of posthumanism. According to Siegert, American posthumanists have tended to ask, "How did we become posthuman?" In contrast, German posthumanists have asked, "How was the human always already mixed with the nonhuman?"[12] The technical recalibration of the human is not some postmodern disruption that burst upon the world as a result of capitalism or digital computing or anything else. On the contrary, Siegert argues, at any point in history, "humans as such do not exist independently of cultural techniques of hominization."[13]

We might draw a preliminary distinction between the terms *people* and *humans* by suggesting that, although people are theoretically possible without culture, humans are by definition acculturated through and through. As Bernard Stiegler has argued (drawing on the work of anthropologist André Leroi-Gourhan), the human has always been externalized in technics, and thus in culture too. Features such as vertical posture, freed hands, opposable thumbs, and even mental capacities did not arise by chance or divine law and then prove convenient for tool use. They evolved *because* of tool use. The tool shaped the body as much as the body shaped its tools. It would be strange, Stiegler points out, if an opposable thumb evolved without any tools to grasp and use. For that reason, he says, both "body and brain are defined by the existence of the tool."[14] Humans and machines might even be imagined as reproductive partners, each of which needs the other to propagate more of its kind, though in the process they get mixed up with each other continuously. Bereft of technologies, nobody would be human anymore, for to be without technology is to be without culture. Similarly, bereft of humans, technologies would not replicate and persist, transmitting forms across space and time. The name of the activity by which people and machines reproduce culturally is not *sex,* but *technique.*

There are methodological implications to this. Thomas Macho has argued that "cultural techniques—such as writing, reading, painting, counting, making music—are always older than the concepts that are generated from them." To put it differently, new practices lead to new theories, not the other way around. "Counting . . . is older than the notion of numbers," Macho suggests.[15] Macho's point translates one of Saussure's core claims about forms of language into an equally compelling claim about forms of culture. Just as language does not simply reflect preexisting concepts but defines or produces those concepts, so too techniques are not just material expressions of preexisting ideas or theories, but the source of ideas and theories. Even so, openness to the idea that techniques precede theories should not become too sweepingly polemical. It is, after all, comparatively easy to show that theories yield new techniques too, as we will see in chapter 7, on the relation of the musical keyboard to Greek harmonic theory. New techniques generate new theories, and those new theories in turn generate still other techniques, all within a reciprocal and ongoing process that makes it rather pointless to insist that one must always come first.

As a result, I am sympathetic to the general point that there is something to be gained by starting with techniques themselves, and giving them pride of place in the analysis. There will be moments when it is important to refer to Martin Heidegger on tools, Jacques Lacan on subjectivity, or Whitehead or Deleuze on process, but these well-known authorities and their influential ideas will not be my starting points. To learn about hammering, one should start with hammers, and not with Heidegger's theory of the *ready-to-hand.* If what we discover about hammering makes Heidegger's analysis seem relevant or clarifying, all the better, but starting with theory can limit our view to what we are used to seeing already. When you're a hammer, as the saying goes, everything looks like a nail, but when you're a Heideggerian everything can start to look like a hammer. By taking seriously even seemingly trivial devices such as switches on children's toys, shutter buttons on cameras, and telegraph keys, we may feel our way toward a new set of questions about the human and its agencies, questions that are not always easily resolved by earlier theories.

There are other sources of overgeneralization to resist. Marshall McLuhan famously compared media to the water that surrounds a fish, entirely invisible and, as a result, the one thing a fish knows nothing about.[16] If that metaphor helpfully cautions against regarding media as transparent, perhaps it also tempts us to think of media as rather too homogenous, an atmosphere that has become practically universal. Much of what follows will prefer a more local, historical, particularist approach that is somewhat less interested in the kinds of mass media that saturate entire societies, and more interested in what Siegert calls "inconspicuous technologies," such as index cards, blackboards, and phonographs. Under the influence of Friedrich Kittler, German media studies has been especially adept at focusing attention on the least obtrusive mechanics of mediation. Precisely because these machines seem trivial, users might not consider that they can have profound effects, but as we shall see, it may be that the television remote control is as consequential as the sound and images transmitted through the set. Accordingly, I will take little interest in macromedia such as mass media broadcasts, the industries behind them, complex technological systems, or sprawling actor-networks, and will attempt instead to drive down into the smallest local details, those forms and functions scaled to the individual human body. This is not to deny the importance of larger systems and structures, but rather to build from a generation of insights about them by attending to the micromediation of the human instead. Others have addressed the systemic effects of telegraphy as a network phenomenon, but very few have thought about different forms of telegraphic input as equally powerful humanizing techniques.

Accordingly, in addition to prying open various black boxes, my goal will be to understand the effects of the black-boxing on those people who cannot see inside, or who prefer not to try. To experience only the inputs and outputs of some switched system is to fancy that distant, amplified, and heavily mediated effects are really one's own. A simple light switch on the wall is a wired remote control, a surprisingly stable relic of late-nineteenth-century home automation. Every adult knows that wires run through the wall from switch to ceiling, and many people have a rough idea of where electricity comes from and how it works, but for these purposes little can be gained by demystifying the entire system. Instead, I will want to ask what the mysteriousness accomplishes and why users are so disinterested in peering inside. Such facts are not to be found inside the black box. They become visible only when we attend to the fantasies that people have about their devices, and that their devices often actively encourage. When Jean Piaget wanted to understand how children conceived of causation, he asked them to explain the functioning of a simple bicycle. Their answers were often ludicrous, as we will see in chapter 11, but they also revealed prevalent assumptions about how people relate to the world. Today, an adult trying to explain a computer keyboard would probably fare even worse than Piaget's children explaining a bicycle, but if we listen carefully to those explanations, alert to the metaphors, analogies, images, and sheer fantasies at play, we may learn a great deal about ourselves.

The more local, situated, particularist sensibility of some media theory has begun to break down the ocularcentric orientation of media studies that derived from the study of visual media, such as film and television. Once smaller and simpler devices receive attention, other parts of the body become relevant, and other senses beyond vision. Caetlin Benson-Allott's recent study of television remote controls shifts attention from what happens on screens to what happens in the living spaces of those watching television, and even to what happens with their hands rather than just their eyes and ears.[17] Similarly, Jocelyn Szczepaniak-Gillece's recent work on seating in twentieth-century movie theaters has recalibrated our sense of the kinds of mediation happening there and of the ways that visual experience relates to the rest of the body.[18] But remote controls and theater seats long went unnoticed because media studies had inherited a visual bias from its origins in the study of film. Given that vision has traditionally stood at the top of the hierarchy of the senses in many Western societies, Benson-Allott's and Szczepaniak-Gillece's work has begun attending to mediated experience through other senses, particularly touch.

In that vein, several other recent books provide crucial historical context for the discussions that follow, and all of them are highly attentive to material conditions and technical forms. David Parisi's *Archaeologies of Touch* is the most thorough media-theoretical account to date of the ways in which haptic technology has refashioned touch itself.[19] Roger Moseley's *Keys to Play* is both an original contribution to the field of organology (the study of musical instruments) and a sophisticated meditation on the digitality of finger-operated musical keyboards.[20] And Rachel Plotnick's *Power Button* examines how push buttons in the late nineteenth and early twentieth centuries organized both power and pleasure in the modern world.[21] Plotnick's recent book is an especially thorough study of related devices. Focusing specifically on push buttons from about 1880 to 1925, it attends to power dynamics within mechanically mediated social contexts. According to Plotnick, the push button is an important but neglected part of the story of the deskilling of manual work, the distancing of management from labor, the naturalizing of inequality, and more. Although some of what follows will extend Plotnick's arguments about objects such as light switches and servant buzzers, this study differs in several important ways. With a longer historical scope and a broader definition of switching, it includes a wider range of devices, especially those arrays of switches on keyboards that require more skillful input and yield more complex results.

Such devices sometimes work very differently, mechanically and socially, from the objects of Plotnick's study, buttons that require "a single push to create an effect."[22] More importantly, when these chapters do attend to social issues, they generally do so to raise broader questions about political subjectivity, individual agency, and even the category of the human. Accordingly, the core questions the following chapters entertain tend to be more theoretical than strictly social, and lead by degrees to this book's analysis of the ways in which technical media may in fact be the source of that kind of liberal subjectivity long thought to be purely and exclusively human.

There will be more to say about these sources in the chapters that follow, along with other important studies of technical media by Thierry Bardini, Matthew Kirschenbaum, Thomas Mullaney, and others. I am fortunate to be able to build on such a rich body of work in the fields of media studies and technology history, fields that no longer seem quite as distinct as they used to. Many of these scholars also have helped reorient critical interest toward small-scale and seemingly mundane interfaces, and also specifically toward the finger techniques they require.[23] Moseley, for instance, is expressly interested in the relationship between musical keyboards and digitality, but even though scores of critics have noted that the term *digital* refers back to a Latin term for the fingers, Moseley is one of the few to have elaborated the matter with theoretical rigor. This book may be considered part of this lively, on-going conversation, which is beginning to understand how the digits and the digital coincide.

My point is not to claim that switching has uniform effects on people, on their political systems, or on their subjectivities, but only that it has certain kinds of effects that nonetheless shake out in a wide range of different and sometimes contradictory ways. These chapters do not amount to a linear history of switching, although they do try to cover some of the most important stages of that history. This is, as the title claims, an "off and on history," one that delves into the past when it matters most, but also frequently turns to other modes of theoretical and critical analysis. Individual chapters often proceed chronologically to chart brief histories of particular switching techniques, such as telegraphy. Overall, however, chapters relate to each other thematically, grouped in five major parts on related kinds of interfaces, switching techniques, or social contexts.

The first part of the book, "Start," addresses the earliest history of actual switches, the terminology traditionally employed for them, and the design processes behind them. Because it is surprisingly difficult to determine what counts as the first switch, the chapters in this part conclude that switching eased into existence over time by a long series of analogies, gradually becoming more binaristic in the process.

With that foundation in place, the second part, "Digital Bodies," attends to the body itself as it comes into contact with the switch. The contact between the finger and the switch it touches is so carefully bounded, so hygienic, that it might be mistaken for a profound separation rather than an intimate encounter. Where they meet, bodies and switches appear to maintain the most decorous relations, one on each side, like Tristan and Isolde sleeping with the sword between them. Then again, everyone knows how that worked out. Sometimes the biggest shows of decorum betray the deepest intimacies. By starting with the body and working back toward the interface of the switch, chapters on the indexical gesture, the mathematizing of the fingers, and the conflation of buttons with human nipples show how bodies and buttons frequently have come to coincide.

The third part, "Keyboard Rationality," turns to the interfaces of keyboarded musical instruments from the Greek water organ to the modern piano, and the legacy of Greek rationalism transmitted through them into the keyboards of today's writing machines. The musical keyboard is an influential protoswitch, for even though its function is not binary in the least, it certainly is digital in that it organizes discrete actions—individual notes—as an array of finger-activated touch surfaces. Accordingly, the chapters in this section trace the many different remediations by which the modern QWERTY keyboard emerged. Against the utilitarian explanations for the rise to dominance of QWERTY, I show that it in fact resulted from a surprising confluence of factors that include medieval Anglo-Irish accounting techniques, European telegraphy, and an American variant of Morse code. Those who prefer to think that people drive technological change may be surprised to find that the keyboard is neither a utilitarian triumph nor even a purposeful design, but a haphazard accumulation of features transmitted over centuries for competing and often contradictory reasons.

The fourth part, "Objects of Play," turns from the rational to the ludic, as it examines the meaning of that activity well known to many parents of small children, the delight of playing with buttons. There have been hundreds, probably thousands of books about play since the early twentieth century, many of which, including the work of Piaget and Donald Winnicott, have identified play as a core means of subject-object differentiation. Still, in these discussions, play often seems to be a human-directed activity, something active people do with passive things. As a result, the materialization of play in those objects termed *toys* tends to be oversimplified. Starting with the fact of children's deep interest in and fascination with buttons and switches, the chapters in this part take toys seriously by considering the subjectivizing effects of a world of binary playthings. Through inquiries into twentieth-century children's toys, the history of pinball, and button-based video game controllers, these chapters suggest that binary switches hasten the onset of subject-object differentiation in some ways, while in other ways temporarily restoring the intense absorption in material things characteristic of very young children. Especially for adults, I argue, playing with buttons can make and unmake the subject at the same time.

Finally, the last part, "Haptic Liberalism," shows how liberal political societies lodge some of their primary values in switching techniques. Through examinations of the mechanization of voting, the role of binary switching in Donald Davidson's influential action philosophy, and technical analogies with shooting firearms, the chapters in this part examine the different ways in which technologies preserve, transmit, and naturalize key political values. This is a darker side of digitality, as the effortless activation of binary switches invests people with an exaggerated model of human agency too often complicit with domination. When routed through switches, the liberal value of self-determination often depends on *other*-determination too.

I began this chapter by switching off my alarm and switching on some lights. Now the morning's preliminaries are over. Press on.

Part I

»» Start

CHAPTER 1

Origin Stories

Beginning at the beginning is always fanciful, because every beginning was born from something earlier. New technologies develop incrementally from old ones, often concealing but never entirely escaping the past. We will have many occasions to note what Richard Grusin and Jay Bolter have referred to as the *remediation* of earlier technical forms in later developments, a historical masquerade in which posterity parades as novelty.[1] As a result, there is no single original moment of binary switching, no sudden instant when everything changed or when some entirely new technology and technique appeared on the scene. However, there are comparatively earlier phases of this history when devices recognizable as switches were less prevalent, less powerful, and less binaristic. As a result, they were also talked about less, and so we may as well start not so much at the beginning of switching itself as at the beginning of its entry into modern languages. Looking at what people call machines can sometimes tell us as much as the devices themselves. For instance, it would be hard to perceive any necessary relationship between computer keyboards and piano keyboards by the form and function of either, but the English reference to the *keys* of writing machines betrays their extensive intermingling with pianos over the course of the nineteenth century. Facts that take some effort to excavate from the history of technology sometimes remain visible on the surface of language.

When new objects appear, old words must be pressed into different kinds of service. As Ralph Waldo Emerson noted, the most well-worn and familiar words started their lives as daring imaginative metaphors. "The poets made all the words," Emerson wrote, "and therefore language is the archives of history, and, if we must say it, a sort of tomb of the muses. For, though the origin of most of our words is forgotten, each word was at first a stroke of genius." Or as Emerson also put it: "Language is fossil poetry."[2] He meant that everyday language is made out of dead metaphors—even the word *literal* is made out of a metaphor—but, over time, these once surprising associations became dully familiar. In recovering the metaphorical origins of words, then, readers can excavate early evidence about how people conceived of new machines and the techniques they required. Language can tell us what those new technologies were *like* before they ossified into tedious fact. To extract the poetry from the fossil is to recover the living history of technical change, that vivid reckoning that everybody undertakes while struggling to align existing assumptions and habitual ways of speaking with new phenomena. The English terms *switches, keys,* and *buttons* that I have relied on so far all turn out to be surprisingly young, but not so young that speakers today can easily perceive their once-fresh resonances.

In German, one name for what English speakers variously and inconsistently refer to as *buttons, keys,* and *switches* is *die Taste,* which derives from Latin *tangere,* to touch. The German verb *tasten* serves as a root for several common German verbs related to touching, but on its own refers more to the gesture of *attempting* to touch, of feeling around for something. Yet this telling conflation of the act of touching and the object to be touched does not appear in equivalent English terms, save for a few descriptive compounds such as *push button* or *touch screen.* Instead, English speakers settled on a motley array of terms that sometimes differentiates devices that are similar, if not identical, and sometimes conflates devices that are very different. One must not expect bright lines between these categories. The *keys* on my computer also qualify as *buttons* by any definition, but one leading manufacturer markets them only as *switches.*

Ivory Buttons

The term *key* is the oldest. Medieval musicians, music theorists, and instrument makers adapted it from the Latin term for an input mechanism on organs, a *clavis.* Early-medieval organ keyboards like those we will encounter in chapter 7 typically featured long, widely spaced, and sometimes hand-sized levers projecting from the vertical face of the instrument. Given their decorative shapes, these must have strongly resembled keys inserted into locks. Later, once keys became narrower and more closely spaced, the name persisted, as it does to this day on writing machines. The term *key* tends to be applied primarily to switches used for registering alphanumeric characters, as on keypads or keyboards. Accordingly, the "mouse button" is not a key because it is not keyed to a single alphanumeric character, an invariant pairing that resembles the one-to-one relationship between a musical key and a given note. That pairing was presumably also reminiscent of the invariant pairing of a lock and key.

The terms *switch* and *button* emerged much more recently. Most nineteenth-century American patents for electrical devices refer cumbrously to a "circuit opener and closer" or some other such descriptive phrase, and that terminology, or rather lack of terminology, persists in some patents until the late nineteenth century. Although it seems reasonable to assume that switches are a broad category of devices for opening and closing circuits, and buttons simply a subcategory of those switches that operate in certain ways, their relationship is more complicated. In fact, the term *button* began to be applied to various electrical or mechanical actuators a decade or so before anyone seems to have used the term *switch* for the same thing. The word *button* derives from romance languages, likely from Old French *bouton,* a bud or a knob, something that *butts* out.[3] The reference is not necessarily to clothing fasteners, but to protrusions for some mechanical purpose, and at a comparatively small scale. Some conjecture that at a prior stage *bouton* may have entered Romance languages from Middle High German, and if this is the case, then the Modern English "button" and Modern French *bouton* may derive from words related to the English verb *butt,* to push or strike.[4]

The *Oxford English Dictionary* dates the first use of the term *button* as a reference to a touch interface or switch to the 1860s in reference to a naval signaling device, but there are earlier examples. Rachel Plotnick notes the early application of the term to the interface for starting and stopping a stopwatch in the 1820s, in what presumably looked like a bud or knob, a *bouton.*[5] Not long after, English telegraph pioneers William Cooke and Charles Wheatstone refer to an electrical *button* in their 1837 English patent and 1842 American patent for the five-needle, six-wire telegraph. In both, Cooke and Wheatstone repeatedly refer to the insulated touch interfaces of the electrical switches used to transmit messages as "buttons or finger keys."[6] The repeated alternation of the terms is telling, suggesting that neither seemed quite adequate for this new apparatus but also that neither was intelligible enough on its own. The American patent distinguishes buttons from keys in its diagram of an interface similar to that shown in Plate 1. In the patent, the term *key* denotes the long, flexible metal strips that bend down to close the circuit by contacting the metal bar beneath, and *button* the round, raised touch surfaces mounted on the ends of each key.[7] The keys were likely named for their resemblance to long, thin piano keys, even if the user does not touch them directly. The buttons, on the other hand, take the form of disks mounted on small stems, protruding upward from a horizontal surface. In shape and scale they resemble the kinds of studs that nineteenth-century men used to fasten their collars. As we will see later, at least one early writing machine did repurpose collar studs for its buttons. As a result, this early appearance of the term *button* in English may have been a descriptive metaphor based on a formal resemblance to an actual clothing fastener, a different kind of small, finger-operated device for opening and closing.

On the other side of the Atlantic at roughly the same time, similar terminology emerged around very different telegraphic instruments. The transmission terminal of the Morse-Vail telegraph was much simpler than the Cooke and Wheatstone system, consisting of a single long lever made of metal, sometimes with an insulated knob or touch surface on the end. It functioned as a simple spring-loaded contact switch. When the operator depressed the round touch surface at the end of the lever with a finger, a screw through the arm contacted the metal underneath, closing a circuit momentarily and transmitting the signal to the receiver. When the operator removed his finger, a leaf spring raised the arm, breaking the circuit. Before the telegraph, there was little need for electrical switches, since there were few electrical devices to control, and none that needed an automated device for opening and closing a circuit. When Joseph Henry first rang a bell at a distance in 1835—an important precursor to telegraphy—he described his device in detail, but he took no notice of the means of closing the circuit.[8] He probably touched a wire to the terminal of the battery by hand. Samuel Morse and Alfred Vail would use a more sophisticated version of Henry's electromagnetic relay in their telegraph, but paired on the other end with a switch sensitive enough that it could articulate clearly distinguishable units of code. Morse's original design from his 1840 patent resembled a toothed lever that encoded the message, but Vail devised what he termed a *correspondent* or *key* like the one depicted in Figure 1.[9] Morse modified his later patents to include a similar design, which he termed a "straight lever or key" in 1846, and a key with a "finger knob" in 1849. By 1850 journalists commonly referred to it as a "telegraph key."[10]

Period descriptions of the device were common in the popular press, and the struggle for a new vocabulary much in evidence. In 1844, one journalist described the transmitting terminal in scant detail, but reported that the operator transmits a signal "by placing a finger upon this key and pressing it down." He described the signal lever itself as "a slim flattened piece of brass . . . with an ivory button upon it . . . of the size of a five cent piece."[11] In 1851, another observer distinguished the entire transmitter key from the touch surface: "This key is fixed upon a pivot axis, to be gently pressed by the operator's fingers on the top of an ivory button."[12] To my knowledge, none of this terminology appears in Morse's own writing or public comments, but Vail did adopt language similar to that of Cooke and Wheatstone. In an 1847 description of early telegraphy, he describes an alphabetical correspondent with thirty-seven keys, in which "each key has its button, with its letter, A, B, C, D, &c., marked upon it."[13] Like Cooke and Wheatstone, Vail also distinguished the touch surface of the raised button from the lever key that makes the electrical contact.[14] Not until the late date of 1909 did *Webster's Dictionary* define the button as "in electrical apparatus, a push button," which seems to conflate the touch surface and the circuit opener/closer, as is common today. We need not share Emerson's sense of poets as "liberating gods" to appreciate that even pedestrian struggles with language are not just adaptive but also fundamentally creative, for they bring new relations into existence, and with them, new ways of thinking and knowing.

Another early use of the term *button* appeared in France. Jan Józef Baranowski, a Polish inventor living in France, devised various meters and calculating machines,

FIGURE 1. Early Morse-Vail telegraph transmitter key, made by Alfred Vail around 1844. Courtesy of Division of Work and Industry, Smithsonian National Museum of American History.

which he patented in both France and the United States, before designing several voting machines around 1849. We will encounter some of these in chapter 13, but for the moment it bears noting that his machines contain input devices that amount to crude binary switches, used to indicate a binary choice between *yea* and *nay* votes, or between one candidate and another. Baranowski's patents contain some of the earliest uses of the term *button* for a nonelectric binary actuator. A promotional pamphlet for his machines similarly refers to "pushing this button" *("en poussant ce bouton"),* at roughly the same time that Americans began using similar language for the Morse-Vail telegraph.[15] Still, Baranowski's reference is ambiguous, as he also uses *bouton* to refer to rotational inputs that, in English, would be termed *knobs* or *dials.* Accordingly, Baranowski also refers to "turning the button" *("en tournant le bouton"),* clearly referring to the handle of a small dial.[16] Even so, one of Baranowski's three voting machines contains a rudimentary switch, also termed a *bouton,* that works by sliding back and forth between two mutually exclusive positions: "By pushing this button to the right, a vote is made for; and to the left a vote against."[17] This is, in practice, a binary switch, but one that Baranowski refers to with a French term entirely appropriate for the protruding knob.

Baranowski seems to have adapted this particular interface at least partly from his machines for reckoning and calculating taxes invented several years earlier, the French patents for which similarly refer to the input device with the term *bouton.*[18] The

American patent for a similar Reckoning Machine refers to the touch interfaces of these machines as "studs or buttons" just a year or two later, another early application of the term *button* to something like a binary switch in English.[19] However, perhaps sensing that English *button* was not quite as capacious as French *bouton,* Baranowski refers consistently to a "stud or button" on the reckoning machine, as if either term were not quite adequate on its own. The use of the term *stud* further suggests the metaphor of finger-sized clothing fasteners, collar studs.[20] And as with Cooke and Wheatstone in England and Vail in the United States, Baranowski's "stud or button" singles out the small round touch surface protruding from a larger mechanism that he otherwise designates as a *slide.*

Switching Terms

The term *switch* never appears in these early telegraph patents or in other sources, even though the input mechanisms of the Cooke and Wheatstone telegraph and the Morse-Vail telegraph clearly employed what would now be categorized as simple electrical-contact switches. Even in the 1860s, scientific and technical books on telegraphy seem not to refer to switches.[21] The history of the term *switch* turns out to be more obscure than that of the *button,* but equally revealing. Samuel Johnson's dictionary of 1755 defines a switch only as "a small flexible twig," the kind that might be used to beat an animal or a child, a similar definition to that found in the 1828 edition of *Webster's* in the United States. By its 1859 edition, *Webster's* had added a new definition for switch: "A moveable part of the rail for transferring a car from one track to another." By 1864 the dictionary added a definition of *switch* as a verb: "To turn from one railway track to another." Only in 1890 did *Webster's* begin referring to a switch as "a mechanical device for shifting an electric current to another circuit." We might extend these examples to other Western languages without changing the fundamental fact that there simply was no word for binary switching before the mid- to late-nineteenth century. And the reason there were no words for such devices was that there were no such devices.

The term *switch* is thus sedimented with a whole series of metaphors layered one on top of the other. The railway switch was probably named metaphorically after a thin stick for beating, which resembles the curved "switch rail" that seems to lash against the "fixed rail" as if it were a long, flexible piece of wood. Early railroad designers might have drawn on different analogies with hydraulic systems, such as irrigation canals, which also redirect a flow from one avenue to another, or with doors, which also function as gates that admit and exclude. Then again, perhaps doors and gates were not quite binary enough, given that they are often partly opened or partly closed, to let in the breeze but not the dog, for example. In those ways, doors function as analog screening devices, continuously variable rather than functionally binary in their operation.

By a further metaphorical extension, the railway switch track eventually provided a name for an automated circuit closing device. How exactly that happened or where it happened first is far from clear, but it seems to have begun intermittently in the 1850s. Early automated circuit openers and closers may have physically resembled railway switch tracks, with exposed contacts that could be manually touched together using an insulated handle that swung the contacts sideways. In an 1853 patent for an early quadruplex telegraph, the inventor refers to "an apparatus, called a 'switch,' used for establishing a metallic connection" between electrical contacts.[22] The term was evidently new enough that it had to be bundled in quotation marks even for a technical audience. Similarly, in 1856 a medical journal reported on a curative electrical bath that could be controlled "with a little piece of apparatus called a 'switch,' from its resemblance to a railroad switch."[23] An 1860 manual of telegraphy casually referred to "a switch, by which [the operator] can throw the battery off the line."[24] Even as late as 1864, *Scientific American* still placed the term *switch* in quotation marks, indicating the relative novelty of the term.[25]

Switches may have derived their name from more than just a resemblance to actual switch tracks. As railway networks grew in sophistication, telegraphy promised a means of coordinating the network's traffic in real time. Cooke and Wheatstone's first commercially successful operation connected Padding station to West Drayton thirteen miles away, both part of the Great Western Railway network.[26] By 1850, between a quarter and a third of Great Western's total twelve thousand kilometers of rail line was accompanied by telegraph line.[27] Telegraphy proved especially useful for coordinating network traffic from a distance, including by communicating the proper position of switch tracks.

This circuitous route by which the name for a thin stick used for beating migrated to a railway network junction and then to an automated circuit opener and closer did not end there, for it eventually continued into more familiar idioms of action and choice. When people talk about *switching* jobs or planes or anything else, they employ a mechanical metaphor that derives from the binary logic of railway systems and electrical networks. Technology provided not just a new vocabulary for talking about action and choice, but a new mechanical model for how action and choice work. When people perform this act of switching, they have exchanged a model of gradual or incremental change for a more binaristic one. Needless to say, people could choose one course of action over another long before the advent of mechanical switching, but English had a less ready vocabulary for it until mechanics provided analogies. Enmeshed in all sorts of mechanical systems, including actual arrays of switches, people readily internalized that organizing structure as a name for a kind of activity that seemed to emanate wholly from themselves.

I am pointing to a certain interoperability of machines and languages, which shape each other persistently and continually. Neither exclusively determines the other. The fossilization of poetic language that Emerson lamented refers to the static conceptualizations that settled terms eventually enforce. But, at any time, another poet can breathe the life of metaphor back into them again. This has real ramifications for how people conceptualize the world and themselves. Technology and terminology feed back into each other. New techniques call forth new ways of talking, and new ways of talking can reshape techniques. All of this has relevance not just for what people say and do, but also for what and how they think. Once everyone refers to a *switch* with a confident sense that they know what it is—once it shakes off the quotation marks that swaddled its infancy—language primes everyone to regard switching as a fact of life, simply the way things work. In forgetting the metaphor, people also forget that things could be entirely different, and recently were. But, by lingering over the slippages, appropriations, and elaborations of this terminology, as we already are beginning to do, it is possible to perceive just how alien switching technologies must have seemed to the people who first encountered them, and how temporary they yet might prove.

Pianos and Guns

At this stage, some readers may feel impatient to know not where words came from or who called things what, but plainly and simply what were the first switches, regardless of what they were labeled. The question is a fair one, but the answer, unfortunately, is not so simple. The closest devices to modern binary switches are probably the various catches, triggers, or releases for stored-energy devices. Edward Topsell's *The Historie of Foure-Footed Beastes,* published in 1607, describes a mousetrap with what may be the earliest use of the word *button* in the context of a machine with moving parts. In a chapter on mice, Topsell describes a trap in which the mouse, "by stirring the iron doth losen the butten, and so her heade is shut fast in the hole."[28] Presumably the mousetrap releases stored energy in the form of a spring or a raised weight when the mouse disturbs some sort of trigger, but the *butten* here is likely just a knob that butts out, or some other form of catch or clasp that releases stored energy when disturbed. This is rather different from a finger-activated switch, but even so, Topsell's mousetrap already has some of the features familiar from modern digital switches, in that a slight touch activates an automated script with only two binary positions, sprung or not sprung. All sorts of traps and snares are important protobuttons, from Native American deadfall traps to seventeenth-century leg-holds. Still, all are closer to passive sensors activated inadvertently, and thus are not as relevant to an analysis of the production of human agency.

Other energy-release devices are more recognizably part of the switch's technical lineage, but like traps, these too are instruments of violence. Long before the modern era in Europe, energy-release triggers on Chinese crossbows from the fifth century BCE functioned much like switches today. Once loaded with a bolt and cocked, the crossbow stored kinetic energy in the bowed wood, which the archer could release all at once and with minimal effort by pulling a trigger beneath the stock. These Chinese triggers often looked remarkably like the triggers on modern firearms. Later European crossbows worked similarly, though their triggers were often squeezed with the whole hand. Crossbows were thus functionally binary weapons that, once cocked and loaded, released stored kinetic energy all at once. There was still plenty of analog work required by such a weapon, not least of all in the matter of aiming, but crossbows required considerably less strength and skill than longbows, and the act of loading them with stored energy could even be delegated to someone else.

This brings us to one of the most important protobuttons of the modern era, the finger activated trigger of a different kind of stored energy weapon, this one chemical rather than kinetic: the modern gun. There will be much more to say about guns in the last section of this book, for the form of the gun appears in a wide range of contexts where buttons proved decisive, including remote control, photography, typewriting, video games, and more. The gun was among the first widespread devices to automate the triggering of action in such a strict binary form, as a nearly instantaneous choice between one of two mutually exclusive alternatives. The gun not only stores energy in the form of the charge but also stores prior decisions about how much powder to load and what kind of shot to use, all of which the user deploys later and all at once with the touch of a finger. Like crossbows, guns also require other kinds of skill, but far less than a sword or a longbow. The success of a shooter derives in part from the gun's much more elaborate mediation of agency.

We have already encountered the other major kind of protobutton, this one rather less lethal: the interfaces of musical instruments such as piano keyboards and the valves of wind instruments such as flutes. No less than the triggers of guns, the touch interfaces of musical instruments represent an important precedent for the finger-activation of automated actions broken down into discrete parts. Unlike the fluidity of the human voice, which all too easily slides between desired pitches, a piano programs each note mechanically in advance. However, unlike the trigger of a gun, a piano key is not a binary switch at all, because it is designed to express fine gradations of pressure, velocity, and duration of touch, and thus to articulate a wide range of qualitative differences between the alternatives of sound and silence. Even so, earlier keyboard instruments were more binary. The harpsichord plucks strings, and so does not have the same velocity sensitivity as the piano. Early organ keyboards opened and closed valves that controlled the flow of air to the pipes, but not in a graduated fashion. One might even suggest that the earliest proto-buttons were the finger holes in the forty-thousand-year-old Paleolithic bone flutes recently discovered in central Europe.[29]

So, we start from this curious fact that, before about 1840, there were, first of all, almost no devices termed either *switches* or *buttons* and, second, a striking paucity of devices that resemble switches or buttons, with the exception of certain kinds of weapons and certain kinds of musical instruments. These two sources—weapons and instruments, aligned respectively with violence and art, power and play—will persist strikingly in many of the switching technologies we will encounter ahead. They indicate two poles of experience that we will have occasion to observe in more local contexts. The first, associated with guns, entails the automation of power in sudden releases of stored energy nearly instantaneously and often at great distance, a compression of time and an extension of space that people often experience as a flattering demonstration of their own innate capacities. The threat of violence is often not far away, for where the legacy of the gun remains apparent, as in the *shooting* of photographs, relations of domination and fantasies of mastery are often apparent too. The second pole of experience, associated with musical instruments, entails the automation of art and aesthetic experience in deeply absorbing activities that often blur the boundary between people and things, but which are also intimately related to the legacy of Greek rationality expressed in harmonic theory. If we must have an origin of the switch, let us have a symbolic origin, not a strictly historical one. The switch was born from a gun and a piano, its violent father and musical mother, whose early modern union tangled two family trees in ways that will take many more pages to sort out.

CHAPTER 2

Designing the Button

When any technology becomes familiar enough, it starts to seem like it dropped from the sky. People take it as a given and forget all of the various factors that went into shaping it, along with most of the alternatives that might have been. The most insidious naturalizing mythology is that of necessary utility, as if machines must have assumed the most productive or efficient forms possible. This is also a covert way of saying that the technology was compelled by circumstance, that it had to come out exactly as it did. This is a mistake. It is true, for instance, that modern bicycles have two equal sized wheels because that works better than the old high-wheel bicycles from the 1880s. But "works better" for what? That phrase smuggles in a set of hidden values seldom examined. If the primary values are convenience, safety, and efficiency, the modern bicycle works better than the older high-wheelers of the 1870s, and those values are in fact the ones that prevailed. If, however, people value self-display over convenience or safety, just as some do in preferring high heels rather than high wheels, then suddenly the high-wheeler "works better" after all.[1] Utilitarian assessments always conceal values, and not just private values but social and political values as well—which is to say, values that are much more difficult for individuals to negotiate entirely on their own.

Many of the values that shape technologies are primarily symbolic and aesthetic. As Beatriz Colomina and Mark Wigley hare recently noted, some of the earliest artifacts attributable to modern humans are ornamental shell beads, which date back one hundred thirty thousand years. Whatever social or economic role beads might have played, such as signaling status, they also seem to have had a surplus of aesthetic value not strictly reducible to function. Even some early stone tools seem to have aesthetic forms, such as the symmetrical profile that distinguishes some *homo erectus* hand tools from much older asymmetrical ones. It might be that making or having symmetrical tools conferred social or sexual advantages, but then again perhaps some tools were designed mainly to be admired. "More often than not," Colomina and Wigley conclude, "what is seen as ornament is doing the real work and what looks like a tool is really for show."[2]

Colomina and Wigley contribute to a growing body of work on design that will inform what follows. Some of this work has focused insightfully on the ways in which the design of environments alters the people who live there. People reshape their environments, of course, but then environments subsequently reshape people too. In truth, people and environments mutually and reciprocally constitute each other continuously. As Larry Busbea has put it: "There can be no environment without the subject to call *it* into being, to topologically confirm its manifold structures, outlines, textures, smells, pheromones, switches, track pads, and touch screens. There is no such thing as a subject free from these interactions. There is no humanity without them."[3] Colomina and Wigley historicize the expansion of modern mid-century design by tracing it to Bauhaus and Walter Gropius, the English "arts and crafts" movement, and other sources who all averred, one way or another, that design had a moral mission. Better design would help make better people, which typically implied that people would be healthier, happier, more productive, more compliant, or more at ease in the modern world. So the stakes of design could not be higher. Accordingly, institutions recognized early that design was more formative than merely decorative. John Harwood's history of Eliot Noyes's redesign of IBM details an astonishingly ambitious project that reached far beyond corporate products. By redesigning everything from IBM's logo to its offices to its machine interfaces, Noyes was in effect prescribing entirely new ways of being not just for IBM but also for its millions of customers around the world.[4]

Like Busbea's and Harwood's work, much of the recent work on design has attended to the mid-twentieth century, when design already had become theoretically elaborate and institutionally extensive. But design theory and practice were both far less developed in the nineteenth century, with the significant exception of the fields of architecture and city planning. However, that is not to suggest that earlier artifacts were free of design, for they all have some amount of intentionality behind them, but rather that the process of making them was more haphazard. The form of a stone hand-axe from the Neolithic period is a reflection of how millions of hominids had concluded a stone hand-axe should be, including the hominid that made it. But that is not to say that those hominids arrived at their consensus through the kind of deliberate and conscious design process that Noyes led at IBM. In many of the machines from the nineteenth century and earlier considered here, there is a striking lack of consciously articulated and managed guiding purpose. Individual designers obviously had ideas about what they wanted to make and why they wanted to make it that way, but they rarely clarified principles up front or articulated, let alone followed, a conscious design process. Though the metaphor of evolution for technical change can be problematic, it rightly expresses the process of trial and error that so often gives rise to even long-lasting technical forms. What people term *design* in the mid-twentieth century, then, is often a reactionary campaign to eradicate Victorian mishmash and the comparatively unguided tinkering that nonetheless gave rise to some of its most successful products.

In the absence of detailed design records, we will have to attend to the forms of machines themselves and to their changes over time, convinced that design must have been involved at some local level, but also that what a machine finally does is just as important as what a designer thought it should be doing. So, although I often will have to speculate about the intentions or values that designers might have had, there is nothing speculative about reading the forms and functions of the devices themselves. In fact, literary scholars like me have learned not to bother too much with the intention of the authors of poems and plays, for what would knowledge of their designs add to the evidence of the work itself? Even if Homer had clearly recorded his intentions in composing the *Odyssey,* would readers be bound to understand it only in light of those stated designs? The best statement of his design is the *Odyssey* itself. Trust the tale not the teller, many scholars have concluded, and I will usually take a similar approach to reading machines. Even when records of design decisions survive, the machines themselves often will have something else to say. In chapter 13, on voting machines, for example, we will see how Jacob Myers thought he was building a machine to safeguard privacy and democratize the masses, but his enormous fully enclosed iron voting chamber seems to this observer to tell a rather different story.

Adventure Travel

Not all switches will tell the same story, because there are so many different kinds applied to so many different purposes: toggle, knife, pressure, leaf, key, pull-chain, and rotary switches, and these are only some that respond to the touch of human hands. Many other automated switches are activated by pressure, moisture, temperature, or other inputs. Utility compels some of this variation, but creativity plays a role as well. In my house, there are at least five different kinds of switches just for turning lamps on and off at the stem, and surely not every one of these is precisely tailored to the demands of each occasion. Their varied forms may be seen as ornamental, an aestheticizing of interface that yields pleasure from variety, which I appreciate in my devices no less than in my dinners or my dress. Even a single kind of switch admits surprising variety. The common push buttons that festoon modern homes, automobiles, computers, telephones, calculators, and more, are themselves exceedingly various. A button is a one-directional, usually spring-loaded touch interface for a binary switch, typically scaled to the finger or the hand and usually situated so that its direction of travel is perpendicular to the surface on which it is mounted. But this still admits a wide range of variety in terms of size, travel distance, color, resistance, sound, force required, and much more. All those factors have been designed into each button, though not always with the same degree of conscious care. Their actuation varies too, such that buttons might be further divided by number of contacts, number of poles and throws, and whether or not the switch is momentary or locked.[5]

To appreciate the design detail in a single switch and the meanings that exceed that design, let us look more closely at one familiar to everyone, a simple computer key. Even though most computer keys look roughly the same on the outside, they can be subcategorized by concealed inner functions, including dome switches, membrane switches, scissor switches, ALPS switches, and torpe switches, among others. All of these have been used successfully on modern keyboards, but each sounds and feels different, and each tends to be valued for different reasons. The sonic and haptic experience compacted into a button's miniscule travel during a fraction of a second is surprisingly rich, so manufacturers produce many varieties with different qualities. The Cherry Company, founded in the United States in 1953 and now owned by ZF Friedrichshafen in Germany, produce dozens of different switches, including twenty-one different versions of the standard computer key. Their popular line of MX switches exhibits a wide range of differences in terms of subtle sensory factors: actuation force, tactile force, noise at bottom-out and top-out, audible click, travel distance, pretravel distance, bounce time, reaction time, and durability, to name just some of the factors designed into every one.[6] Most of these qualities are not essential for typing alphanumeric text, but operate at the boundary between aesthetics and utility. Laptops designed for unobtrusive use in meetings or libraries tend to use quiet switches with little travel. Keyboards designed for gaming often have keys with more resistance and an audible click, to give immediate feedback that an input has been accomplished.

At the level of user experience, then, pressing a button turns out not to be so binary after all. It may occur in a fraction of a second, but users experience it as a process unfolding in time, even if subliminally, and modern designers have carefully scripted that unfolding. Industrial designers describe a button press with a simple graph, a force–displacement function that charts the force required to press the key on the vertical axis and the travel distance on the horizontal axis. A button with a strictly linear function would have a horizontal line: as the key travels downward it requires the same amount of pressure at all moments. But this is rare, and most graphs are both curved and inclined, as Figure 2 shows. This graph of the force–displacement function for the Cherry MX Blue switch illustrates the characteristic bump that a user would feel just before the switch closes the circuit. The completion of that tactile bump communicates back to the user that the operation had been completed. At roughly the same moment, the key also makes an audible click.

When we correlate available information about design with observations of the

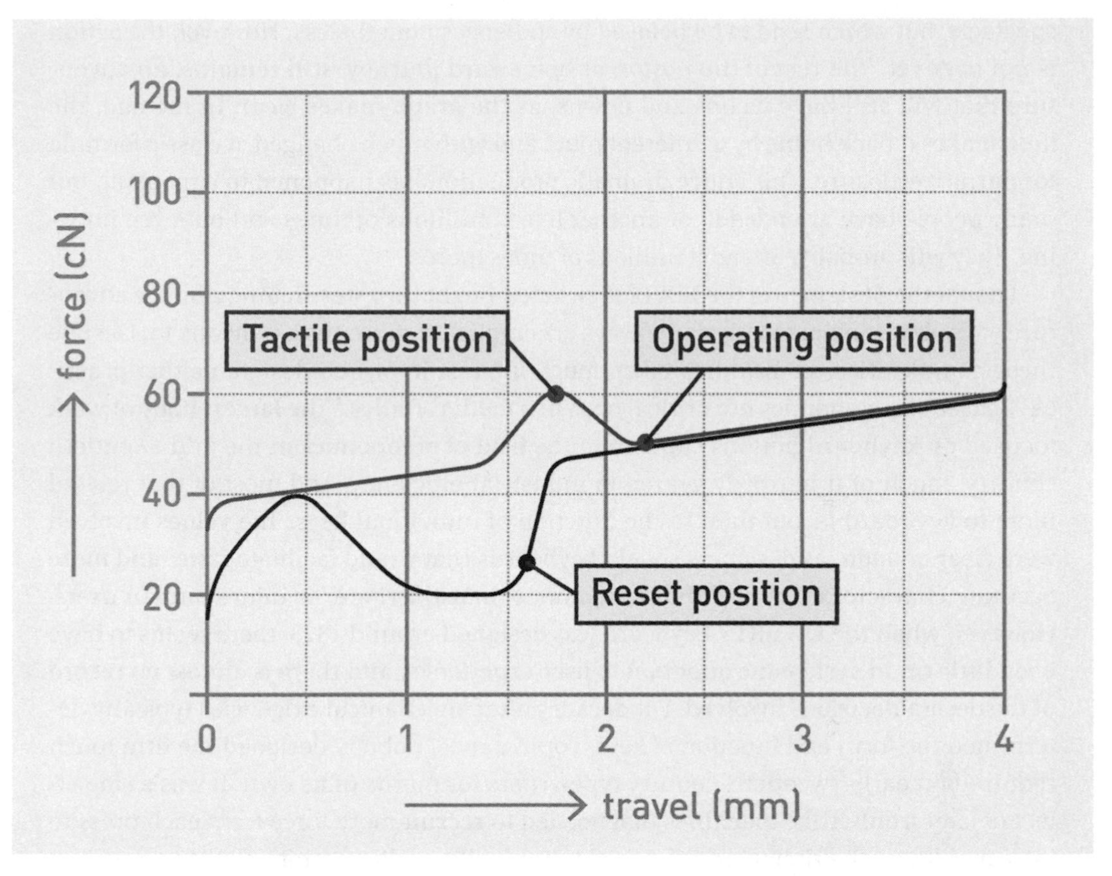

FIGURE 2. The force-displacement function for the MX Cherry Blue keyboard switch. Courtesy of Cherry AG.

machine itself, we gain access to the fullest range of stories that any device can tell. When it comes to telling stories, the MX Blue switch turns out to have considerable literary ambition. In a fraction of a second, it scripts a small dramatic narrative, an adventure story for the fingers. The graph above summarizes its plot. Like many traditional stories, this is a circular tale of departure and return with plenty of adventure along the way, an *Odyssey* compressed to less than a tenth of a second. In the first moments of this epic, there is stasis. The hero sits on Cyprus with his head in his hands, waiting for something to happen. But heroes are made for action, so before long he gets going on a journey that is smooth only at first. But as in any good story, conflicts quickly intrude and the hero must overcome resistance. More force is needed. These efforts prevail, and the hero surges onward. At that precise moment, an audible click celebrates the hero's success. These are the special effects, those trappings that Aristotle denounced as mere spectacle, but which tend to be beloved by audiences nonetheless. However, the action is not over yet. The rest of the *nostos,* or homeward journey, still remains, an adventure that will still have its ups and downs, as the graph makes clear. In the end, the hero makes it back home by a different route and with much changed, a classic formula for narrative closure. This entire dramatic production has happened in a moment, but many people have attended it, or another like it, millions of times without even noticing. They will probably attend it millions of times more.

I doubt the designers of the MX Blue switch thought they were telling a heroic adventure story, but then again, design always accomplishes more than it means to. Despite these complexities, no field has taken much interest in switch design, neither practical fields like ergonomics nor critical ones like media studies.[7] The largest body of work focused on keyboard buttons came from the field of ergonomics in the mid-twentieth century, much of it narrowly geared to industrial efficiency, and most of that related more to keyboard layout than to the function of individual keys. The values involved were clear enough, as designers sought keyboards that would facilitate faster and more accurate character entry and that would not exhaust, irritate, or injure human users. However, when the QWERTY keyboard was designed around 1873, there seems to have been little or no systematic attention to user experience, and there is almost no record of the design decisions involved. For decades after, mechanical exigencies typically determined the form and function of keys. For instance, nobody designed the firm touch required on early-twentieth-century typewriters for merits of its own. It was a side effect of later front-strike machines that needed to recruit more force from each press to raise and lower the ribbon and the shield. Ironically, earlier bottom-strike typewriters did not need to recruit that extra energy, so many had significantly softer keyboards with less travel for each key. Eventually, electrification allowed designers to separate such strictly mechanical functions from matters like key travel, resistance, and audible clicks, which permitted more thorough shaping of user experience through intentional design. By the 1980s, designers were looking at the shapes of force–displacement curves like the one in Figure 2 and even contemplated designing different curves for different fingers.[8]

Even so, the archived papers of the Association for Computing Machinery reveal surprisingly little interest in keyboard design until the 1980s, when personal computers began to become more popular, but even then we find almost no interest in individual keys. Eventually that changed, but for a surprising reason. Computer engineers—as opposed to experts in ergonomic efficiency—seem to have grown interested in buttons not when electronic keyboards first became the standard computer interface in the 1960s, but decades later, when graphical user interfaces (GUIs) began representing virtual buttons on screens. From the Xerox PARC system to Apple's Macintosh, new GUIs compelled new attention to the form and function of mechanical buttons in order to imitate them better virtually. Virtual buttons brought switching into the realm of software design, which eliminated all the mechanical engineering that went into making a material interface: no ANSI (American National Standards Institute) standards to observe, no Underwriters Laboratory specifications to navigate, no limitations on materials, voltage, environmental conditions, or ergonomic factors, and not even the fundamental problem of creating a label that would not rub off. Because programmers could design a virtual button however they wanted, they suddenly had to reckon with far more details of design, without relying on mechanical exigency to handle so many of the decisions. As utility ceded ground to aesthetics, a host of new questions arose. What should a virtual button imitate? Should it simulate a mechanical delay, audible clicks, or visual simulations of travel? How large should virtual buttons be, and what shape? Or, should virtual buttons be animated in ways that physical buttons never could? As buttons first became electrical and then virtual, they could be thoroughly shaped by negotiable design decisions after all. Once those decisions were free from previous material constraints, they also could reflect a wider range of values.[9]

Ironically, then, virtual buttons compelled new interest in mechanical buttons too, a tendency that extended far beyond keyboards. One revealing study from 2014 inventoried the formal and tactile properties of more than 1,500 mechanical buttons on 112 different domestic appliances in order to "recreate the experience/familiarity of physical buttons" in virtual settings.[10] This ecology of domestic buttons was no exercise in ergonomic design, no attempt to improve mechanical interfaces, but rather an effort to better understand the thousands of devices operating in modern life already and the habits and expectations that their users had developed. This effort was also surprisingly late, for by this stage, four or five generations of people had already learned to take mechanical buttons for granted. As Rachel Plotnick has shown, buttons had been proliferating wildly since about 1880, and with surprisingly extensive social and political consequences. According to Plotnick, buttons became part of a "management style," a form of "digital command" that relocated communication and control to the fingers. As such, the push button became an important source of manual deskilling, but also, and as a result, a means of distancing management from labor physically and symbolically. On one side of a servant buzzer, there is a commanding master who exerts an absolute minimum of effort; on the other side, an obedient servant meant to comply instantly, unfailingly, and from a distance. Although it is possible that an industrial designer in the 1970s might have considered such far-reaching consequences, there is little evidence that anyone did so in the 1870s. What switches do is never strictly limited to what designers thought they should.

Master Switch

In the four-millimeter travel of a computer key that I fancifully compared to the journey of Odysseus, we encounter an apparatus that does more than mediate between servants and masters; it also confers something like mastery itself. To trip a switch is to project the full force of one's will powerfully and at a distance, as if in a micro-epic of technically mediated heroism. The hero in this tale, of course, is the person pressing the switch, for the switch confirms that they are among that special class of people who are active rather than acted on. The swift and effortless results affirm them as free, empowered, individualized agents, sources rather than targets of action. To be otherwise is to be less than fully human, and to drift toward categories of the nonhuman, such as animals or machines, but also toward those kinds of people traditionally accorded a lower level of rights and responsibilities: children, nonwhites, the disabled, and women, among others. So even as switches mediate between masters and servants, they also help constitute mastery itself: they confer that ideal standard of personhood that never has been equally afforded to all.

This helps explain the special appeal of mechanical feedback, such as the tactile bump and audible click of the MX Blue switch. People receive unavoidable feedback from analog tools such as hammers, which transmits tactile, audible, and visual confirmations of each blow. So too does a mechanical typewriter, which transmits sound and vibration as side effects of the stroke. But with many electrical buttons and all virtual buttons, this feedback must be scripted and recorded. Those that fail to provide some form of feedback yield telling results. Older crosswalk buttons in the United States often make no click and provide no audible or tactile confirmation that they have registered an input. As a result, users often give them repeated angry jabs, seeking some confirmation that they truly have acted as intended. An elevator button that fails to illuminate provokes a similar sort of frustration. When switches withhold feedback, people experience something rare in analog interactions, which is deep uncertainty about whether they truly have acted after all. In the drama of the key-press I described above, every user is cast as the hero in an infinitesimal ordeal. But, without feedback, they rightly fear that they might be less heroic than supposed. And because the effects of switches often happen far away, microscopically, instantaneously, or after some delay, scripted feedback creates an illusion of immediacy, a flattering confirmation that intention and action perfectly coincide.

In that way, the kinds of feedback designed into switches anticipates and alleviates anxiety about agency itself. The disproportionate emotional response that some people feel when an elevator button fails, a key sticks on the computer, or the shutter button on a digital camera lags is difficult to explain in purely practical terms. The stakes seem so low. But "button rage" makes more sense when considered not as a response to some trivial practical impediment, but as a response to a more profound threat to self. To doubt whether one has acted is to doubt in a more serious way whether one is an agent after all.

There will be much more to say about how switching organizes and even constitutes so-called human agency, but for the moment we can summarize four general ways that will matter most. First, binary switching has made a great deal of action functionally instantaneous. When I flip a light switch, I condense the process of illuminating a room down to an imperceptible instant. In previous centuries, illuminating a room would have involved the more protracted analog process of lighting a candle or lamp, or of summoning a servant to do it. Those actions not only unfold gradually over much more time but also involve many different private decisions sometimes compounded with social meaning: where to strike the match, how high to turn on the gas, what to use for kindling, or whom to ask for help. In contrast, when I flip the light switch on the wall, the room appears to illuminate instantaneously. Even switches that are not functionally instantaneous tend to compress action down to drastically smaller timescales, such as the keys on even the earliest writing machines.

Second, switches reconfigure the relationship of people to distance. This is most obvious in remote controls, but also evident in my computer keyboard, where the action of my fingers produces writing on a simulated page ten to twelve inches away. Switches distance action from the body, whether by inches or miles. Light switches, elevator buttons, servant buzzers, library bells, thermostat inputs, camera shutters, dashboard controls, and more all mediate between *here* and *there,* which sometimes takes the form of mediating between *inside* and *outside* as well. This has complicated effects on the spatial logic of causation. People have become accustomed to compelling action at a distance from their bodies, a flattering arrangement that had far fewer precedents before the nineteenth century. The phrase "action at a distance" comes from classical physics, most conspicuously in Aristotle's claim that every motion requires a contiguous mover, which means that there cannot really be action at a distance after all. As John Durham Peters, Florian Sprenger, and Christina Vagt have recently noted, this classical and (later) early modern view has held that action can be transmitted only through a medium, and anything else can register only as miraculous.[11] As a result, they argue, media studies must reckon with this persistent fantasy of action at a distance and with the historical injunctions against it. But switching persistently conjures the illusion of action at a distance, the flattering sense that my touch *here* can instantly accomplish some result *there,* with no apparent mediating stages. The wires for the light switch are hidden in the wall. The light from the remote control is beyond the visible spectrum. Even though media studies often has dispelled the illusion of action at a distance, it might inquire more deeply into why its simulation has been so appealing and what the broader effects of that simulation might be.

Third, the button makes human skill or strength far less significant for the course of any action. The minimizing of labor largely eliminates questions of trying from the contemplation of action, given that to try to turn on a light with a switch is, in the vast majority of cases, to succeed in doing so. Where trying still matters, switches typically involve either intricate sequencing or precise timing, as in the operation of a pinball machine. Actions appear less as intricate entanglements of social and material factors that might take long practice to master, and more as practical expressions of some individual and internal resolve, desire, or intention. Switches mass-distribute mastery. The sheer simplicity and effortlessness of input eliminates so many other factors from consideration that most of what remains appears to emanate from a single person's magisterial will.

Fourth, and finally, most switches conceal intermediate stages of action, black-boxing automated processes so that little appears to come between the input of the finger and the output of the result. As I type these words, I am largely unconscious of the technical processes taking place inside my computer. Even when I try to be conscious of them, I struggle to put them into words. If I had to explain exactly how the movements of my fingers make Roman letters appear on the screen, I am sure I would find it humbling. No wonder those subroutines that operate invisibly in the background of modern computers are called *daemons,* as Wendy Hui Kyong Chun has pointed out, given the mysteriousness of what goes on in there.[12] Just as action at a distance is miraculous, so action within the black box can seem magical too. The agent at the switch thus encounters action not as a process, still less as a complex multistage interactive process, and more as the immediate translation of cause into an effect.

For all these reasons, all switches enlarge and aggrandize, like a mirror that reflects a person's image back at twice its normal size. But switches return that magnified version haptically rather than optically, through feeling rather than seeing. Fantastically empowered, that individual is nonetheless pitifully isolated too, autonomous but alienated, and thus all the more dependent on mediating technologies. Little wonder that designers invest so much ingenuity in subtle feedback mechanisms that confirm to a human user that the button has been pressed. The clicking keyboard, the glowing elevator button, the subtle haptic bump—these are gentle reassurances that individual human agency is secure after all. As a result, although the button magnifies, it diminishes too. Actions are larger, faster, and more powerful than ever, but movements of the body smaller and more trivial. This is a precarious arrangement. So much depends upon so little. If the agency of a being this powerful and puny were to fall into doubt, what would be left?

CHAPTER 3

Analogs and Analogies

The binarism of the button requires some further comment, for the closer we look at the materiality of binarism, the harder it is to perceive. The term *binary* denotes any paired system, but Gottfried Leibniz developed the fullest, if not quite the first, base-2 mathematical binarism in the late seventeenth century, which represented all numbers with only the symbols 1 and 0. Long before Leibniz, many other philosophers organized the world around different oppositional pairings, especially the various dualists from Plato to Descartes who insisted on strict and foundational distinctions between *real* and *ideal,* or *substance* and *accident,* or *mind* and *body.* Hegelian dialectics made such dualism more critical and dynamic by positing the mutual interdependence of such opposed terms while also insisting on their synthesis. Later, structuralist linguistics and anthropology made binary opposition integral to their theories of language and culture, although sometimes (but not always) shedding the Hegelian determination to synthesize opposition into higher unity. For Ferdinand de Saussure, Claude Lévi-Strauss, and the generations of structuralists they inspired, linguistic and cultural meaning derives from a relationship of difference within binary oppositions.[1]

The prevalence of binaristic thinking in Western societies is hard to overstate, but not many have considered how this philosophical and mathematical legacy relates to the mechanics of everyday life. The matter is perhaps most conspicuous in the history of early computing. Starting in 1945, the computer ENIAC (Electronic Numerical Integrator and Computer) used decimal rather than binary number storage, requiring ten vacuum tubes for each decimal order, such as ones, tens, or hundreds. ENIAC stored the digit 5 by turning on five of the ten vacuum tubes in any given place order. ENIAC's successor EDVAC (Electronic Discrete Variable Automatic Computer) used binary number storage instead. In binary code, the decimal number 5 is 101, which requires more numerals but fewer vacuum tubes—one for each digit of 101—and only two of which are turned on. The efficiency gains from combing the binary mechanics of vacuum tubes with the numerical code of ones and zeros were thus considerable.[2]

But I am not fundamentally interested in the internal architecture of computers and the switches that encode numerical values; instead, I am interested in those switches that interface with the body, a more practical and pedestrian binarism of input that includes those banks of switches and plugs on ENIAC, the punch cards used to instruct it and later computers, and eventually the electrical push buttons of modern keyboards. Such humble materializations of the binary have received far less attention from historians of computing, but I want to suggest that their development across many technologies has been as profound a development as Western dualist philosophy, base-2 numerical code, Hegelian dialectics, structuralist binary opposition, or any of the other prominent conceptual and symbolic binarisms of modern life.

Perhaps structuralist binarism is part of a culminating efflorescence of dualism that is related to the material binarism of switching, for it emerged at precisely the moment when techniques of binary input and control were rapidly colonizing much of everyday life. I make no attempt here to show that Saussure, Lévi-Strauss, or any other early structuralist modeled their theories of binary opposition consciously or unconsciously on switching technology. Their objects of study generally were not technological, and there appears to be little evidence of any such narrow and one-directional influence. However, I will conjecture that twentieth-century structuralism's preoccupation with binary opposition across many scholarly fields was part of a broader cultural development in which theories and practices mutually informed and reinforced each other. A world convinced that all knowledge is structured as binary opposition may reproduce that logic at the level of technique; conversely, a society dependent on binary switching techniques has every reason to assume that the rest of the world works similarly. To hazard an analogy: when people conceived of the universe in terms of the mechanics of rotation, as spheres and later orbits of planets, they eventually imitated those motions in their rotating technologies, such as clocks and gears and wheels. Having represented a whirling world at the level of technology, they also came to perceive rotation and revolution elsewhere too, such as the fluctuations of their fortunes, the aging of their bodies, or their relationships to God. It is not so much that one of these domains determined the others, but rather that they are part of a self-reinforcing whole, an elaboration of complementary practical and theoretical logics across very different contexts.

Similarly, when the known universe seems to be made up of binary oppositions (not celestially but conceptually), new techniques arise to represent that state of affairs: switching. Subsequently covering the globe in billions of switches has further repercussions for how people conceive of the world and themselves. By the time Saussure delivered his *Course in General Linguistics* in Geneva between 1906 and 1911, the functional binarism of new switching techniques would have seemed more familiar, more omnipresent, more essential to everyday conduct than at any other time in human history, even though it would expand even further in the decades ahead. In the twentieth and twenty-first centuries, then, binarism has not been limited exclusively to either theory or practice, to either ideas or instruments. Binarism has rather united theory and practice. It is a form that both ideas and instruments have assumed, and thus represents one of the grandest and most inescapable of all mythologies.

In recent critical theory, however, binarism has enjoyed rather less prestige. A wide range of poststructuralist theorists and their successors have rebelled against binary opposition and its allegedly essential role in human language and thought by showing how easily any given binary might be reversed, or by demonstrating how poorly any two terms account for more sprawling networks of relation. Binaries might be potentially useful heuristics, some concede, but they are also confining, reductive, and potentially oppressive. As Gilles Deleuze and Félix Guattari put it: "One becomes two: whenever we encounter this formula, . . . what we have before us is the most classical and well reflected, oldest, and weariest kind of thought."[3] Accordingly, the general temper of more than a half century of continental philosophy, literary and art criticism, and even sociology and anthropology has been tremendously skeptical of binarism, which many deem complicit with ideologies of domination and the most familiar forms of oppression. This is what Jacques Derrida meant by the "violent hierarchy" of all binaries, which privileges one of the two terms over the other, thereby wiring power disparities into the available conceptualizations of the world.[4] Deconstruction was one of many subsequent attempts to subvert binary logics, and thus to upend the settled structures of power they subtend. But this is easier said than done, for reversing a binary to destabilize it is not quite the same as escaping the cultural logic of binary opposition altogether, and Derrida's prose is one index of how hard one must work even to make a start.

I take it for granted that habits of thought and cultures of inquiry are so suffused with the legacy of Western dualism that nobody can hope to escape it altogether. Nor should they try, for the evident power of binarism is hard to surrender willingly, and some of the most powerful methods of reasoning depend on it. Even the most adventurous critics of the binary have a hard time escaping it. Deleuze and Guattari's *rhizomatic* model opposes the totalizing logic of what they call the *arborescent,* a term of opprobrium for all that is rigidly hierarchical, and thus aligned with binarism. Yet this opposition of the rhizomatic to the arborescent is itself a binary opposition, which its authors surely recognize. If Deleuze and Guattari cannot assail binarism without also employing it, I hold little hope of doing so either. Nor is it quite clear how one might think and argue within this culture without relying on even the most traditional paleo-binaristic pairs, such as *right* and *wrong,* or *here* and *there,* or *past* and *future.* The basic structuralist insight that concepts and culture derive from binary opposition does not seem to this critic to have been invalidated, though we are right to question the alleged naturalness and permanence of that arrangement.

Digital and Analog

At this stage, we also must reckon with a broader issue, the fact that the binary has a more historically specific relation to the digital. There will be much more to say about this later, but for the moment, suffice it to say that the use of the term *digital* as a byword for computation is a comparatively recent development, devised only in the mid-twentieth century to distinguish those computers that performed calculations using binary code from analog computers that performed calculations using continuously variable quantities, such as the rate of rotation of a spinning wheel. The sprawling mid-century field of cybernetics did more than any other to define the digital and distinguish it from the analog, for it was a pressing matter for them to understand just how systems of information feedback and control were organized. Even so, at the famous Macy Conferences on cybernetics in the late 1940s and early 1950s, John von Neumann made a comment that remains as true today as it was then: "Present use of the words 'analogical' and 'digital' in science is not completely uniform."[5] They never did achieve a clear consensus about the application of those terms, though most eventually agreed that *analog* refers to processes of continuous gradation while *digital,* as Lev Manovich has claimed, refers to the breaking up of continuities into series of discontinuous bits.[6] Accordingly, the digital is not just computational, nor is it strictly binary, and still less to be associated exclusively with binary code. In the broadest sense, the *digital* refers to the dividing up of continuities into any discrete units, regardless of whether we describe those units with base-2 or base-10 numbers. Benjamin Peters has urged us to associate the digital even with the numerical distinctions of finger counting and insists that "we foreclose against a fuller understand of the limits of our digital condition . . . when we understand digits only computationally."[7] According to Peters, the digital is a much broader ensemble of ways of anchoring abstract mathematics in material practices, especially those practices that transform continuities into discrete enumerable units. The binary is thus the minimum quantity of the digital, the differentiation of unity into the smallest multiple, or as Alexander Galloway puts it, "of dividing one into two."[8]

Needless to say, the emergence of the term *digital* in the mid-twentieth century installed a new binary in our lexicon, *digital–analog,* a staple of decades of media studies. One problem with this particular binary is that it appears to be clearly affiliated with the digital side. As McKenzie Wark has argued, "if the relation between the analog and the digital is digital, an absolute boundary, then the domain of the digital can be perfected" as a "locus of command and control of the analog remainder."[9] To accept the binary *digital–analog* is already to have accepted the governing logic of digitality. One way to mitigate this problem is to adopt an analog view of the *digital–analog* binary, by referring to features that are *more* or *less* digital or analog. This retains a dualist polarity that is not always welcome—a basic structure of opposed extremes—but it insists on the porousness of the boundary between them and the gradual transition of one into the other. Many of those who first articulated the distinction between analog and digital at the Macy Conferences regarded the terms similarly, remarking in their papers and discussions that the analog and digital are generally mixed.[10]

There are dangers from privileging the analog too much as well. If we side resolutely with the analog against the digital, we run the risk of lapsing into nostalgic naturalism, for such assumptions can blithely align the analog with nature and the digital with culture. Von Neumann made the point explicitly at the Macy Conferences: "In almost all parts of physics the underlying reality is analogical, that is, the true physical variables are in almost all cases continuous." In contrast, he continued, "the digital procedure is usually a human artifact."[11] This is a seductive idea, but as Jonathan Sterne has noted, the analog only acquired this organic, romantic allure once it could be defined as *not-digital.* It is not foundational or natural, but for much of its history a concept that expressly designates the digital's antidote.[12] For example, Sterne notes, critics of digital audio often denigrate it as less "live" or "natural" than analog because they perceive that "digitisation kills audio bit by bit," chopping it up, leaving gaps between the digits, and taking it ever further away from living voice and unmediated reality.[13] However, Sterne cautions, "the meanings we commonly attribute to the word *analog* did not even fully exist in the so-called analog era"; the warmth and naturalness that seem to be essential to the analog are just as much cultural artifacts as the digital, and indeed developed at mid-century in relation to the digital.[14]

Perhaps it bears mentioning that naturalizing the analog can both privilege it and denigrate it. Sterne is concerned that the analog would be privileged as the form of authenticity, intimacy, and immediacy, while the digital would be linked with derivation, alienation, and mediation. But naturalizing the analog also can denigrate it as unformed, irrational, and primitive, while distinguishing the digital as structured, rational, and civilized. Even the term *nature* is not exclusively one of approbation. It will not do to generalize too much at this stage, for different circumstances exhibit different values, but I will try to resist any instinct to stage a moral drama between the analog and the digital in any all-encompassing way. I prefer a messy interpenetration of the digital and the analog that in no way dispenses with the binary of *digital–analog* but does attempt to trouble it productively. If I cannot escape the confines of the binary *digital–analog,* at least I can make a racket on the bars of the cage. Accordingly, against those pioneers of cybernetics who sought to evacuate all value and meaning from the "forbidden ground" of analog gradation, I will want to attend to the entanglements of the analog with the digital, especially in material manifestations that are hard to characterize as solely one or the other.

For instance, it might be tempting to imagine a simple electrical switch as the very epitome of binary function, in that it turns current on and off instantly, with no analog middle ground in between. Although this may appear to be true to human perception, the process of switching is replete with analog properties nonetheless. As Florian Sprenger has recently shown, the apparent instantaneity of electrical transmission has long been a source of its "phantasms of immediacy," the illusion that electricity could somehow be in two different places at the same time.[15] Later, science established that electricity does travel through different media at different speeds, including copper wires, even though imperceptibly to human senses. Electricity actually consists of two separate functions, drift and wave. When a circuit closes, the electrons themselves drift slowly through space, just a few millimeters an hour. But the wave of energy that accounts for what most people understand as electricity moves much faster, though slightly slower in copper wire. In systems that use alternating current, the wave reverses direction many times per second as well. Alternating current at 60 Hz changes direction sixty times a second. Given that the speed of the wave is not more than three hundred thousand kilometers per second (roughly the speed of light), the length of the wave of energy in a single-hertz cycle can be measured as roughly one sixtieth of that, or five thousand kilometers long, just a little longer than the distance from New York to Los Angeles. What seems to be an instantaneous and hence entirely binary function of *on–off* at human time scales is instead a great wave of energy sloshing back and forth across a continent. With that in mind, it becomes harder to think of opening and closing an electrical circuit as a strictly binary operation after all.[16]

For that reason, in all practical and material functions, the binary is an approximation, not an essence, and binarism may be fully realizable only in symbols such as code. This is why later chapters include devices that might not seem to qualify as switches, such as early typewriter keys. An electrical switch is more binary than a typewriter key, but neither one eradicates the analog entirely. Given that there is no bright line between the digital and the analog, there is no good reason to attend only to switches that control electrical energy and to exclude those that control chemical or kinetic energy, such as gun triggers or shutter buttons. It might be better to think about digital technologies as concealing the analog, rather than as eradicating it. Nobody thinks of a door as digital, because even though it has two opposite states, open and closed, it also has a visible range of useful positions utterly unconcealed in between. A door is not an all-or-nothing affair. In contrast, while railroad switch tracks also pass gradually through a middle range between open and closed, these continuous gradations have no functional value on their own, and if maintained will be regarded as malfunction. So even though the analog might be impossible to eradicate from any material system, I will concentrate on those devices that have attempted to minimize it or conceal it, and that have succeeded to a considerable degree. This amounts to a focus on devices in which the output is mostly binary, even if the input and subsequent internal processes often remain stubbornly analog.

Historical Analogies

The analog is present in binary instruments in another way, not in the function of individual machines but in the changing forms of machines over time. Binary interfaces spread through culture in the most analog ways, by a series of striking analogies that we already have begun to document: telegraphs borrowed from pianos, typewriters from telegraphs, computers from typewriters, and so on. We noted in chapter 1 that the earliest buttons may have been made from clothing fasteners, that keyboards cribbed from musical instruments, and that switches were named after railroad junctions. The role of analogy was obvious enough in the terminology of switching, but we should now note that technologies themselves change analogically, as relations of similarity unite otherwise different technical domains. Considered as a species of logic, analogy is not a method for arriving at certain and permanent conclusions, but for managing uncertainty in the absence of better options. That is analogy's greatest weakness but also its greatest strength. Arguments from analogy never deserve the confidence associated with formal logic, for they belong to that class of inferential processes that can advance knowledge without necessarily generating certainty.[17] Analogy in this sense is itself analog, in that it always operates somewhere between the poles of true and false.

Yet those plausible conjectures have great power, for as Keith Holyoak and Paul Thagard have shown, an impressive list of scientific discoveries developed from analogical thinking: in the first century CE, the Roman Architect Vitruvius explained sound with the analogy of water waves and designed amphitheaters accordingly. In the 1770s, Antoine Lavoisier understood animal respiration using the analogy of combustion. In the 1940s, Salvador Luria won the Nobel Prize for his work on bacterial gene mutations, inspired in part by the analogy of the irregular payouts of a slot machine.[18] There are many more examples. The great power of analogy lies in its ability to shuttle between otherwise separate fields of knowledge, proceeding not by necessity, but by intuitions of similarity. Research on analogy in philosophy and psychology has tended to focus on better ways of making useful or valid analogies, on improving analogical reasoning in practical endeavors, or on the ways in which people build up domains of knowledge analogically from a young age. Most philosophers of analogy agree that the greater the similarity between two items, the more likely the analogy will prove useful.[19] But the fact that analogy often resembles literary metaphor confirms that it is often most powerful when it dares to make more radical associations, arranging surprising conjunctions of the otherwise estranged. Analogy challenges people to think that a snake might be like a molecule, a book like a child, or a moral lesson like a meal, and that real knowledge might result from supposing so. As a mode of radical antiparticularism, analogy refuses to observe difference, and indeed translates difference back into similarity in order to transport knowledge from one domain to the next.

The perils of that approach are clear enough. Because analogy insists on similarity, it can easily become reductive or leveling. It can make everything the same, obliterating necessary distinctions. As Nelson Goodman once argued, "similarity, ever ready to solve philosophical problems and overcome obstacles, is a pretender, an imposter, a quack."[20] Everything is similar to everything else, Goodman says, depending on the criteria applied. My spleen is similar to the Taj Mahal, for both are about the same distance from the planet Venus. Goodman's point is that the frames of reference that determine similarity are relativistic, given that similarities are not properties of things but properties of common ways of talking about things. If this is dangerous for philosophy, Goodman acknowledges that it can be "serviceable in the streets," by which he means as a method for solving practical problems in the absence of stricter forms of logic.

I think analogy is serviceable in more than just the streets, especially because the poetic qualities of surprising analogies can disclose new possibilities that settled habits obscure, and this can create social and even political opportunities. Analogy mixes things up, violates established orders, and shuffles the old into something new. How any given analogy will serve in the long run is impossible to say and depends on the trial of actual experience, but one great value of analogy is that its components are so radically mobile, so effortlessly interchangeable, that it militates against the rigid permanence of binary opposition. Analogy does not pretend to assert the everlasting validity even of itself. If the analog is a state of continuous variability, that fact expresses something about the variability of analogies too. Every analogy will expire, because it either fossilizes into fact or degrades into incoherence. The changeable states of analogy are strikingly different from the stubborn durability of binary opposites, which is why philosophers have never needed to deconstruct it. Analogy continually collapses on itself, discredits itself. Never purporting to be entirely true or false, the analogy is by definition a creature of convenience, a provisional arrangement, a mere mortal trespassing in logic's eternal temple.

Another peril is that analogy can transfer values with great speed from one domain to another, often before anyone has time to think through the implications. Wendy Hui Kyong Chun has argued that modern genetics and modern software developed analogically from one another, and with lines of influence that ran in both directions. DNA was not just a biological model of people as programmable computers, but computers themselves may have been analogically influenced by even earlier Mandelian genetics. For Chun, genetics and software both exhibit "a tightly prescribed logic of programmability"; neither precisely came first, and she associates both with neoliberal discipline.[21] If values slide between domains analogically, attached surreptitiously to technical theories of how things work, imagine how widely and rapidly the values associated with switching must have spread, as switches colonized one area of life after another during and after the nineteenth century.

Let me conclude by acknowledging that the purely analog is as fanciful as the purely binary. Wherever we find one, we find the other. Analogies even have their own internal binary structure, for just as a binary opposition pairs two different terms, an analogy pairs two similar ones. Even though the analogy converts *two* back into *one,* to reverse Galloway's definition of the digital, a logic of pairing remains common to both. Nor is the conversion of two into one ever truly complete, for analogy ensures that difference always remains visible beneath the surface of purported similarity. Every analogy is also a relationship of difference, a difference that the analogy only partly overcomes. This is to say that the *binary* of *analog–binary* might also be expressed as the *analogy* of *analog::binary.* What looks like distinction from the perspective of the binary looks like identity from the perspective of the analog. The terms themselves enjoy relationships both of difference and of similarity. These two structures of comparison are so entangled that one would want to resist siding polemically with either one. Although the task here is to understand with some degree of historical detail just how the advent of binary switching changed the relation of the analog to the digital, we should apply critical pressure to both.

An analogical account of the history of switching techniques must fully acknowledge what we have already started to apprehend, which is that switches are themselves analogies, vehicles that transfer techniques from one part of culture to another. Accordingly, an analogical account of switching techniques must dispense with the hope for a neat, linear progression of forms through time, and with any notion of a gradual approach to technical perfection. Instead, it must seek out the lateral routes of exchange by which drastically different devices swapped forms and functions across different social and technical contexts, in ways that are as unpredictable as analogy itself, but also sometimes just as delightful.

Part II

»» »» Digital Bodies

CHAPTER 4

The Point of Touch

Let us begin with the body itself and the kinds of gestures associated with extending a finger to operate a switch or button. To recognize technical interactions with machine interfaces as gestural is to recover the latent symbolic meaning that persists within them, which is not reducible only to utilitarian operations, and which did not arise exclusively in technical contexts. It is also to step back from mechanical interactions that have become so routine that they no longer seem to be socially and semiotically complex. Whenever people extend their fingers to activate certain kinds of switches, they are performing one especially prominent gesture, the indexical gesture by which people point to one object among many. Pressing an elevator button is thus conceivable as technically mediated pointing, a gesture of indication detected not by another person but by a machine interface designed to recruit and register it. Buttons might even be thought of as recording devices for pointing. Like more recent machines that recognize human faces or human voices, buttons recognize and respond to one of the most prevalent gestures of the human hand.

The sociology and semiotics of pointing are complex, but precisely for that reason they can disclose new aspects of ordinary actions otherwise all too easy to overlook. One such ordinary action is the simple act of operating an elevator with its wall-mounted push button interfaces. To summon an elevator or to direct it to the proper floor typically involves pointing a finger at one among many buttons. This is hardly the only interface that recruits the indexical gesture, for it is common on many summoning devices such as doorbells or servant buzzers, some nineteenth-century voting machines, early-twentieth-century light switches, and twenty-first-century touch screens, to name a few. But elevator buttons provide an especially rich context for thinking about indexical interfaces, for they are familiar across cultures, have been formally and functionally stable for decades, serve a wide range of people, and have hardly ever been considered in any detail. The elevator interface is also far simpler than others already mentioned, such as keyboards, but it still mediates a complex technical and social environment. By putting a little pressure on elevator buttons now, we may deliver ourselves to higher levels of analysis later.

Paleodigitality

Benjamin Peters has correlated modern computational digitality with the indexical function of the fingers. Peters's account is the most ambitious attempt to take seriously the reference to the fingers resonant in the etymology of the term *digital,* and thus to locate a source of the digital not just within technology but also within people. For Peters, the digits do a number of things, including counting numerically and pointing to the real.[1] We will turn to counting in the next chapter, but for now Peters's argument that the digits point to the real suggests that the digital might be defined in part as the automation of a semiotic function, the keying of a referent to a certain kind of sign. "Digital media thus have meaning insofar as they index the world," Peters says.[2] His point is that the digitality of computers derives more fundamentally from the indexical reference itself, a kind of paleodigitality associated with the fingers.

Philosophers have long debated the logical and semiotic nature of indication.[3] An *indexical* is a sign that points to one of several objects in a context-dependent way. For instance, you and I both refer to ourselves with the pronoun *I,* yet we understand that the same pronoun is indexed to different referents, depending on who voices it. In social groups, other pronouns sometimes require gestural support to be indexed correctly. A teacher might point to a student while saying *you* to denote the proper referent. The semiotics of indexicality derive in significant measure from Charles Sanders Peirce, who organized his system not around a binary pair of *signifier* and *signified,* as Ferdinand de Saussure would later do, but around a distinction between three different kinds of signs.[4] For Peirce, the first of these, a *symbol,* has a purely conventional connection to its referent. It is effectively arbitrary. In contrast, an *icon* shares some quality with its referent, such as visual resemblance. But an *index* "signifies its object solely by virtue of being really connected with it," Peirce says, as in causal relationships. Thus, smoke may be an index of a particular fire. By that logic, all symptoms are indexical, and so too are other noncausal indicators, such as the labels on a geometrical diagram, demonstrative and relative pronouns, and of course, "the pointing finger," which Peirce calls "the type of the class."[5]

Peters's insights are primarily semiotic, and Peirce's entirely so, but such semiotic functions have social meaning too. In some cultures the indexical gesture is innocuous, but in many others it is deeply offensive. In one, it must be performed with an index finger, in another with the middle finger, and in still another with two fingers or the whole hand. Peirce might have thought that the pointed finger was the "type of the class," but he imagined so only because people in his society had developed the custom of indicating that way. But different ways of indicating are significant, because all gestures acquire a surplus of social meaning that exceeds their strictly referential function. Vilem Flusser has suggested that a gesture is "a movement of the body or of a tool attached to the body, for which there is no satisfactory causal explanation."[6] He means that a gesture has some additional meaning or value beyond the strictly utilitarian purpose for which the action is performed. A gesture is socially dense, symbolically resonant, replete with values, and never reducible to purely semiotic or utilitarian functions. If I raise my arm and extend my middle finger with the back of my hand facing away from me, almost everyone I know will perceive that as an aggressive insult. Yet my youngest child is baffled by the fact that one of his classmates types with his extended middle fingers, and I have not been able to explain to his satisfaction why extending the middle finger in one context is a forbidden insult while the same configuration is meaningless while typing. He is right to be suspicious of my rigid distinctions, for Flusser's point is that technical interactions not typically categorized as gestures, such as photographing or telephoning, really do have gestural meaning too. To read technical interactions as fully gestural is, for Flusser, to recover that social meaning from the illusion of mechanical necessity.

That is fundamentally the goal here. The physical act of pressing an elevator button is so minimal, so brief, and so common that anyone could be forgiven for assuming that it has no real significance beyond its mechanical or semiotic function. And yet, if we regard that simple act not solely as a mechanical exigency but as the elaboration of a much older gesture of pointing, we can discern complications worth addressing. Sociologists, anthropologists, and behavioral psychologists have studied the indexical gesture intensively for almost a century, and it does seem to be the case that human infants in all cultures indicate gesturally. However, that is not to say that they indicate in the same way or that their gestures of indication have the same social meaning. For instance, in some cultures, pointing with the index finger is so taboo that children learn early to indicate with the bottom lip instead, and some indigenous cultures were baffled when Western anthropologists attempted to indicate people or things with what was, to them, the indecipherable oddity of a pointing finger.[7] Moreover, even in Western societies, there are significant differences in how people indicate and what they indicate for. Adam Kendon identified seven different kinds of pointing in Naples, Italy, some with the open hand, some with the index finger, and one with the thumb.[8] One gesture might be used to indicate an object, another a person; one to call attention to something, another to indicate an object while talking about it. Despite these cultural differences in social meaning, most research on the topic supposes that the capacity for indication is as innate as any symbolic activity could be, and many have noted that its onset in children coincides with the onset of spoken language.[9]

The fact that the indexical gesture can give offense in some societies attests to one important aspect of its social meaning. Peirce himself recognized that there is something a little imperious about indexicality. "The index asserts nothing," Peirce says. "It only says 'There!' It takes hold of our eyes, as it were, and forcibly directs them to a particular object, and there it stops."[10] This taking hold of the eyes, as Peirce describes it, disrupts the Western hierarchy of the senses, such that the organs of touch can command the organs of sight, rather than the other way around. Perhaps this challenge to ocularcentric norms accounts in some measure for the gesture's modest taboo. Not only does it take the eyes in hand, but it implies considerable authority for the person doing so. For that reason, pointing is most acceptable in contexts where power imbalances are obvious, such as a speaker gesturing from the rostrum or a professor calling on students in a class discussion. One of the most famous representations of the indexical gesture in Western cultures that regard it as mildly taboo appeared in an aggressive recruiting effort by the British military in World War I. In Alfred Leete's poster, the British secretary of state for war, Lord Kitchener, indicates the viewer with his raised finger above the slogan "Your country needs you!" (see Figure 3).[11] Lord Kitchener's pointed hand is in an orthodox Western form, palm down and fingers tucked beneath. With a martial glare, he sights down the finger as if it were a weapon. When deployed socially in that configuration and directed at another person, the gesture can connote a wide range of meanings most latent within Leete's design: "I see you"; "Come here"; "Stop what you're doing"; "I want you."

The indexical gesture of Lord Kitchener is thus an imperative summons, which is one of its two most common meanings: the first is the imperative ("Bring it here!"), and the second the declarative ("There it is!"). It is not entirely clear whether young children's pointing is fundamentally protoimperative or protodeclarative, whether they first develop the gesture as a way of demanding possession or as a way of sharing attention. The stakes of this basic question are high. A major debate in the study of pointing involves a fundamental disagreement about whether children develop pointing as an elaboration of reaching for and grasping objects (imperative: "Bring it here!") or as an attempt to communicate through what amounts to sign language (declarative: "There it is!"). These are the *prehensile* and the *communicative* theories, respectively, which correspond to imperative and declarative meanings of the gesture.

An early and influential proponent of the prehensile theory, Lev Vygotsky, claimed that "the indicatory gesture is simply an unsuccessful grasping gesture directed at an

FIGURE 3. Alfred Leete's illustration of Lord Kitchener for the cover of the *London Opinion*, September 5, 1914. Courtesy of the Library of Congress, Prints and Photographs Division, LC-DIG-ppmsca-37468.

object and designated a forthcoming action. The child tries to grasp an object that is too far away."[12] According to Vygotsky, as children better learn to assess their capacities over time, they come to understand that they can more economically attain a distant object by pointing to it, which recruits their mothers' attention and aid.[13] In contrast, the communicative theory holds that pointing gestures are not failed attempts at touching or grasping at all, but from the start direct the attention of others to an object of shared regard. Rather than trying and failing to grasp the object, such gestures orient everyone's attention to something that remains at a distance.[14] A child who points will often check to see whether the mother is watching, and before long will check to see if she is watching before even beginning to point.[15] Because most researchers now seem to agree that pointing is closely related to the onset of language, many have concluded that finger pointing is fundamentally and not just derivatively communicative.

The communicative theory of pointing has mostly triumphed over the prehensile theory today, but that consensus should not imply that the two different functions of pointing are mutually exclusive or that they are utterly unrelated. George Butterworth argues that pointing may not be an elaboration of prehensile gestures at all, but rather antithetically related to them. In Charles Darwin's work on the expression of emotions in people and animals, Darwin identifies the *antithesis* as an opposite and corresponding gesture to some emotional expression. Darwin suggests that, in dogs, the low posture of friendliness or submission is the antithesis of upright aggression. Similarly, he says, in human cultures that signal indignation or attack by "squaring the elbows and clenching the fists," the antithesis is shrugging the shoulders with palms held out and up, signaling "a helpless or apologetic frame of mind."[16] Butterworth suggests that pointing may be antithetical to a particular kind of grasping, the two-finger pincer grip that infants develop by about eight months of age to pick up small things. It is a failed grasp, or the gesture of a finger that cannot or does not want to grasp something for itself. As such, according to Butterworth, pointing does not develop as an elaboration of an actual grasping gesture, as Vygotsky argued, but as its antithesis, which communicates precisely the opposite of the original gesture.[17]

Butterworth's argument helpfully keeps the prehensile and the communicative in meaningful relation. There might be other forms of relation between them still. Indeed, much of the research on the indexical gesture from Vygotsky to Butterworth confirms that pointing and possessing are, at the least, hard to separate, which may be because they remain so entangled socially thereafter. Even if pointing originates purely from either communication or possession, it rapidly acquires the ambiguous property of being both communicative and possessive at the same time. Whatever it arose from—if indeed it developed from only one source—the indexical gesture developed in Western societies as a sign with two referents, one of which means "There it is!" and the other of which means "Bring it here!"

The difference between a prehensile and a communicative conception of pointing is also the difference between a solitary and social understanding of pointing. All pointing is communicative in some way, for even Vygotsky was arguing only that it developed socially from what was initially an individual act of possession. Pointing is thus a way of making reaching and grasping social, of converting the prehensile to the communicative, where it operates socially through the assistance of others. But are there other ways in which the reverse might happen, in which the communicative is converted back into the prehensile? The primary instruments for accomplishing this conversion of the communicative back into the prehensile are certain kinds of switches and buttons.

Mothers of Invention

Many kinds of interfaces have practically individualizing effects. Bruno Latour argues that the residual sociality of some technologies amounts to the displaced labor of another person. In his famous example of the *groom,* the French name for an automated, pneumatic door closing device, a machine stands in for the porter it has now replaced. Latour's point is that the social does not disappear just because its function has been delegated to the nonhuman. On the contrary, door users continue to have a properly social interaction with it, for it preserves and enforces values, including such simple values as "It is good to close the door."[18] Still, most users are oblivious to this residual sociality unless they have been reading deeply in science and technology studies. By replacing a human *groom* with a mechanical one, the technical system creates the flattering impression that door users are acting entirely on their own, as individual agents rather than social participants. Even though the mechanical *groom* might be better for all sorts of utilitarian ends, such as reliability and regularity, as Latour suggests, it might be best for one conspicuously political reason, which is that it systematically conceals sociality by technical means.

In purely practical terms, elevator buttons similarly conceal the displaced labor of what used to be a human operator. Until the first decade of the twentieth century, all elevators had human operators, and in public and commercial buildings human operators persisted even longer. Riders prized operators who could drive the car quickly and smoothly without overshooting or undershooting the desired floor. Before long, automated systems could achieve better results more reliably and with lower costs. At the most practical level, these systems automate the value that "It is good to travel quickly and smoothly." But this is only their most superficial social function. They also promote more overtly political values of Western individualism, such as "It is good to do things for yourself," or even "It is bad to depend on others." Just as the human mother watches dutifully to see what her child might be pointing at, so too, behind the elevator panel, an electronic mother (or motherboard) checks and rechecks thousands of times per second to see whether anyone is commanding her attention. What she lacks in versatility she more than makes up in vigilance, for though she attends only to those who point at certain spots designed to detect that gesture (the buttons themselves), she never wavers or sleeps, and almost never fails to deliver the desired result. Indeed, she is so quiet and reliable that it is little wonder that the people who depend on her begin to imagine that they have accomplished these actions by themselves. They know that there are intermediate layers of control, but because they physically press the button, they can easily imagine that they are producing the result themselves, rather than communicating a desire to another. That is, they forget that this is a communicative operation and imagine that it is a prehensile one.

Let us think a little more from the infant's perspective about the kind of communication pointing entails. When children point to an object across a room, they are not typically attempting to grasp it physically. Their gesture is communicative, but it communicates in one of two modes, either the declarative ("There it is!") or the imperative ("Bring it here!"). Children commonly use both modes, often conflate them, and leave adults to sort out which one they mean. When a child appears to summon a toy from across a room with the indexical gesture, it may not seem quite as communicative to the child as it does to the adult observer. Researchers observe that the child often checks to see whether the mother is attending before or after pointing, but the question is to what extent the child experiences the mother as a separate object. The child knows that the presence of the mother is required, but at the same time, the mother of a one-year-old child is still gradually detaching from the child's subjectivity. Unlike a newborn baby, the one-year-old is learning that the mother's breast is not part of its own body, but it still perceives the mother as a partial extension of self. Invested with this egotism, the child surely does not perceive a social relation in the same way the mother does, but merely understands that pointing at a thing in the presence of the mother will bring it to hand. In that way, although pointing is communicative, it might also register as prehensile to the child, for to point at certain things really is to summon them from a distance. Through the indexical gesture, the incapacities of the infant translate into colossal strength.

The residual presence of the indexical gesture in technical interfaces approximates something like this infantile egotism in which the communicative turns into the prehensile. Even for adults presumably possessed of more stably bounded subjectivity, elevator buttons foster the illusion that pointing really is a form grasping. The infant's insensibility about the mother's separateness has, as its technical equivalent, the adult's minimal consciousness of the mediation of machines. And these technical systems accomplish this not just by automating the provision of satisfaction, but also by recruiting one of the earliest and most powerful childhood gestures associated with the child's expansive ego. The imperiousness of the pointing child is not so different from that of Lord Kitchener, for both command those around them with more than common authority. It might be better to say that Lord Kitchener has recovered one of the core gestures of infantile egotism, and with it fostered the illusion that social interaction really is just individual possession after all.

It is not hard to imagine elevator interfaces that might go even further, such as cell phone apps that control the elevator, and allow each passenger to select a floor from anywhere in the car, no matter how crowded. After all, sometimes there are still outbreaks of sociability in crowded elevators when one person is appointed button pusher for all new entrants. Yet technology can and is eliminating this interpersonal dynamic too. The Otis elevator company provides an app called eCall that summons an elevator from a smartphone and allows users to preselect its destination. Its British marketing language refers to it as "your own personal lift button," as if a single control panel were too difficult to share.[19] Marketing photographs for the Otis eCall system show a woman with her hand configured in the familiar northern European and American position for pointing—index finger extended, others retracted, and thumb tucked beneath—but instead of looking up and reaching for the familiar elevator button she touches her iPhone screen instead. She does so alone, outside the elevator, and apparently before she even arrives at its doors. Her gesture is a more intensely private version of the gesture that early-twentieth-century elevator buttons had already organized, as they turned pointing into pushing, and with it, interaction into activation. The gesture of pushing the button is thus a demonstration of private mastery, an affirmation of individual power over other people and things, and thus a symbolic investment of the person making the gesture with extraordinary powers, even approaching those of a young child.

If pressing certain kinds of buttons really is an elaboration of the indexical gesture, and if it is more specifically a rehearsal of infantile egotism, that can explain two other curious elevator phenomena: children's intense desire to press elevator buttons, and adults' intense frustration when the same buttons do not work. There seems to be no serious study of either of these matters in sociology or developmental psychology, but I have not yet found a parent of young children who reports that their children are indifferent to elevator buttons. Perhaps pressing the elevator button is itself a powerful surrogacy for the infantile egotism that children must, at some level, be aware is waning day by day. As the world constantly disappoints children by failing to regard them as its center, the elevator can still summon a room at the touch of a finger. The very hiddenness of the operation reinforces the impression of magical power. And what's more, while the whole affair is not communicative in any evident way to children, it truly does seem to consist of commanding touch. It may be that elevator buttons are one of the earliest and most powerful instruments of modern liberal individualism, as they literalize infantile egotism in the form of new, technologically mediated illusions of autonomy and self-control.

But if children are delighted by the simple function of the elevator controls, many adults are exasperated by their failure, as noted briefly in a previous chapter. On those rare occasions when an elevator button fails to illuminate, as one particular button frequently does in my elevator at work, I feel an outsized frustration that does not at all correspond to inconvenience. Such strong feelings rarely characterize my more analog interactions with the material world, even though those are full of more and greater disappointments. When I wash dishes, for instance, I lose the brush, splash water on my clothes, topple glasses, scald or freeze my hands, and drop slippery glasses, but this haplessness is just part of doing the dishes, or so I have come to accept. Unless I break something, I hardly even notice it. Yet when I press the elevator button and it does not immediately illuminate, I jab at it angrily, have muttered curses, and have seen others pound it furiously with the heel of a fist. The moment the button illuminates, my fragile emotional equilibrium is restored. The mother in the machine is watching where I point once more, and she is off to fetch what I want. It is painful to admit but impossible to avoid the conclusion that my response does not just resemble the tantrum of a child, but truly is one, having been fostered by technical systems that preserve infantile egotism as a permanent norm.

Button rage signals far more than just the *unreadiness-to-hand* that Heidegger associated with nonfunctioning tools, which sends the tool reeling back to its position as a separate, obdurate object. That may happen too, but more fundamentally button rage confirms what all adults know but prefer to forget, which is that the indexical gesture they use to press the button is communicative after all, and that they are neither reaching out and grasping the elevator directly nor summoning it magically into their presence. Although the button's failure damages its human users' sense of agency, it also, and more importantly, damages their sense of self, especially in societies that prize values such as privacy, autonomy, self-reliance, and independence. When a button fails, the problem for adults is not just that a loving algorithmic mother who dotes on their every gesture has been revealed as indifferent, but that they are dependent on those mothers in the first place. The illusion of autonomy is dispelled. The ego contracts to its proper proportions. When the indexical gesture is revealed to be communicative, the person who makes it must concede that their actions are interactions after all.

We will return to the techniques of individualism in far more detail in the chapters in the last part, "Haptic Liberalism," but for the moment let me simply suggest that the persistence of the indexical gesture in some switched interfaces embodies the fullest range of the digital. If the digital really is inflected by indexicality, then surely it is replete with indexicality's contradictions too. For, although the indexical gesture often involves a pretense of mastery, it is also a confession of dependence. Although it connotes presence, it is occasioned by absence. Although it facilitates touching, it does so at visual range. When people point their index fingers at buttons, then, they regress to a digital infancy that is both dependent and despotic, both social and solitary. The very point of the digital may be to hold these contradictions in tense but stable relation.

CHAPTER 5

Counting on the Body

The elevator buttons discussed in the previous chapter have one feature that we ignored entirely: the fact that the buttons inside most elevators are keyed to numbers. Elevator buttons are just one among many different numerical buttons in the world today, including those on telephones, computer keyboards, calculators, ATMs, credit-card terminals, cash registers, microwave ovens, photocopiers, media remotes, automobile dashboards, and many other familiar devices. If the previous chapter focused on a certain kind of gestural digitality organized around the social act of indication, this chapter turns to a somewhat different kind of gestural digitality, the act of counting on and with the fingers.

As noted in the previous chapter, Benjamin Peters is one of the few who has expressly associated the digits with "symbolic counters" and attempted to sketch a line of development from finger counting to digital computing. But what does it mean to take the digital several next steps, not just from fingers to numbers, not just from numbers to computers, but from there to all the things that computers digitize, including music, photography, writing, and most other forms of media today? Given that forms of digital media largely operate microscopically, instantaneously, automatically, and within tightly sealed black boxes, they hardly touch the body at all, so one might well ask whether the fingers really are relevant to the digital anymore. In what follows, I argue that they are. The fingers still are in the digital, but more importantly, and in ways that people have rarely acknowledged and perhaps prefer to ignore, the digital has long been in the fingers.

This chapter addresses just one limited mode of embodied digitality, which is the relation between techniques of enumeration and calculation that involve the fingers, including recent techniques of mechanical calculation. Jabbing at numbers on a keypad with an extended finger is neither natural nor inevitable, for it derives from centuries of earlier techniques, and from the technologies they involved. Today, finger-based technologies of enumeration and calculation are hard to avoid. Their ubiquitous keypads are largely uniform, enshrined in international standards that govern their form and function across much of the planet.[1] Obviously there are other forms of digital media that involve the fingers but that are not expressly numerical, and other finger-based gestures, such as pointing, that might be conceived as digital for other reasons. For these purposes, however, a narrower scope must suffice, for the embodied digitality of counting and calculation has had especially extensive effects.

This chapter is thus part of the larger story of the body's adaptation to digital environments that Mark B. N. Hansen and others have already told, though typically in relation to much more recent media technologies.[2] But by tracing the emergence of the digital through earlier and simpler techniques, we can see that the fingers themselves were, at one time, "new media" too. The embodied digitality that this chapter examines is thus not only, and not even primarily, about modern digital computing, but rather about the ways in which the fingers became digital in specific historical contexts, mostly before the twentieth century. To think of these developments as techniques is to refuse to privilege either people or machines, and to attend instead to the mutual and reciprocal processes by which human practices and technical affordances have shaped and reshaped each other. Culture, as Bernhard Siegert has put it, is always "technologically constituted," inextricably bound up with the devices used for writing, plowing, moving, telling time, and just about everything else. People make technical devices to do those things, but the technical devices also make people, or rather organize the conduct of people in ways that then count toward the criteria of the human. As a result, according to Siegert, "humans *as such* do not exist independently of cultural techniques of hominization," just as "space *as such* does not exist independently of cultural techniques of spatial control."[3] That is to say, the category of the human is not natural or given, nor is it a matter of the theories people have about themselves, but rather gets generated through technical operations.[4] When those operations are mathematical, and when the techniques involved map numbers onto the extremities of the hand, the embodied digitality that results becomes a key piece of evidence for humanity. Just as tears connote an allegedly human capacity for feeling localized in the eyes, so embodied digitality connotes an allegedly human capacity for mathematical reason localized in the fingers.

A number of recent scholars have returned attention to the fingers and the machines they operate, including Caetlin Benson-Allott, Till Heilmann, Roger Moseley, and Rachel Plotnick. They generally share the sense that scholars across many fields have rather neglected the fingers and the devices operated by them, from remote controls to piano keys to push buttons, devices that operate powerfully but inconspicuously within most of the developed world today. Moseley has been especially interested in the theoretical ramifications of thinking about digitality in relation to the fingers and, with Peters, is among a small number of scholars who have made the digits central to the digital.[5] However, in pursuing that line of thinking, we must guard against the temptation to assume that the digital is somehow grounded in the naturalness of the human body, or even that the human fundamentally is digital. In fact, techniques of enumeration and calculation digitized human fingers differently and at different times, and in some places not at all. The irony at the center of this chapter is the simple fact that, although bodies are in no way naturally digital, as they became more digital over time they also became more human.

Continuous People

In chapter 3, we considered some of the early articulations of the terms *digital* and *analog,* primarily at the Macy Conferences on cybernetics in the late 1940s and early 1950s. We noted that one prevalent definition of the digital relates to those practices that transform continuities into discrete enumerable units, chopping up wholes into parts, or as Alexander Galloway puts it, "dividing one into two."[6] We can call this process of transforming the analog into the digital *discretion,* the process of making discrete units out of otherwise continuous wholes. Nothing is more routine than discretion, even when it is not functionally mathematical. People do it every time they see the trees rather than the forest. But modern digital culture is characterized by a marked intensification of discretion processes that have been increasingly automated, mediated by computer technologies, embedded in powerful institutions, and enumerated. Value turns into dollars, space into miles, time into minutes, and any supposedly unified essence into so many quantifiable local instances, often mediated by the forbidding language of institutional experts, or concealed within automated processes that operate below the threshold of human perception. Although there is more discretion than ever, much of it is getting harder to perceive.

This has led more than a few people to castigate the digital as inhumane or unnatural, especially when it is automated, miniaturized, encoded, and accelerated by modern computers.[7] Even those who are not nostalgic for vinyl records or film photography frequently see the analog as foundational and the digital as an imposition upon it. Brian Massumi has argued for the "superiority of the analog" and Bernard Stiegler's Marxist critique of digitality associates it with the most dehumanizing symptoms of capitalism, such as assembly-line work and commodification.[8] The assembly line chops up time, labor, and materials into discrete and alienating units in roughly the same way that computers turn continuously variable quantities into ones and zeroes. As we saw in chapter 3, John von Neumann made the point explicitly that the analog is more natural than the digital: "The digital procedure is usually a human artifact," he said.[9]

Although von Neumann also supplied some careful qualifiers, he was entirely clear that the analog equates to nature and the digital to culture.

However, a serious difficulty arose when some attempted to define the digital as natural too. Von Neumann was responding to a paper by neurophysiologist Ralph Gerard, who had attempted to discern digital operations at the level of human neural function. If basic neural operations are properly digital, as Gerard suggested, then the digital would seem no less natural than the analog. Gerard's claim proved contentious at the conference, but it must be acknowledged that there have been at least as many claims for the naturalness of some conceptions of the digital as there have been for the naturalness of the analog. Pythagoras and Democritus thought the world was pre-divided into discrete units already and that what appear to be continuities are in fact invisibly assembled from small parts. Modern atomic theory generally agrees. Had DNA been fully described in 1950, the Macy participants would have had an even more difficult example, one involving discrete biochemical units sequenced like code.

My point is not that either the digital or the analog is more natural than the other, and still less that both are natural, but that naturalizing claims have been made on behalf of both, and that all such claims should be regarded with healthy skepticism. As noted previously, the distinction between the digital and the analog is itself a digital demarcation of two opposed positions, binary in its structure, which suggests that both categories took shape squarely in the realm of culture. Rather than worrying about whether either the digital or the analog is more natural, it seems more useful to study how these categories emerged historically, how they ended up serving various social and political goals, and how, in the process, both came to be naturalized in the service of competing humanist values.

We should take stock of some of the forms of analog humanity that the digital allegedly disrupts. In discussing analog humanity at some length, I mean only to show how prevalent such ideas have become and how deeply ingrained they are in popular conceptions of what it means to be human. My point is not to affirm these claims, but to recognize analog humanity as a historical formation that is parallel to, rather than prior to, the emergence of the digital. Some of the most prevalent modes of analog humanity emerged from the legacy of Enlightenment humanism. For instance, there are analog assumptions behind the basic belief that, in spite of changing continuously through time and across space, a person remains the same person. This idea is absolutely essential to liberal political theory, despite its own complex philosophical heritage.[10] Without such assumptions, contracts could not work. When I sign a contract, I cannot usually object later that a different, noncontinuous *I* signed it instead, one who has no claim on a different and discontinuous *me* here or now. Still, there are exceptions. Evidence of coercion or mental illness, for instance, can indeed register as personal discontinuity, and in those cases, a judge might rule that I truly was "not myself" at the time I signed the contract.[11]

Another common version of analog humanity involves the prevalent metaphorizing of emotions as continuously flowing liquids, such as *surging* anger, *overflowing* love, *brimming* joy, or *melting* sympathy. There are many others. Metaphors of liquid feeling cast the body as a container of continuously variable flows and pressures, and they have precedents as ancient as Aristotle's tragic *katharsis* and as recent as Freud's hydraulic model of the psyche.[12] Crucially, emotional flows are not just metaphors for abstract inner life, but correlate with the body itself, specifically with its leaky fluids: not just surging with anger but "spitting mad"; not just brimming joy but actual tears; not just an undercurrent of fear but a cold sweat; not just overflowing love but the damp business of sex. Even those rare mechanical metaphors for the emotions connote the management of liquids within continuously variable analog systems. For instance, the sexual "turn on" implies the properties of an analog dial, which further suggests operating a valve to manage the graduated flow of gas, oil, or electrical current, rather than suddenly flipping a switch.

To the extent that other aspects of thought count as feeling, they also tend to dissolve into liquids. William James saw consciousness itself as a matter of feeling.[13] He even claimed logical operators like *if* and *and* are fundamentally affective. "We ought to say a feeling of *and,* a feeling of *if,* a feeling of *but,* and a feeling of *by,* quite as readily as we say a feeling of *blue* or a feeling of *cold.*" For James, consciousness that works like this is, in his famous formulation, a "stream of thought," a continuous liquid flow. And he specifically emphasized that it does not take the form of discrete units. "Consciousness . . . does not appear to itself chopped up in bits," James claims. "Such words as 'chain' or 'train' do not describe it fitly as it presents itself in the first instance. It is nothing jointed; it flows. A 'river' or a 'stream' are the metaphors by which it is most naturally described."[14] The "jointed" and the "chopped up in bits" are the markers of digital discretion, that which has been divided into units. To designate thought as a *stream* rather than a *train* is to naturalize human consciousness as analog against the implied threat of acculturated digital alternatives.

However, we should remember that these popular conceptions of analog humanity make sense only in relation to the burgeoning digitality of the modern era. To insist on a *stream* of consciousness, as James does, is to differentiate human thought from the competing technological models he also names, the *chain* and *train.* The analog human thus emerges at least partly as a response to the increasing pressures of the digital, which is why the digital and the analog should be seen as complementary developments. Moreover, even though digital and analog properties are always complexly blended in actual practice, the construction of both terms as binary opposites entails a marked acquiescence to the overall logic of digitality. It might even be said that analog humanity became conceivable as pure, natural, or essential only once a more all-encompassing digital model emerged as a possible alternative.

Embodied Digitality

If body fluids materialize feeling as liquid, and thus testify to the allegedly analog nature of humanity, other parts of the body naturalize digital humanity instead, particularly the fingers. At a distance from the wet and pumping heart, the fingers indicate both a competing and complementary digital humanity no less intimately associated with the body. For Peters, as we have seen, the digits manage the world through symbolic gestures such as counting and pointing, gestures that are the prototypes of all symbolic reference, including those increasingly automated by computers. "Digital media do what fingers do," Peters concludes.[15] However, in presuming so, we must guard against the temptation to naturalize these techniques simply by sourcing them from the body. Seeing the fingers as a source of digitality, rather than as another symptom of digitality, can lead to a familiar impasse. As with shedding tears, so too with counting fingers: the embodiment of any activity can make it seem to proceed straight from nature, never mind that people both cry and count in drastically different ways, for different reasons, and with different meanings across cultures. So, although I agree that the digital is an embodied condition that long predates electronic binary computation, I feel more reservations about claiming, with Peters, that "the human species has always already been born digital."[16] It might be better to say that it has *sometimes* been born digital, for the transformation was historical. Rather than a projection of human bodies into mediating machines, the digital is better understood as a more sweeping reconfiguration of both bodies and machines, a reconfiguration that often amounts to a conflation.

Some developmental psychologists and anthropologists considered finger counting a universal human activity, until (as with so many supposed human universals) they began looking more carefully at just how people do it.[17] It turns out that entire cultures have no finger counting methods whatsoever, such as the Pirahã people of the Amazon.[18] Others count on the body but not with—or not only with—the fingers, such as those in New Guinea who count joints up the arm to the shoulders, then across facial features such as ears and eyes.[19] Even where finger counting prevails, it takes highly variable forms. In much of Europe, simple one-handed finger counting starts with a closed hand and raises the thumb for 1, then proceeds across the hand to a fully open hand for 5. In China and the United States, in contrast, the index finger typically signifies the number 1 and the counting then proceeds to the pinkie for 4 and culminates in raising the thumb for 5. Still others differentiate between gestures for counting privately to oneself and gestures used for displaying a count to others.[20] Even I would raise my thumb for 1 if counting to myself but if communicating that count by gesture to others would raise my index finger instead. Other more complexly symbolic systems of counting on fingers such as the one depicted in Figure 4 date back to the Roman Empire, flourished in Europe through the Middle Ages, and survive today in different forms in the last open outcry trading pits of some American financial exchanges.[21]

One might object that, even though finger counting is culturally variable, still there is something essential or foundational about the human brain's need to employ it in techniques of embodied cognition, even if only as a matter of utility. Yet some recent studies have suggested not only that finger counting is not necessary, but that young children can count and calculate better without the fingers, a sign that such techniques might persist in culture for reasons other than utility, or for utilities other than mathematical efficiency.[22] Aristotle called the human hand the "tool of tools," but all tools are products of culture rather than nature, even when they are made of flesh and blood.[23] We must be especially careful, then, to avoid presuming that, just because finger counting is common, it is also natural or universal. In fact, finger counting is just the most rudimentary digital technique, culturally variable, and by no means guaranteed to spread everywhere or persist forever. Our task should be to see how that technique has both persisted and changed over time, especially once it began to involve more complexly mediating technologies.

The fingers became digital as cultures mapped numerical values and mathematical operations onto the body, thereby enrolling them in techniques of embodied mathematical cognition. They became a form of embodied digitality, a haptic fact, incorporated into the extremities of the hand, and could then stand in for all that is multiple, enumerable, and discrete. In the process, the human became digital as well, as certain parts of the body came to represent not the singularity of the continuous self, but the multiplicity of so many discrete units. The fingers became digital markers of mathematical reason in roughly the same way that tears and other body fluids became analog markers of feeling. Both materialized certain functions that then registered as naturally and nonnegotiably human, though in drastically different ways. Accordingly, the individuality of the analog human has, as its less acknowledged complement, the *dividuality* of the digital human, signified by the division of the fingers of the hands and their involvement in numerical techniques. The digits are that part of the body where one divides into many, where the unity of the self is equally and simultaneously comprehensible as multiplicity. At the tips of the fingers, people are made out of math. Or rather, they were made into math, as they gradually developed techniques of digital

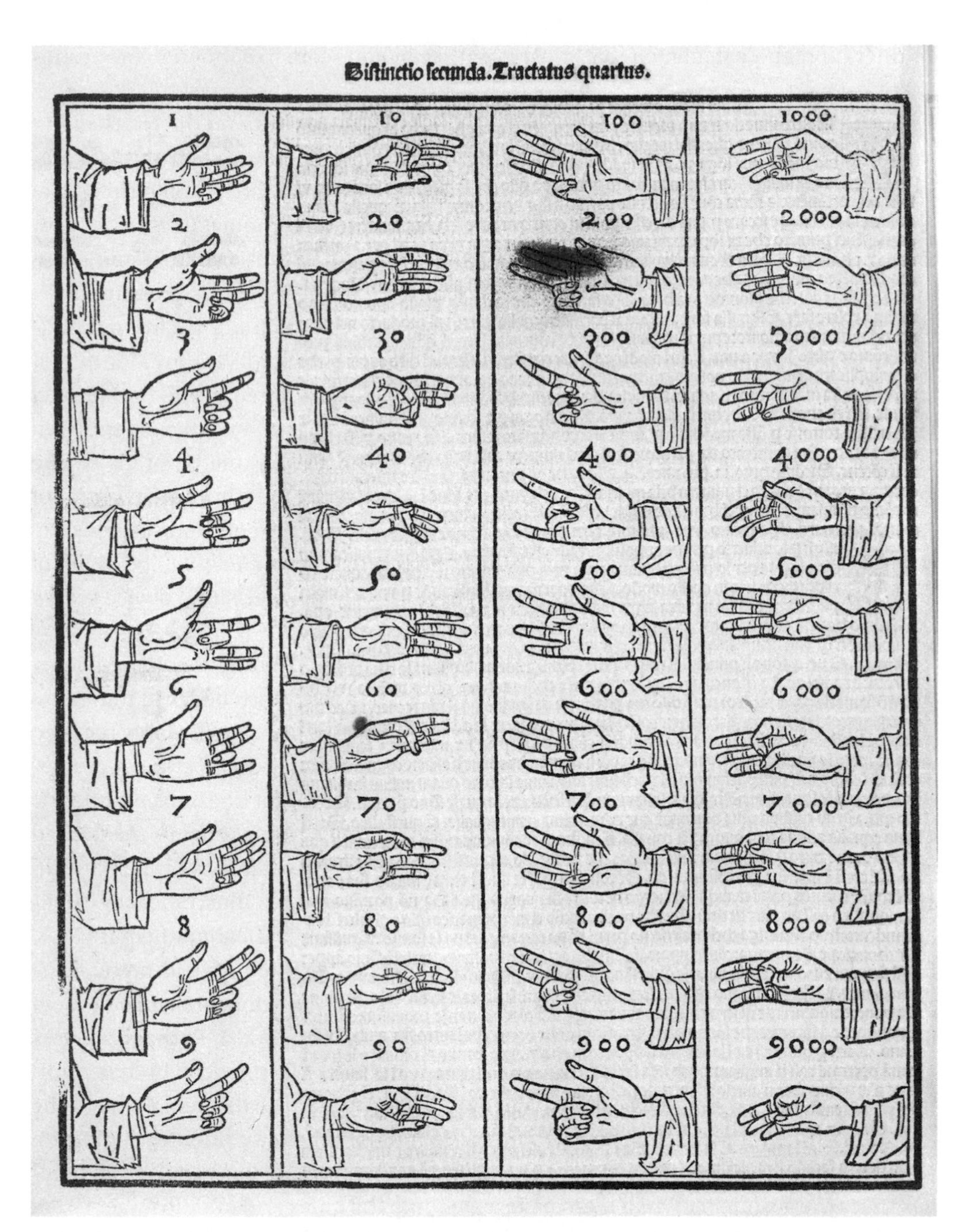

FIGURE 4. An early illustration of finger counting employing symbolic gestures that involve both hands from Luca Pacioli's 1494 *Summa de Arithmetica*. Courtesy of the Library of Congress, Rare Book and Special Collections Division.

enumeration and calculation that subsequently came to seem like inherent properties of themselves.

Such claims may seem counterintuitive in light of recent work on the irrationality of bodies, haptics, and touch. Mark Paterson has argued that, over centuries, geometry forgot the body, abstracting mathematical reason into primarily visual forms while repressing "the intuitive epistemology of grasping and tactile contact."[24] By this view, touching is primitive or nonmodern, and in many Western cultures, mature adults are expected to dispense with childish finger counting.[25] Yet, despite this pressure away from the haptic and toward the ocular and the abstract, which are traditional hallmarks of reason, many people still kept calculation close at hand. As Karl Menninger has shown in detail, many cultures developed technologies for counting that proved far more powerful than the fingers alone, such as the classical and medieval European counting tables that used small stones, chips, or disks of ivory, metal, or glass. The Romans referred to such counters as *calculi* (little pebbles), the diminutive plural of *calx,* Latin for *limestone.* The term *calculate* derives from these counters, as does the study of continuous change involving infinitesimal differences, *calculus.* The name of advanced Western mathematics thus derives from this simple mediating device scaled to and operated by the fingers.[26]

However, even as various counters detached from the body, functionally replacing the fingers, the fingers remained digital in Western cultures, both because they were still continually employed for simpler mathematics and also because the body recalled this earlier installation of the digital. Paterson has called for a *re-membering* of geometry that would concretize abstract math by restoring long-forgotten limbs and haptic experience, but my point is that, in other contexts, numbers were being re-membered already, as they took up different forms of residence in the extremities of the hands.[27] To take an early and telling example, in one sixteenth-century illustration of calculating techniques reproduced as Figure 5, the two greatest classical mathematicians, Pythagoras and Boethius, appear with the female personification of Arithmetic, each performing calculations with methods he was commonly, though mistakenly, thought to have invented. Boethius calculates with written numbers, while Pythagoras uses a medieval European counting table. In contrast to Lady Arithmetic, who has numbers cascading down her flowing analog gown, the men manage discrete digits in the form of written numerals on one side and *calculi* on the other. (Arithmetic might be a woman, but men alone seem to have the agency to manipulate her.) Especially noteworthy is the gesture that both men make, extending the index and middle fingers from an otherwise closed hand. Pythagoras might need to extend his fingers to move his *calculi,* but for Boethius the gesture can be only symbolic, for it plays no role in the act of writing, nor is he grasping a pen. In fact, their identically pointed fingers seem to be

FIGURE 5. Illustration of Boethius and Pythagoras calculating with written figures and a counting table, respectively, and accompanied by the figure of Arithmetica. From a 1508 edition of Gregor Reisch's *Margarita Philosophica,* originally published in 1503. Courtesy of the Library of Congress, Rare Book and Special Collections Division.

indexical gestures, as they point with the fingers to their new modes of calculation. But this unnecessary pointing perhaps confirms Peters's supposition that the digital subtly combines finger gestures of both indication and enumeration.[28]

My point is that digitality continues to reside in the fingers even after new techniques have replaced the fingers with other kinds of counters. Cultural techniques endure in the fixed forms of machines, which require people to adapt to them generation after generation, but techniques also endure in the body itself, in habits and conventions, practices and methods, and these can prove more durable still. A limitation of some recent work on cultural techniques is that the field's trenchant antihumanism sometimes borders on technological determinism, a charge its adherents vigorously protest but one that some invite by giving more attention to machinery than to people. In any event, anyone fleeing retrograde anthropocentrism always risks running aground on the shoals of an equally backward determinism. But to take seriously the full meaning of *technique,* which combines technical forms with human practices, we need to leave a place for the acts, desires, and habits of people too, and for the ways in which people preserve and transmit culture with a minimum of technical mediation. Sometimes the body alone is medium enough. Once embodied, the digitality of Boethius's pointlessly pointed finger could and did persist. And it persisted for much longer than this. In fact, these early forms of embodied digitality persisted even as new kinds of mechanical calculators threatened to remove human digits entirely from the equation.

Alternatives to the embodied digitality of the finger are evident in the earliest automated calculating machines, including Wilhelm Schickard's Rechenuhr (1623), Blaise Pascal's Pascaline (1642), and Gottfried Wilhelm Leibniz's Stepped Reckoner (1672), to name only three of the most well-known. All employed dials that work on a rotational principle and that required more complex modes of manipulation, such as grasping and twisting with multiple fingers. Pascal's machine even required users to employ a stylus to operate the dials, which functioned a little like the dials on mid-twentieth-century rotary telephones. Grasped like a pen, the stylus substitutes for the single finger, which now plays no role on its own. If the stylus preserves something like the indexical gesture in technical form, it also compels the body to develop new and more complex manual gestures that bear little resemblance to those of Pythagoras and Boethius in Figure 5.

Most of the work on calculating machines has focused on their inner function, such as the impressive conjunction of mathematics and engineering in Charles Babbage's analytical and difference engines. But historians have paid much less attention to the interfaces of simpler and more widely used calculating machines, and to the techniques they required. Given that many societies are saturated by key-based calculators and other numerical keypads today, it might be surprising to learn that calculating machines did not employ buttons or keys in any successful machine for roughly two centuries after Pascal's first device. The first calculating machine in widespread use employed slides to input numbers, which the user grasped between two fingers and carefully positioned next to the desired digit. First patented and demonstrated in France in 1820 by Charles Xavier Thomas de Colmar, the Arithmometre was a stepped reckoner that resembled Leibniz's design and was capable of all four arithmetical operations. It was a two-stage device: the user first entered a value and then performed the desired calculation with a second operation, twisting a small crank. Thomas's Arithmometre saw little use from the 1820s to the 1840s, but around 1848 he returned to work on it in earnest and filed several new patents, at which point he began manufacturing his Arithmometres in larger numbers and selling them to government bureaus and businesses such as life insurance companies.[29] This was mass production, but it was a modest mass to be sure. Until the late 1870s, Thomas never sold more than one hundred per year, even though it remained in production until World War I.[30]

Similar but more sophisticated machines emerged later in the nineteenth century and enjoyed far more commercial success. In the 1870s, the Swedish engineer Willgodt Theophil Odhner developed a sophisticated pinwheel calculator, which he marketed successfully in the next decade.[31] His earliest versions used knobs to input numbers, while later ones used sliding tabs on a curved metal housing. Clones of the Odhner Arithmometer proliferated in England, Sweden, Germany and Japan, as well as in the Soviet Union well into the 1970s. In much of Europe, then, calculation looked like slides and cranks, not buttons and keys, well into the age of computers. Just as people can count on parts of the body other than the fingers, so too they might employ calculating interfaces that slide and rotate. The fact that this is virtually unfathomable to most people born after the decline of the logarithmic slide rule only confirms how closely aligned numeracy now is to button pushing. The keypads most people take for granted today are not natural, just naturalized—the products of a different historical process that finally did realign embodied digitality with calculating techniques.

Remembering the Digital

When the first key-based machines appeared, there seems to have been little demand for anything faster than Thomas's Arithmometre, and certainly there was no demand for machines that were more complex and expensive to make. When English users suggested improvements to Thomas's device, they wanted more reliability and quieter operation, not greater speed.[32] Moreover, key-based calculators proved far more complex and difficult to design and build than any earlier devices. Their interfaces simplified human action considerably, but only by massively complicating the machine. As a result, although designs for the earliest key-based machines date to the 1820s, for decades the vast majority were extremely crude, and almost all were unreliable.

So daunting was the project of creating a workable key-based calculator, and so disheartening were the results over decades of attempts, that one wonders why so many people stuck with it for so long. After all, these many inventors were attempting something at first hard to fathom. They wanted to make machines that were substantially less capable than ones already widely available, for purposes that were usually obscure, in contexts where no demand existed, and with results that were decidedly discouraging. The only real advantage these machines could claim was some possible improvement in the human experience of using them. The overriding goal appears not to have been to calculate better, faster, or more accurately, then, but to calculate in ways that felt more intuitive or comfortable to people, apparently without much thought to overall efficiency. It is hard to find evidence that bears directly on the matter, but it seems plausible that so many inventors were so devoted to building key-based machines because forms of embodied digitality persisted through that long period when earlier calculators dispensed with anything that looked like finger counting. Put differently, bodies remembered the digital, for they both recalled an earlier stage of finger-based numeracy and then re-embodied it through new techniques keyed to different kinds of machines.

There seem to have been only three key-based machines before 1850, by James White in England, Jean-Baptiste Schwilgué in France, and Luigi Torchi in Italy.[33] White published the first description of a key-based adding machine in 1822, which may be the first proposal for any kind of key-based device other than a musical instrument. White wrote, "This machine is not, generally, an arithmetical Machine. It points lower: and therefore promises more general utility."[34] It was, he claimed, "a Machine fit for the daily operations of the counting house; by which to favour the thinking faculty, by easing it of this ungrateful and uncertain labour."[35] As White makes clear, the goal was not improved calculation, but an improved experience of calculating for the human half of the assemblage.

By the 1850s, key-based inputs had become increasingly common in patents, especially in the United States, which would dominate the industry of mechanical computation in the twentieth century. Apparently working independently, and sometimes in comparative isolation, otherwise obscure inventors, including DuBois Parmalee, Thomas Hill, Leonard Nutz, and Caroline Winter, patented key-based calculators. Most of these devices were simple adding machines, many with a similar design in which the user depressed a key on a tall stem downward toward the case or along its front side. As the stem moved down, it rotated geared wheels within. The dial that used to be at the interface now lay concealed inside the black box, operated at one further remove by spring-loaded keys. Because keys for different numbers had to rotate the dial different distances, they typically stood on shorter and taller stems. The 8 key had to travel twice as far as the 4 key, and four times as far as the 2 key, so the lengths of the stems were proportioned accordingly, a feature that survived even in much later commercial machines such as the one pictured in Figure 6.

The gesture required to operate those keys was supposed to be simple and even effortless, as White suggested, but two enormous technical challenges intruded from the start. First, when pressing a key to enter a number, the force applied could easily impart extra momentum to the wheel within, causing it to spin past the desired position. The most successful solution to this problem employed a mechanical detent to lock the wheel at precisely the right instant, but as one early-twentieth-century inventor and historian of mechanical calculators noted, a proficient operator might make five hundred and fifty strokes per minute, or just over nine per second.[36] This estimate surely presumes the simultaneous entry permitted on later machines, but let us assume that earlier machines permitting only sequential entry would have had to function for brief bursts at a rate about as fast as that of a highly proficient typist, about ninety strokes per minute. At that rate, each stroke would occupy roughly two thirds of a second. But the detent had to work in only half of that time, about one third of a second, because the key acts on the mechanism only during half of its travel, either when going down or when coming back up (machines differed). Even worse, the wheel for the 9 key had to be stopped at precisely the right instant between the wheel's rotation past the ninth point, but before extra momentum carried it past the tenth point. The detent mechanism thus would not have the luxury of an entire one third of a second to operate (half of a keystroke), but had to operate in one ninth of one third of a second, or less than .04 seconds. This is rough reckoning, but close enough to illustrate the challenge designers faced: in such a brief instant, the wheel had to complete its rotation and some mechanism had to arrest it from rotating even one tooth further.

The second problem was related. Many inventors sought machines in which a single key stroke both entered the digit and performed the calculation all in one stage. In these single-stage machines, to touch really was to calculate all in one gesture, a marvelous accomplishment, for no other manual operation intervened. Even modern calculators are two-stage devices, with an "equals sign" button where the crank on Thomas's Arithmometre used to be. However, single-stage operation required considerable energy to carry values from one place to the next, and those additional wheels also had to be prevented from overrotating too. Some early and successful machines, such as the Felt and Tarrant Comptometers of the 1890s, pictured in Figure 6, recruited that energy from springs wound during the pressing of keys, but those springs had to be firm if they were to act within .04 seconds. As a result, depending on the numbers calculated and the number of place values employed, it could take up to 5.4 lbs. of force on a single key

FIGURE 6. The Felt and Tarrant Manufacturing Company's Comptometer, one of the first and most successful mass-produced key-based calculators. This early model was produced around 1900. Copyright by Science Museum Group.

to advance the mechanism and compress the springs needed for carrying values and activating the detents, a rather far cry from the fantasy of effortless operation.[37]

So the bold and arduous project of key-based calculation led to these two related problems, first that the fingers moved so quickly that the machines could scarcely keep up, and second, that solving the first problem required every ounce of force a finger could manage. No wonder dials and slides predominated for decades, slow and cumbersome as they were. And yet, a widespread desire for some form of push-button interface justified the enormous investment necessary to overcome obstacles like these. And once those obstacles were overcome, machines like the Comptometer and the similar Burroughs Arithmometer rapidly took over the market, especially in the United States, where they first appeared. Within just decades, hundreds of thousands of

Position of Hands for Division.

In dividing keep the eyes on the register, not on the keys. There is no occasion to look at the keys after the fingers are once placed. After once understanding division and obtaining a little practice, strike the keys in rapid succession. Where one hesitates it is because of habit acquired in learning, not because of any necessity of proceeding carefully.

FIGURE 7. Illustration of "Position of Hands for Division" from page 18 of an 1895 instruction manual titled *Methods of Operating the Comptometer.*

people had organized their mathematical activity around key-based techniques. In 1895 the Burroughs company sold 286 units; by 1909, it was selling almost sixteen thousand each year.[38] There were other competitors as well, including the slide-based Odhner machines still popular in Europe, but over time the trend was clear. In a span of about two hundred and fifty years, the primitive embodied digitality of finger counting first disappeared from mechanical calculating techniques and then returned again, having been reconstituted in different forms.

More might be said about the methods for operating machines like the Comptometer or Burroughs Arithmometer, and about the kinds of people who ended up using them, but for these purposes, the point is that these machines organized techniques of embodied digitality that persist to this day. Embodied digitality is not some timeless fact, but a continual, ongoing adaptation of people to machines and of machines to people. Figure 7 depicts the kind of digital humanity these machines have now made ubiquitous, a sort of *homo digitalis* whose rationality is most evident in the activity of the fingers. To be reasonable—which is in significant measure to be human for millions and probably billions of people today—is not only to think in certain ways, or even to complete certain tasks successfully, but to embody reason at the extremities of the hands, where fingers and numbers have become the same thing.

Let me emphasize that I do not see embodied digitality as either the foundation or the negation of the human, for it is just one especially prominent configuration of the human, durable enough to persist even through long periods of technical neglect. My point has been that, over centuries, digitality lodged so securely in the body that it persisted there even in the absence of direct interactions with complementary machines, for it became part and parcel of what counts as human. Indeed, the massive proliferation of digital interfaces today, from binary switches to keypads to touch screens, may be seen as technical accommodations of an embodied digitality that long predates them. After all, how long can most readers of this chapter go without pressing a numbered key on a phone, computer, ATM, or automobile? Not long. When people do those things, they manifest one kind of humanity, just as they manifest another with empathetic tears. Against those who see digitality primarily as alienating or even dehumanizing, I would suggest that, for roughly a century and a half, people have surrounded themselves with a digital infrastructure in order to render themselves all the more human.

CHAPTER 6

Darth Vader's Nipples

In Roger Vadim's 1968 film *Barbarella,* the title character played by Jane Fonda falls into the clutches of the villain Durand Durand, who proposes to execute her in his "Excessive Machine." The machine contains an inclined bed on which the victim lies, covered with undulating strips that run from shoulders to feet, and that move when the operator presses keys on an organ-like keyboard. The machine kills with pleasure, Durand tells Barbarella, and even within the parameters of its PG rating, it is clear enough that Barbarella faces death by orgasm. As Durand begins to press the keys, Barbarella remarks that it is rather pleasant, and soon after the machine begins to eject articles of her clothing from vents on its sides. As the intensity increases both she and Durand grow disheveled and sweaty, and Barbarella moans ecstatically. Instead of killing Barbarella, however, the machine itself combusts, overpowered by her sexual vigor as it meets its own climactic end. Durand, distraught, calls Barbarella a "shameful girl," though she appears to be mostly satisfied with the procedure.

Although the film plays the scene for laughs, this is sexual violence nonetheless, a machine-mediated rape and attempted murder, despite the campy pretense that Barbarella prevails. For these purposes, however, the scene reveals a different version of embodied digitality than we encountered in the previous two chapters, for it shows one person operating electrical switches or buttons and another person operated by them. The gendered power disparities could not be clearer. He stands while she lies; he is free while she is constrained; he operates the machine while the machine operates on her. And further, he derives a purely scopophilic pleasure from the mechanically mediated sexual operation of a woman's body, while she endures sexual stimulation that she can neither decline nor control. By the logic of this arrangement, men are the kinds of people who operate switches like those on Durand's keyboard, and women are the kinds of people who are operated by them.

This technical mediation of gender extends well beyond a single film. *Barbarella* is symptomatic of an entire culture of mechanized women, for while Durand is decorously separate from the machine, Barbarella is enclosed within it and subject to it.

Durand's command over the machine is parallel to his command over Barbarella, for the scene configures both as mere technologies for the satisfaction of human masters, which really means male masters. Even though Barbarella reasserts her humanity and autonomy by destroying the machine in the end, the entire scene at first rehearses a much more conventional power arrangement in which autonomous men operate mechanical or mechanically mediated women.

Similar power dynamics may be more familiar in analogous labor relations, such as the dehumanizing discourse of slaves, serfs, or any other kind of alienated laborers as mere machines.[1] As Despina Kakoudaki has recently shown, robotic slaves "embody ethnic and racial otherness despite their nonhumanity," which is really to say that their mechanical nonhumanity is the marker of their ethnic and racial otherness.[2] Moreover, the same mechanical qualities associated with slaves—"repetition, relentlessness, inevitability, or single minded focus"—frequently translate into the realm of gender as a fantasy of feminized sexual automation.[3] Thus does the slave's mechanical devotion to labor become the woman's mechanical devotion to sex, and their bodies end up similarly configured as instruments for satisfying others. Versions of this fantasy can be traced back to Pygmalion creating his erotically compliant Galatea, or even to the second-order creation of Eve for Adam's satisfaction. Many other eroticized *gyndroids* in this long tradition include E. T. A. Hoffmann's Olympia, Auguste Villiers de l'Isle-Adam's Hadaly, Fritz Lang's Maschinenmensch, Ira Levin's Stepford wives, the seductive cylon named simply Number 6 from the *Battlestar Galactica* reboot (2003–2009), and Barbarella herself. All of them, and many others are, in different ways, sex machines.

The sexual gendering of androids and robots raises complex questions, however. For, ever since Donna Haraway first argued for the feminist potential of the cyborg, it has not always been clear whether mechanical women are more of an antidote to or an instrument of patriarchy. Haraway argues that the cyborg disrupts the allegedly natural status of the human, a privileged condition claimed primarily by and for white men. By blurring the boundaries between people and machines, the cyborg also, and more importantly, challenges binarism itself, including gender binarism and its attendant hierarchy. Even so, it can be hard to separate the disruptive potential of the cyborg from the kinds of mechanical dehumanization that have been applied selectively to women, to nonwhites, and indeed to almost everyone except white men. Mechanization might even be a version of feminization, and thus not an escape from gender hierarchy but a doubling down on it. The more mechanical women appear to be, the more dutiful and servile they are presumed to be, which may be why almost all voice-activated digital assistants such as Apple's Siri or Amazon's Alexa have been voiced by default as women. As Susan Squier, Mary Ann Doane, and others have argued, the problem with taking the cyborg as an emancipatory alternative to traditional humanism is precisely that women have long been excessively mechanized, especially in relation to sex and reproduction.[4]

There is some truth to both positions. Mechanization can disrupt and affirm patriarchy depending on the circumstances, and much depends on the precise way in which people are mechanized in any given rehearsal. However, the mechanical people considered in the pages that follow (various dolls, automata, robots, androids, and cyborgs) proceed largely from those traditions in which mechanical people do not disrupt the naturalness of the human, but rather define a privileged version of it by way of opposition, as Barbarella initially does in relation to Durand. I will refer to such beings as "mechanical people" rather than as *androids* or *cyborgs* in order to indicate that their nonhumanity typically affirms rather than disrupts traditional humanism. The first part of what follows will trace the history of these mechanical people back to Enlightenment automata, which were frequently represented as women, children, nonwhites, and even human–animal hybrids such as fauns. Adult white men appear to have been less often simulated. Period theories placed heavy emphasis on the allegedly surplus mechanism of women's reproductive systems, which were thought to interfere with higher mental functions. In Cartesian terms, women consisted of too much mechanical body, which inhibited the reasoning mind. But this account of women's reproductive systems persists in a curious form that has become part of the Western iconography of mechanical people ever since: a bank of buttons or switches on the front of their chests roughly where nipples should be, and sometimes also lower on the torso near the womb. If a moment's reflection does not confirm the preponderance of this android form in Western culture over the last century, the many images reproduced in the pages that follow probably will. To render switches as nipples in film and television, on children's toys and in books, and on almost a century of working automata of various kinds is to preserve much earlier assumptions about the relationship between gender and mechanization. And not just gender, but race too, and indeed any form of humanity regarded as inferior to that of adult white men. In fact, when switches appear expressly as male nipples, as they often do, the mechanical person is usually rendered as a bare-chested savage, more brawn than brain, and accordingly, sometimes expressly marked as nonwhite. Already an embodiment of surplus mechanization, the nipples of automata, now transformed into switches, define a category of subhumans who wear their operable mechanical parts on their chests.

This chapter is not an analysis of practical machines like those encountered in the previous chapter, but a consideration of fantastic machines as represented in popular cinema, avant-garde art, toys, and mass market journalism and advertisements. As imagined technologies—as science fictions—these androids contain a denser concentration of ideas and values, unhindered by the demands of utility. They are purer projections of operative values, and thus sometimes more openly symbolic, and more candid too. The study of imagined technologies can reveal more about social and symbolic meaning than technologies in widespread practical use, especially when imagined technical forms get repeated across many different contexts. If any individual exhibit seems excessively idiosyncratic in what follows, or if the variations between exhibits seem too pronounced, I ask readers to focus not on any instance on its own, but on the mythological consistency across so many different examples.

Despite its basis in fantasy, this is a story of historical development too. The sources in the first two sections generally reflect traditional gender binarism and heterosexual norms. If anything, their mechanical people seem designed to affirm and preserve gender hierarchies, organized around the presumed inevitability of the heterosexual dyad. And yet, by the mid-twentieth century, a new generation of much more androgynous mechanical people conflates what were originally separate submasculine and hypermasculine figures (women and monsters) into a single ambivalent figure. This returns us to the mechanical person named in the title of this chapter, Darth Vader. Vader is identifiable not just by his black garb and wizard-like powers, but also by the bank of switches situated on his chest. More important, he is not just permitted but also required to operate those switches himself. As a result, he is a far more ambiguous and ambivalent figure than either Barbarella or Durand. Vader is both Barbarella and Durand at the same time, the one who operates the switches and the one operated by them. The switches on his chest confirm that he is simultaneously the one in control and the one being controlled, both the machine-mediated master and the machine-dominated slave. Vader is trapped in his own excessive machine, but he also plays its keys. By preserving the power disparities of Durand and Barbarella intact, and by impacting them into a single body, Darth Vader became a figure for everyone whose apparent humanity was already technologically derived.

Sex Machines

To peel back the layers of meaning behind that control panel on Vader's chest, we should begin a little closer to the beginning, for *Star Wars* viewers inherited centuries of representations of mechanical *others,* especially mechanical women. As Monica Casper, Anne Balsamo, Mary Ann Doane, Allison Muri, and others have shown, women's sexual and reproductive systems were especially prone to mechanical description.[5] As Muri argues, seventeenth and eighteenth-century mechanistic theories of the body may have applied equally to men and women, but the main question for men was a Cartesian one, the nature of the mind, whereas the main question for women was how the reproductive machinery interfered with what otherwise might have been equal mental capacities.[6] So it was not necessarily that women were more mechanical than men as a whole, but that their mechanical nature was somehow more pronounced, disruptive, and decisive.[7] As a result, according to Muri, we can trace a more or less direct line from seventeenth-century accounts of automatic wombs to contemporary accounts of the fetus as a kind of astronaut protected in the capsule of the mother's body.[8]

Seventeenth- and eighteenth-century mechanical automata thus propose different versions of mechanical personhood for men compared to everybody else. Among the most celebrated automata by Jacques de Vaucanson and Pierre Jaquet-Droz, there seems to be a preponderance of women, children, and animals. Vaucanson's famous musical automaton of 1737 could play a flute by blowing air through its mouth and operating the valves of an actual flute with its fingers. This remarkable android did not simulate flute playing, but played an actual flute, a remarkable feat that required soft, moveable lips and tongue.[9] Although Vaucanson's flutist appears to be a man, Jessica Riskin has shown that it is based on Antoine Coysevox's statue "Shepherd Playing a Flute," which depicts a faun, half man and half goat.[10] Similarly, Jaquet-Droz created three complex automata in 1774, one of which consisted of a young woman or girl playing a small organ, her chest rising and falling as she appears to breathe, and her body swaying slightly in time to the music. His two other automata were small children, one a writer and the other a draughtsman. Many other musical automata were also women, children, or nonwhite men. Jaquet-Droz made an even more sophisticated female harpsichord player ten years after his organist, and his collaborator Henri Maillardet produced several other women musicians.[11] Marie Antoinette owned a German-made female automoton that played a dulcimer.[12] Although Wolfgang von Kempelen's chess-playing automaton was eventually exposed as a fraud, it was nonetheless racialized as The Turk. Much later, Jean Roullet imitated Vaucanson's flute player, but recast the faun as an African man.[13] There are many other examples. Although there were some simulations of adult white men too, a large number of automata represent other kinds of people. Mechanical children, women, animals, mythological hybrids, and nonwhite people ensured that, however successful a mechanical simulation might be, it would not encroach too far into the category of the human.

There is another relevant feature of these automata also evident in the examples mentioned above, which is that they all operate other machines, instead of exclusively performing other recognizable gestures of the human, such as breathing or speaking. The androids by Vaucanson and Jaquet-Droz that were among the most celebrated in Europe are uniformly engaged with the operation of other devices, such as drawing implements or musical instruments. This must have raised fundamental question for thoughtful viewers of these "philosophical toys": is the realistic android who plays an organ more like the organ itself or more like its human creator?[14] On that question, Cartesians and materialists would probably disagree, but the operation of a machine by a machine seems to have amounted to a special claim on humanness beyond the verisimilitude of moving bodies. To operate a simpler machine was to affirm one's humanity by way of contrast. Even so, not even the most dogmatic materialist really thought that these machines were human, but because they conjured the illusion of personhood so successfully it was more conceivable that actual people might be just slightly more sophisticated devices. These Enlightenment androids thus maintain relations of similarity and difference in both directions. The android who plays the organ is like her human creator, even though she is just his mechanical slave. Conversely, she is also like her musical instrument, even though she also appears to be its master.

These eighteenth-century automata do not have the kinds of operable breasts that only became fixtures of androids in the twentieth century, but they do establish an association between mechanical people and those kinds of humans deemed to be creatures more of body than mind. The persistence of these ideas, and the sexual stakes, can be neatly demonstrated in an early twentieth-century artistic elaboration, Hans Bellmer's 1934 photography collection *Die Puppe* (The doll). *Die Puppe* consists of ten surrealistic photographs of a life-sized female doll with articulated joints, and seems to have been directly inspired by Jacques Offenbach's opera *The Tales of Hoffmann,* the first act of which dramatizes E. T. A. Hoffmann's 1817 short story "The Sandman."[15] In that story, a man falls in love with a beautiful but eerily dull android named Olympia, who drives him to madness. Significantly, "dumb, stiff Olympia" enraptures Nathanael at her first public exhibition when, like Jaquet-Droz's musician, she plays a keyboard instrument, a harpsichord. Nothing, it seems, makes one appear more alive than operating another machine. Bellmer recreates Olympia in a more overtly eroticized form, consisting of strangely displaced body parts that appear in his photographs in various stages of assembly and disassembly, dress and undress. Some of the photographs are conventionally aestheticized with flowers and lace, others are flatly documentary, but most are grotesque, even hideous.

The result is more than usually disturbing. Great jumbles of disassembled plaster flesh, mounded and rounded but also clearly mechanical, appear undressed as if for pleasure yet posed as if in pain. In his introduction, Bellmer calls this "the salt of deformation," a violent defamiliarizing conducted with "a hint of vengeance."[16] He alleged that his dolls amount to a decadent protest against the Nazi cult of the nuclear family and of healthy athletic vigor, but he is entirely clear that the doll also stands in for "young girls" in general, who stoke sexual desires that (he insists without a great deal of conviction) must be satisfied only through art.[17]

Bellmer is trafficking in familiar aestheticist doctrines of art for art's sake, by which the highest reality exists on the plane of the artificial. As health, virtue, decorum, and good taste came to seem complicit with banal naturalness, and even fascist dogma, only aestheticized violations of decency registered as authentic. In that way, Bellmer's art derives significantly from the French decadence of Charles Baudelaire's *Les Fleurs du Mal (The Flowers of Evil)* which wrenches the conventional aesthetic value of flowers into the idiom of "evil." Accordingly, Baudelaire's love is not like a red, red rose, but in one famous poem more like the colorful, maggot-infested carcass of a rotting horse. Only disgust can awaken aesthetic sensibilities dulled too long by conventionality. Defiance of sexual norms was an especially common rehearsal of decadence, perhaps most visible in the performance art that was the life of Oscar Wilde, but for Bellmer, such defiance is not homosexual but sadistic and pedophiliac. This results in the dubious elevation of artificial girls as sex objects, which renders them at once allegedly superior to the human because of their artifice but also drastically inferior to it because of their base materials.

Crucially, even the doll's thoughts are mechanical. Just visible in its exposed abdomen in Figure 8 are the rudiments of a mechanism for displaying images. We do not know exactly what images it contained, if it ever contained any, but according to Bellmer, they were to represent something like a sexual unconscious. He insisted that "one must not stop short of the interior, of stripping away coy girlish thoughts so that their foundations become visible, best of all through the navel, deep within the belly in the form of a panorama electrically illuminated by colored lights."[18] The six small panoramas he intended to include "were constructed out of small objects, materials and colored pictures distinguished by bad taste."[19] To the extent that this female figure has thoughts, then, her thoughts are sexual and mechanical all the way down. She thinks with her reproductive system. But further, even her own thoughts amount to nothing but a projection of male sexual fantasy, which she gestates for Bellmer in her mechanical womb.

In an illustration of the working doll included in *Die Puppe,* Bellmer (a superb draughtsman) illustrated a woman's torso with the side cut away to reveal the mechanical interior. This illustration in Figure 9 proves as illuminating as the doll itself. A large human eye peers into the navel where the scenes would be displayed. Above it, an oddly blackened hand, perhaps gloved and clearly garbed in a man's sleeve, extends a single finger to touch the nipple of the doll's left breast, which was to contain a button for swapping scenes. The woman is thus all torso, while the male operator is defined by

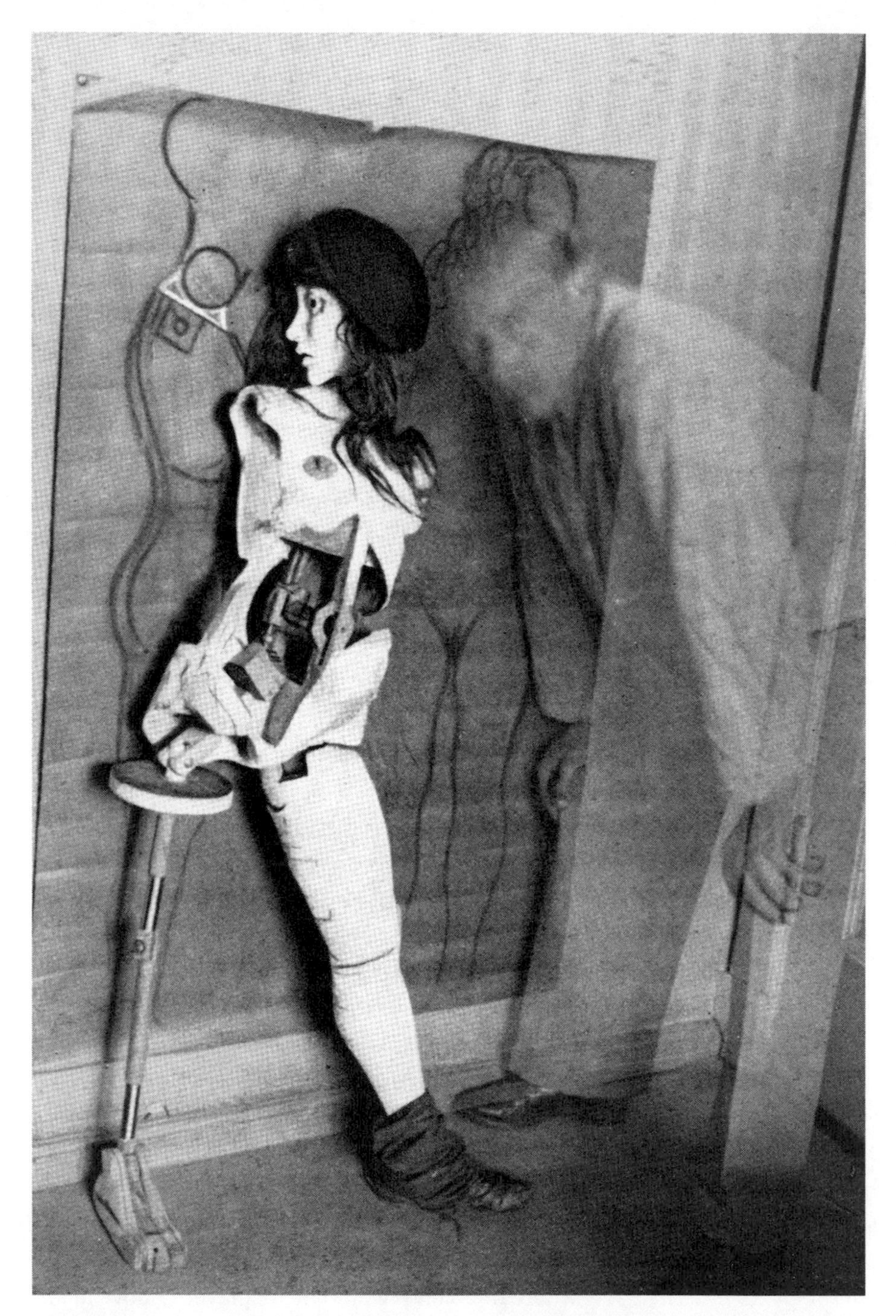

FIGURE 8. Photograph from Hans Bellmer's *Die Puppe,* included as Plate 3, depicting his partly constructed doll, including the mechanism concealed in the abdomen. The double exposure captures Bellmer himself standing next to his creation. Copyright 2023 by Artists Rights Society (ARS), New York, and ADAGP, Paris.

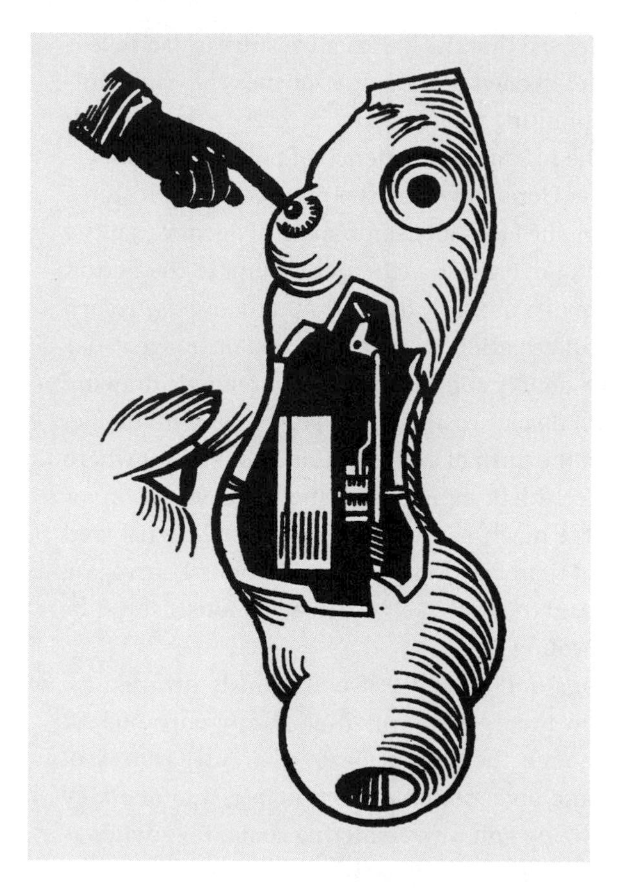

FIGURE 9. Illustration accompanying Hans Bellmer's preface to *Die Puppe.* Copyright 2023 by Artists Rights Society (ARS), New York, and ADAGP, Paris.

the things she lacks, extremities and the sensory organs of head and hands. It is hard to imagine a clearer allegory of men as discerning agents and women as passive sexual objects. Bellmer's doll allows us to see what was already implicit in several centuries of mechanical women, which is that the feminine and the mechanical were becoming interchangeable within the realm of male sexual fantasy. To lack the rationality, autonomy, and agency ascribed to adult men was thus to be a woman or a machine, for as Bellmer makes clear, they amount to the same thing.

Bellmer himself shows up in these two images in curious forms. In the photograph in Figure 8, he is a ghostly figure accomplished by double exposure, apparently more spirit than substance. This is man as mind or soul encountering woman as grossly material body. But when we return to his illustration of the doll from *Die Puppe,* we find something else: the representation of male agency in the form of the indexical gesture, in the disembodied hand pointing toward the button on the doll's breast. In chapter 4 (the first in this part on digital bodies), I argued that the indexical gesture is itself a legible symbol of power and authority, which I likened to the imperiousness of small children just learning to summon things by pointing their fingers. To press a button with a pointed finger, I argued, is to recover the infantile experience of pointing to objects that a compliant parent rushes to retrieve. Here, however, the pointed finger restores the phallic qualities that the doll lacks, for she lacks any semblance of agency that the phallus stands for symbolically. Moreover, not only does the hand point to the button on the doll, but it also does so in the form of a quaintly archaic marker, a *manicule,* a traditional graphical indicator used to call attention to parts of a text or image as far back as the Middle Ages. The hand is thus doubly commanding. As a realistic drawing of button pushing, it commands the visual display inside the torso. But, as a manicule, it commands the viewer's attention too in the form of an imperious order about where to look. The entire image is layered with fetishistic substitutes. The finger stands in for the absent phallus, but it reaches toward the machinery of the breast, itself a fetishized substitute for the unrepresented organs of reproduction.[20] The navel too is a surrogate vagina, an alternate route to the womb. And the doll itself is already a fetishistic substitute for the forbidden sexual objects of young girls.

The extravagant fetishism of Bellmer's doll seems self-consciously attuned to, perhaps even directly driven by, Freudian theory, which Bellmer likely encountered through other surrealist artists. Even though the factual basis of Freud's theory of sexual fetishism seems dubious at best, and objectionable on principle, it undeniably influenced generations of people, naturalizing and universalizing social inequality as psychological fact. The sexual fetish, according to Freud, symbolically restores an absent phallus to the woman's body. An entire generation of feminist critics have amply critiqued the patriarchal implications of this, but it can lead to an important question for these purposes: where is the phallus of the machine?[21] Is it not the case that these androids, both male and female, all appear to lack what was for Freud the totem of agency itself? They are missing something, so to speak, and against the dread of that possibility, according to Freud, men will project a substitute in the idiom of desire. The inferiority of mechanical people thus shows up not just as a surplus of body, but also as the absence of that one body part that Freud and those he influenced associated with agency and authority, with full personhood. After all, even mechanical people who are evidently gendered male rarely have anything between their legs, as generations of children exploring Barbie's boyfriend Ken have discovered. Nor are other familiar male robots equipped below the waist, including C3PO from the *Star Wars* films and every other male android considered in the rest of this chapter.

If machines and women had become figures for one another, perhaps even machines gendered male also received fetishistic substitutes for what they also allegedly lack. In other words, if technology has been feminized in precisely the same way that women have been mechanized, certain privileged locations on the machine might also acquire something like an extra erotic charge, a fetishistic substitute for the lacking phallus. In Freud's terms, these amount to projections from those people observing such mechanical eunuchs, people who might fear that a similar loss might befall them too—or worse, that it already has. Foremost among those locations, I am suggesting, is the apparatus of the binary switch, a substitute phallus that is precisely not a phallus, but a supplement for an alleged lack. The switch is where the action is, even though that action also has to come from without, from the pointed finger in Bellmer's drawing, or from Durand's fingers on the keys of the Excessive Machine.

The fetishism of switching is clear enough in the cases of Bellmer and Durand, but how much remains in less explicitly sexual contexts? Perhaps more than we think. Ali Na has recently argued that the modern computer mouse is a symbolic vulva, a receptive organ that affirms the user's phallic power by constituting interactions as masculine control.[22] For Na, the clicking finger is a pantomime of sexual agency. Perhaps it makes even more sense to see the mouse as more like the interface of Bellmer's doll, as breast rather than genitals, the displacement of sexual energy onto a prefetishized bodily equivalent. Bellmer's doll might even be considered the first draft of internet pornography: a technical system in which clicking buttons automates the display of erotic images. Given that about 20 percent of all Google XHTML mobile web searches are for pornography, making it the largest single topical category in a recent study, perhaps we should wonder whether the touch interfaces of modern computers are just as eroticized as the images displayed.[23] What if internet eroticism has as much to do with touching switches as with looking at screens, with haptic as with scopic pleasure, as Bellmer's drawing so strongly suggests? What if, that is, the pleasure of internet pornography involves not just looking at but also commanding the realm of images with the touch of a finger? There might even be some residual eroticism within switches no matter how they are being used, if, as with Durand, their operation entails a sexualized fantasy of mastery.

Tin Men and Metal Hearts

By the early 1930s, a number of functional androids were being exhibited as popular amusements and commercial promotions, almost always electrical and typically remote controlled. Some of these continued the Enlightenment tradition of high-fidelity simulations of human forms and movements, but a rather larger number abandoned any such pretense.[24] These *tin man* robots are named after the Tin Woodsman of Frank Baum's *The Wonderful Wizard of Oz,* who was born violently as he accidentally

FIGURE 10. Gene Autry presses a control button on the chest of a tin-man robot in a promotional photograph for *The Phantom Empire* (1935), a twelve-part serial science-fictional cowboy musical directed by Otto Brower and B. Reeves Eason. Courtesy of the Everett Collection.

amputated limb after limb with his axe, which a tinsmith replaced with metal prostheses. When he finally cut his torso in half, the smith restored his body but forgot to supply a heart. We might expect these tin men to exhibit this trope of the missing or automated heart, yet few actually do, even though many continue to place special emphasis on the breast and the lower abdomen.[25] For instance, before the Maschinenmensch in *Metropolis* turns into a convincing simulation of a human woman, her gender is defined almost exclusively by her metallic breasts. After 1930, dozens of remote-controlled tin men adopted a similar metallic form, though now gendered male, and instead of breasts begin to feature various kinds of control interfaces in the same location.

In her history of this shift to metal androids, Kakoudaki shows how the performance of what she calls *metalface* rapidly became more common after about 1920.[26] For instance, the 1921 and 1922 productions of Karel Capek's *R.U.R.* (the play that

FIGURE 11. An advertisement for a public exhibition of the Westinghouse Electric & Manufacturing Company's demonstration robot Katrina Van Televox, from the *St. Louis Dispatch,* June 23, 1930. The image shows the rack of controls on her chest, and the text affirms that "Katrina loves housework!"

coined the term *robot*) cast the robots as men in regular clothing, but by 1928 another staging covered them with simulated sheet metal. In the later production, the robots also had square panels on their chests, which Kakoudaki likens to the front grilles of automobiles, but which appear to me to be more like a hatch or access panel covering the inner works or hidden controls. If so, the *R.U.R.* robots would be well-aligned with a rapidly emerging style of tin men featuring operable breasts or windows into their abdomens. For instance, the body of the Westinghouse Electric & Manufacturing Company's 1927 robot Televox consisted of only the crudest two-dimensional human form sometimes cut from cardboard, but its torso contained a panel displaying bulbs, automatic relays, and on later models what appears to be a row of control switches near the top.[27] Similarly, Earl Kent's tin-man 1930 robot named Mr. Ohm Kilowatt has two switches of some sort higher on his chest and what appears to be a round gauge centered below them, roughly configured as nipples and navel.[28] There are many other examples: Arthur Wilson's 1936 Robie, with buttons high on its chest[29]; Yasutaro Mitsui's Steel Humanoid from around 1930, with a clunky pectoral control panel;[30] and Westinghouse's 1930 Katrina Van Televox (Figure 11), a gendered elaboration of the original and more androgynous Televox, now cast as a voice-controlled female robotic housekeeper displaying an exposed control panel for a torso, with switches located at the top. A period newspaper reported that she "does only what she is told."[31]

In some examples, such as Katrina van Televox, while the controls on the upper chest were not configured anatomically as nipples, the arrangement was still suggestive. On others, however, they are unmistakable, including Swiss engineer August Huber's Sabor models, which he produced as Sabor II in 1930, followed by Sabor IV in 1938, the latter of which is depicted in Figure 12.[32] Sabor IV was an imposing seven-foot-tall tin man, though more sophisticated than many others. It walked on wheeled feet and could talk, sing, and yodel, capabilities mostly managed by a person operating it remotely. But Sabor IV also displayed something that period commentators sometimes mention, the two prominent nipples on its torso. Each conical nipple was hinged, allowing a human operator to open it in order to adjust concealed controls. An assistant at Sabor IV's exhibitions during a tour of the United States later reported that they would open the nipple to adjust, or pretend to adjust, the controls beneath, which amused audiences apparently alert to some amount of sexual innuendo.[33] A curving line of vents descends below each nipple, which are ambiguously legible as ribs, abdominal muscles, or even the lower line of pectoral muscles or breasts.

Similar and somewhat later robots also displayed nipples, such as Jean Dussailly's 1954 remote-controlled Anatole, an improvement of an earlier version named Gustave from 1947 by a French engineer recorded only by the last name of Koralek, and depicted in Plate 2. Some depictions of Anatole also show a small trap door with a transparent window in the lower abdomen, through which people could view his inner machinery. Westinghouse's 1939 Electro also had a lower abdominal window through which inner workings could be seen. Like Mr. Ohm Kilowatt, whose nipples were controls and whose navel was a visual display, many of these androids reproduced with surprising regularity the basic configuration of Bellmer's doll, with switches on the breast and some sort of visual access to the interior on the lower abdomen. Crucially, none of these robots displayed mechanisms inside the head, through windows or transparent domes, as would become common in the 1960s, for at this stage, their electronic controls were associated with body rather than mind, and thus located in the torso.

Moreover, although these male androids were often minimally clothed (with the exception of their massive boots), they evidently lacked more than a heart. They lacked organs of reproduction too. Decorum alone is more than enough to explain this, but in any event, their masculinity had to derive from other performances of gender, such as their hulking size, or from publicity photos in which they sometimes pose in various seduction scenes with women models and actors.[34] But if these robots were legible as men, what does it mean that they displayed nipples so prominently? The disarmingly

FIGURE 12. Photograph of August Hubor's Sabor IV on the cover of the Swiss magazine *Zürcher Illustríerte*, December 9, 1938, one of many prominent depictions of this widely exhibited robot. The nipples on Sabor's chest open to allow some access to controls beneath.

simple answer is that the exposed nipples confirmed their masculinity, for if these robots were women, prevalent taboos would require their nipples to be concealed, as were those of the Maschinenmensch. Their nipples signaled a particular kind of masculinity akin to naked virility, the mark of "uncivilized" or "savage" people, rather than the other kind, who wear clothes. To display male nipples was thus to rescue these androids from feminization by means of an equally baleful racialization, for as giant naked savages they remained too much body and not enough mind. Nipples thus connoted either submasculinity or hypermasculinity, either that allegedly inferior variant of humans called *women* or that other allegedly inferior variant considered brutes or monsters. The racial content of these bare-chested robots appears clearly enough in the color illustrations of Anatole in Plate 2, whose bright red lips and bared white teeth also belong to the long racist history of European and American caricatures of Africans.

The double duty of nipples to signify the excessive machinery of women and savage men is clearly visible in mid-century children's toys, which must have reached millions of impressionable children. Plate 3 shows a wide range of toy robots from Japanese manufacturers, some of which were marketed in Europe and America as well. Some, like the toy robot named Roberta, were overtly feminized and sexualized. The dials on Roberta's chest are clearly breasts, and she appears to have a suggestive inverted triangle low on the center of her torso. Others were more gender-ambiguous, such as the Radical Robot, which may be wearing a metal-clad dress but which certainly has two large red nipples. In contrast, the Changeman Robot was a hypermasculine monster, with the head of a lizard but with mammalian male nipples on its mechanical chest. Still others tended toward later-century abstraction, such as the Blue Robot Hack, with a full bank of switches or buttons high on both sides of its chest, and with additional machinery in the region of its navel. In all these toys, we find the same conflation of mechanical nipples with mechanical personhood, legible code for subhumanity over many decades and across many different Western and Westernized cultures.

How strange that a fetishized emblem of excessive femininity could function at the same time as an equally clear emblem of excessive masculinity. To be too much of a man is, ironically, to display vestigial lactation equipment shared with female bodies, and which is prevalently fetishized in Western societies.[35] It is as if the excessively mechanical nature of these mechanical people registers as *both* racial and gender difference: racialized because they are tantamount to naked savages, and gendered because their nakedness is defined by the excessive machinery of nipples and sometimes wombs. Then again, perhaps this should not be so surprising, given that early-twentieth-century racist pseudoscience supposed that part of the problem with so-called "primitive" peoples was that they had inadequate gender differentiation: the women were too physically strong and active and the men were too emotional. For white racists already conflating strict gender binarism with racial superiority, male nipples must have been a scandalous fact, best kept out of sight. Perhaps for that reason, many American municipalities outlawed the display of both male and female nipples well into the 1930s. When male nipples first started appearing in Hollywood films, they were most often enrolled in primitivist fantasies such as Tarzan. On both male and female bodies, nipples had become the excessive machinery that signaled estrangement from the ranks of the fully human. Especially when figured as switches and buttons, robot nipples further signaled an appropriate subordination to those humans authorized to direct them.[36]

Operating the Self

As mechanical people grew more abstract, their gender became more indistinct and their chest controls became far less obvious imitations of human anatomy. Two of the most popular mid-century androids in film and television, Robby the Robot from *Forbidden Planet* (1956) and B-9 from *Lost in Space* (1965), both display control interfaces prominently on their upper torsos, despite the fact that their heads have been reduced to large, transparent domes. Instead of peering into their lower abdomens, as with so many earlier robots, observers now peered into their heads. Perhaps prevalent metaphors of early computers as mechanical brains relocated mechanical personhood from the abdomen to the cranium, but this mostly eliminated conspicuous references to uterine machinery. The clear binary gendering evident in most of the exhibits so far, and the clear implication of heterosexual commitments, became more complicated in some exhibits after the mid-twentieth century. Although Robby's voice is clearly gendered male, the chest controls take the form of two circles next to each other, like cartoonish nipples or breasts. However, other characters wonder openly about the robot's gender. When Robby overhears them, the robot replies: "In my case the question is totally without meaning."[37]

Other mid-century androids were also shedding strict gender binarism, and in doing so beginning to trouble the clear distinction between operable machines and operative people. A 1955 cover of *Time* magazine featured a portrait of Thomas Watson Jr., CEO of IBM, in front of an androgynous android with a bank of buttons high on its chest, reproduced as Plate 4. Drawn by Ukrainian-born artist Boris Artzybasheff, well known by this point for his drawings of anthropomorphic machines, the robot's eyes are spools of computer tape and its form vaguely if crudely human.[38] As Watson smiles complacently in the foreground, his lips tight, his left eye drooping, the robot behind him raises one index finger to its lips in a gesture requesting silent complicity from onlookers. As it does so, it reaches to press another button on a separate device on the wall. Although the figure itself has no clearly discernible marks of gender, the history discussed so far suggests that its buttons and button-pushing were legible in more complexly gendered and racial terms. Its own buttons connote submission and obedience. Its indexical gesture connotes power and control. The robot is thus both operable and operative, both Barbarella and Durand, both doll and Bellmer. No longer safely quarantined solely in the category of *woman* or *savage,* the *Time* robot spans the full range of possibilities, embodying mid-century hopes and fears about modern computing. The simple iconography of buttons and fingers thus could connote both power and powerlessness to mass audiences that had long been sensitized to earlier connotations of button-operated mechanical people.

Despite its complexity, the *Time* magazine robot also exhibits a key limitation of every mechanical person considered so far, which is that none of them appear able to operate the switches on their own bodies. But that possibility was also being entertained by mid-century. Artzybasheff himself had drawn a similar robot for *Esquire* magazine four years earlier that did operate its own switches. Titled "Executive of the Future," the illustration depicts a male robot with four arms, a panel of controls on his chest, and a square cranium full of vacuum tubes. With two of his arms he operates a push-button control console before him, but with one of the other arms he operates a toggle switch on his own forehead.[39] So even as the explicit gendering of mechanical people became more muted, the significance of button-pushing seems to have become more consequential, and more connotative of both mastery and servility.

And it was not limited to the fantastic imagination of Artzybasheff. At roughly the same time, Americans were encountering yet another figure that could press its own buttons, through mass market advertising for a new instrument of home automation, the remote-controlled garage door opener. The Alliance Manufacturing Company first marketed this wonder under the brand name Genie in 1954, the year before *Time* published the illustration of the android in Plate 4. That brand name persists in the United States to this day, openly associating automation with both supernatural power and Orientalized servitude. Racial connotations clearly remained. But during the 1950s, the original Genie did not resemble the brawny djinn that the company features today.[40] Then, Genie appeared as an androgynous space-age android with mechanical ears and nose and a fantastic antenna projecting from its head, reminiscent of Sabor IV, Anatole, and other mid-century tin men, but much more diminutive. Its body was the capital letter *A* for Alliance, with the open triangle in the top of the capital *A* rendered as a round circle, clearly a button, which Genie presses with one hand while raising the other to its mouth as if to bellow a command. Sometimes, as in Figure 13, Genie also extends a magic wand to compel action at a distance. The evident sexual symbolism of the extended digit pointed at the navel may be significantly blunted by the cartoonish form,

FIGURE 13. The Alliance Manufacturing Company's original Genie logo from product packaging for a garage door opener from the late 1950s. Reprinted with permission of The Genie Company.

but it is clear enough that the navel has been substantially fetishized too. No doubt the advertising wits had the English idiom *belly button* in mind as well. Androgynous in form, Genie is nonetheless clearly childlike, with an oversized head and something like a child's overalls. Like the *Time* robot, Genie is legible not only as an autonomous agent endowed with considerable power, but also as an inferior that both requires and is amenable to control.

So the Alliance Genie marks a further stage of the collapse of Bellmer and his doll, or of Durand and Barbarella, into one another. The *Time* robot might have occupied both positions at the same time, operator and operative, but it evidently did not operate itself. But the Genie presses its own buttons, and indeed cannot perform its most powerful actions any other way. Genie is Jaquet-Droz's musician adjusting her own clockwork; it is Bellmer's doll pushing her own button; it is Barbarella playing the keyboard from inside the machine. Accordingly, Genie is both a tool that people can use and a kind of person they might be. It collapses the human into the technical, specifically rendered as a being that is both operator and operative, and that might need to be both of those things simultaneously in order to act at all. As such, it reconfigures the sexist arrangement of active men and passive women into a gender-neutral feedback loop in which the agent has both a great surplus of agency and a serious deficit. This being that is both a person and a thing, a subject and an object, a source and a target of agency, is in fact a fair model of what happens to everyone when binary switches mediate so many routine interactions in the modern world. The buttons on which billions of people today depend might not be on their chests, but they are part of the person nonetheless, excessively mechanical organs housed outside the body and through which that mysterious humanizing property known as *agency* must flow.

This returns us to Darth Vader at last, the descendent of the many operable androids with buttons and switches on their chests considered so far, but finally more like the Alliance Genie than any of the rest. Although the interface on his chest changes slightly from film to film, it always includes three indicator lights, two buttons, and four binary rocker switches.[41] The other defining feature of Darth Vader is that he can employ that mystical power that the films refer to as "the force," which expresses itself as clairvoyance, telepathy, psychokinesis, and the projection of electrical current. Like Genie, he also shoots lightning from his fingers. But it is not exactly clear why a being who can control the material world with his mind must also sustain life using switches that appear to have been borrowed from 1970s home appliances. Then again, to object along those lines is to ignore the symbolics of switching evident in mechanical people. Vader is a parody of Cartesian dualism, both its highest development and its most contradictory expression. For Vader has the most powerful kind of mind housed in the feeblest kind of body, but an augmented body that is excessively mechanical too. His split identity as both operator and operative renders him at once both fantastically sovereign and fantastically vulnerable, both superhuman and subhuman, both technology's master and its slave.

To claim that Vader's switches are nipples, as the title of this chapter does, is thus not to insist on a literal reduction of his switches to actual body parts, either male or female. It is rather to read them as containing residual significance derived from a much longer sexist and racist history of how bodies have been conflated with machines. He

FIGURE 14. Promotional still photograph of Darth Vader, used to promote the *Star Wars* sequels *The Empire Strikes Back* (1980) and *The Return of the Jedi* (1983). Courtesy of the Everett Collection.

is in some ways submasculine, symbolically castrated by his injuries and, as some have observed, associated especially by the shape of his helmet with Lang's feminized Maschinenmensch. But he is more evidently hypermasculine in his violence, strength, and overassociation with paternity (*vader* is Dutch for *father*). More importantly, like Sabor IV and Anatole, Vader is a brutal savage racialized as Black. Defined by the color black, voiced by a famous African American actor, and committed to the "dark side," Vader's hypermasculinity suggests at least some reference to the color line.

I confess that I find little in the cyborg that is Darth Vader that might militate against naturalized humanity and the inequalities associated with it. Still less does he disrupt the binarism that Haraway identifies as the source of sexist and racist hierarchies. In fact, Vader seems to be as binary as the switches that control his life-support suit, a creature of equal parts power and powerlessness. What meaningful middle ground is there between these two extremes? Switchable people end up exaggerating the poles of power and powerless, eradicating anything in between. It may be that other kinds of cyborgs do erode the allegedly natural status of the human in more productive ways, and thus unmask the politics that sustain inequalities, but in the examples I have collected here, the switch pulls all manner of cyborgs and androids back into its crude binary order. Fears and fantasies accumulate at interfaces like these, by people who have no partial agency—only mastery and servitude with nothing in between.

Not all of this is lamentable, however. The drastic simplification of power as a binary alternative also opened up new opportunities for some kinds of people. As we turn to women bookkeepers, autistic typists, and working-class voters in the chapters ahead, we will see that the mastery promised by the binary switch could, and sometimes did, restore some measure of agency to those all too accustomed to being controlled by others.

Part III

»» »» Keyboard Rationality

CHAPTER 7

The Keyboard's Checkered Past

There may be no clearer example of the analog nature of technological change than the gradual development of that interface known today as the *keyboard,* first across a wide range of musical instruments, and later across an even wider range of communications technologies. Billions of people worldwide use keyboards of roughly the same size, shape, and arrangement, and practice input techniques that have been standardized to an extraordinary degree. Despite attempts to devise different and better interfaces, the form of the keyboard on writing machines has persisted mostly unchanged for a century and a half. And yet this durability belies a longer history of gradual change, one that began over two millennia ago, starting with the interfaces of Greek water organs, persisting in similar musical instruments in Rome, spreading east to the Byzantine Empire, returning to Europe in the Middle Ages, growing in variety and sophistication during the early modern period, and continuing to adapt to changing demands up to the present. As we shall see, musical keyboards are still in modern writing machines, as are some of the meanings and values originally invested in musical techniques. The main task of the chapters in this part is not just to understand any single kind of keyboard (though we will have occasions to linger over many specific examples), but to begin to understand an entire global keyboard culture that developed around this most prolific and versatile interface.

Today, keyboard culture is most prominent in the interfaces of modern computers, which mediate not just writing but also programming, shopping, gaming, communicating, calculating, accounting, filing, and much more. The adaptability of the interface has been astounding. But its derivation from musical interfaces should not be taken as evidence that typing necessarily preserves the aesthetic pleasures of music. Though that is true in some cases, such as video gaming, as Roger Moseley has recently shown, still aesthetic pleasure and playfulness seem rather remote from many other keyboarding techniques.[1] For instance, the keyboards of modern writing machines have thoroughly preserved the legacy of mathematical rationality associated with music, especially as expressed in the classical Greek theory of harmonic intervals, which has, with some modifications, defined Western music ever since. In an earlier chapter on calculating machines I argued that new kinds of touch interfaces mathematized the body at its extremities, turning fingers into digits. Through discretizing procedures that transformed the continuous gradations of value into discrete numerable units and then mapped them onto the fingers, calculating techniques digitized the human body. Something similar happened much earlier with the musical keyboard, which discretized—and so rationalized—the analog continuum of pitch. By fixing each key to the production of a specific pitch and by permitting the sounding of no pitches in between keys, the keyboard was one of the earliest and most rigid digital technologies. As such, the keyboard not only mathematized music but also mathematized the people who used it. The reasons for that have everything to do with the keyboard's status as one of the most important, and most neglected, sources of Western rationality since the late Middle Ages.

In fact, the keyboard should be considered one of the primary scientific instruments of the late medieval and early modern eras, no less central to the subsequent development of Western scientific culture than the mechanical clock. As musicologist and (fittingly) IBM employee Edmund Bowles has argued, it should come as no surprise that the fifteenth-century polymath physician, astronomer, and clockmaker Henri Arnaut de Zwolle also wrote the earliest technical description of keyboard instruments. Arnaut also may have designed the astronomical clock of Bourges cathedral. It seems to me no accident that clocks and organs both occupied central symbolic space in late medieval cathedrals. Just as the cathedral clock aligned mechanics with the divinely appointed nature of the heavenly spheres, so the cathedral organ aligned mechanics with the physical and spiritual nature of human beings. The clock pointed up toward the heavens; the organ pointed back toward the body. The clock testified that the works of people might express a divinely appointed cosmic order. The keyboard testified that people might suit themselves to that order, fingers first. By mapping the mathematical relations of harmony onto the human body, the keyboard proposed that this rigid alignment was in fact proper, natural, and sacred. To recover earlier configurations of the keyboard is thus really to recover neglected aspects of the early modern reconfiguration of the human.

Looked at this way, early musical keyboards were also calculating machines, devices for automating the complex mathematical relations of harmony and making them readily accessible to the human body. We might expect that musical keyboards might have directly influenced the kinds of calculators we encountered in chapter 5, but that was not the case. Very few if any early calculators imitated musical interfaces, despite the fact that dozens of early writing machines did. If we look back further, however, we find a surprising fact: earlier calculating technologies may have influenced the form and function of musical keyboards instead. Early modern music theorists writing in

Latin sometimes referred to the keyboard as an *abacus,* which clearly betrays a sense of its calculative functions. And as we shall see, the keyboard may have drawn more directly from even simpler calculative techniques, such as counting tables. Instead of assisting with arithmetic, however, the keyboard assisted with the mathematical ratios of Pythagorean harmony, which it stored precisely and made available for rapid retrieval at the touch of a finger. The keyboard is a sonic calculator that outputs its results not in symbols but in sounds. Its graphs are harmonic relations, its proofs aesthetic pleasures.

With its long row of equally sized, equally spaced rectangular units, the keyboard is also the audible complement of a rationalizing grid, that graphic form that has so pervasively influenced early modern visual culture, from cartography to topography to perspectival drawing. The mathematically precise rectilinearity of the keyboard links the mathematics of sound to the mathematics of space. It translates regular patterns that people can hear into regular patterns that people can touch and see. The keyboard grids sound. Or it might be more accurate to say that the mathematics of music project the rectilinear form of the keyboard as harmony's spatial and visual equivalent. In referring to *gridded* sounds, I would not want to be understood as advancing only a banal metaphor. The keyboard is not *like* a grid (an obvious point), but rather *is* a grid. It derived substantially from rationalizing grids also dating to medieval and early modern periods, transported those logics from mathematics to music, and thereby introduced an underappreciated vector of Western rationality that has only grown more prominent in the information age. To discern the grid within the musical keyboard is to begin to appreciate the full extent of the rationalizing power of all those other keyboards in use today.

Gridded Sounds

Before extending the concept of the grid to sounds, let us clarify the properties of the grid in more familiar visual examples. Much and perhaps most recent theoretical work on the grid has regarded it as an instrument of optical discipline and control. Rosalind Krauss has lamented the "antinatural" intrusion of the grid into modernist visual art, and Curt Meine has objected that the topographical grid "has dictated our system of land use and our way of thinking about land—the natural, the wild, the human, the civilized."[2] Brian Massumi has opposed the "gridlock" of "an oppositional framework of culturally constructed significations" with a more process-oriented and expressly analog pliability.[3] And Bernhard Siegert has claimed that visual art, topography, cartography, and architecture all can "be described in terms of a growing totalitarianism of the grid."[4] For all of these critics and others, the grid imposes its rigid order on lives that might be better organized in more fluid and flexible ways.

Those other ways of organizing and understanding life might generally be identified as analog, as Massumi claims. As discussed in chapter 5, one prevalent definition of the digital holds that it defines that which has been made discrete: separated or distinguished, like the fingers of the hand. The analog, in contrast, denotes fluid forms that blur at the boundaries, continuous gradations, or processes rather than discrete products. But it must be noted once again that the opposition of the digital to the analog binarizes the world into two opposed categories, which is a very digital way of thinking about things. Materially, almost everything turns out to exhibit features of both conceptual categories. The pixels on my computer screen are digital in the sense that they compose visual forms with a grid of equally sized and spaced illuminated squares, but the light they emit has evident analog properties too. Digital and analog are thus not properties of things in themselves, but ways of talking about things, categorizing them, and assigning them values. As a result, the distinction between digital and analog can be both digital and analog. It is digital in the sense that it can solidify into a restrictive binary. It is analog in the sense that the digital and the analog are always blurring at their borders.

Just as the digital and the analog are inconceivable without each other, so too the grid is most prominent when juxtaposed with some opposing form that it both defines as other and purports to organize and rationalize. To take one early and famous example, Albrecht Dürer's engraving of a male draftsman drawing a nude woman depicts a reference grid that operates in two directions, with rather obvious gender dynamics.[5] As shown in Figure 15, the woman on one side of the grid is ostentatiously curvy, a fluid sprawl of flowing fabric and rounded flesh. On the other side, the draftsman uses the organizing power of the gridded frame to translate those flowing forms onto his paper. Accordingly, he appears as an emblem of rigid rationality and self-control, stiffly upright and invested with fixed purpose, and thus equated with the highly phallic visual guide protruding from the region of his lap. She, in contrast, appears as yielding and flowing liquids, veritably poured upon the mounded cushions on the table, recumbent in her posture and presumably—by prevalent sexist standards—labile in her character. In these terms, the analog is not just continuous but also passive and natural, and hence feminine. The digital, in contrast, is not just discrete but also active and cultural, and hence masculine.[6] The grid is everything he has and everything she lacks: reason, culture, creativity, activity. He is the digital made manifest; she is the analog being digitized by male operations.

Dürer's draftsman looks through a grid, but he also draws on a grid. His posture is not unlike my own at this moment, seated before an array of small squares just beneath my fingers, my gaze directed at an array of even smaller squares before my eyes, the

FIGURE 15. Illustration of a draftsman drawing a nude through a perspectival grid, which first appeared in the posthumous edition of Albrecht Dürer's 1538 *Underweysung der Messung*. This version from about 1600 courtesy of the Metropolitan Museum of Art.

pixels of my screen. Swap the pen for a mouse, and he looks more than a little like any office worker, the pixelated screen before his face and the keyboard under his fingers. Or, to be less anachronistic, he looks like a person seated at a piano, hands on the rectangular keys and his eyes lifted to the grid of the musical staff. By the early sixteenth century, that arrangement was prevalent across Europe with instruments including organs, clavichords, and harpsichords, some of which were used in domestic settings like this. But beyond this formal resemblance of the draftsman to a keyboard musician, what deeper relation might there be between the grid of the keyboard and the music it controls? What, in other words, is there about music itself that might be comprehended through the logic of the grid, and that might even have projected the grid of the keyboard as an appropriate interface? To answer that question, we must look back to an even earlier stage when Western music theory first took shape.

We encountered the Greek philosopher Pythagoras in chapter 5, on calculating machines, but his reputation for mathematical knowledge was closely tied to his reputation as the most influential musical theorist of the classical world. The early modern illustration reproduced in Figure 5 shows him reckoning on a counting table, which many mistakenly presumed he invented. In truth, Pythagoras reckoned as much by ear as by eye. For him and for his followers, the proportions of musical intervals, expressed as ratios, suggested essential and unchanging structures of the universe. Music was math, and math was metaphysics. In Aristotle's account of the Pythagoreans, he says that they saw "in harmonies attributes and ratios that are found in numbers—since, then, the other things seemed to have been made like numbers in the whole of their nature, and numbers were primary in the whole of nature, they took the elements of numbers to be the elements of all beings, and the whole heaven to be harmony and number."[7] Boethius, a major source of Pythagorean thinking given that none of Pythagoras's own writings survive, concurred: "There can be no doubt that the order of our soul and body seems to be related somehow through those same ratios."[8] Even after Pythagorean musical tuning gave way to more modern versions, these ideas persisted and proved complementary to emerging scientific thought. In the early seventeenth century, Johannes Kepler faulted Pythagoras for too narrowly mathematizing music, but he went even further in his universalizing claims, supposing an entire harmonic cosmology amounting to a musical language of God: "Which planet sings soprano, which alto, which tenor, and which bass?"[9]

Pythagoras famously arrived at his harmonic theory by hearing blacksmiths working at their hammers, producing different pitches with different sized tools.[10] The difference in size or weight of one hammer from another explained the difference of pitch mathematically. The interval between any two pitches determined whether or not they were consonant when sounded together, or pleasing to the ear. The relation between one pitch and another could be expressed as a ratio, and those ratios could be measured physically on musical instruments. For instance, on the simple one-stringed instrument called a monochord, the player varies the string's pitch by adjusting the length allowed to vibrate, typically by sliding a bridge back and forth beneath it. If the player removes the bridge entirely and plucks the string, it will produce a certain pitch, called a *unison,* or a 1:1 ratio of the length that is plucked to the length of the whole string. When bridged at the midpoint, so divided in half, the string will produce a different pitch when plucked. This interval can also be expressed mathematically as the ratio of the length that is plucked to the length of the whole string, or 1:2. That interval is known as an *octave* because it spans from the first to the eighth natural note in the scale. Stripped down to its barest form, this basic structure is the foundation of all Western polyphonic music.

All other notes fit within the octave at intervals also defined just as precisely by mathematical ratios. Bridging the string at one quarter of its length and plucking the longer side would yield a ratio 3:4, known as a *fourth.* Similarly, dividing it into three sections and plucking the longer side is the interval 2:3, known as a *fifth.* This is why music for Pythagoras was not simply a matter of aesthetic pleasure or cultural tradition, but rather, as Daniel Heller-Roarzen has put it, a veritable transcription of the natural world, "not, to be sure, by the letters of the alphabets, . . . but by those various collections of unity that the ancients understood to be 'numbers.'"[11]

Even in the Pythagorean tuning that prized small numbers and simple ratios, these intervals can be forbidding. On a monochord, the player would have to position the bridge manually to produce each note. This might be possible when playing music, but would require no small amount of precision and skill, and would surely introduce delays between selecting different notes. For that reason, monochords were probably teaching rather than performance instruments. More sophisticated machines turn out to be better at automating the production of specific pitches and of preventing the production of any others. Foremost among such machines are modern musical keyboards. The modern piano keyboard records eighty-eight predetermined pitches during the process of tuning and forbids the production of any pitch between adjacent keys. The keys make it possible for a player to produce any one of those pitches precisely, rapidly, and with minimal effort. Relieved of the responsibility of placing each note on an analog continuum of pitch, as a vocalist would have to do with the body, keyboard players can devote their limited attention to selecting notes with unparalleled speed, which in turn allows them to combine notes with unparalleled complexity. Even so, this much automation and mechanization has seemed alienating and excessively rationalizing to some. Max Weber once objected, "The organ is an instrument strongly bearing the character of a machine. The person who operates it is rigidly bound by the technical aspects of tone formation, providing him with little liberty to speak his personal language."[12] Perhaps so, but the keyboard also freed players from having to superintend the accurate production of each and every note, and allowed them to sound multiple notes rapidly and simultaneously.[13] Automating note formation made new kinds of music possible, and Weber acknowledges that Beethoven's later works, among many others, simply could not be played on earlier instruments.[14]

The gridded form of the keyboard is entirely evident in some of the earliest drawings of fifteenth-century clavichords, which predate the earliest surviving instruments by roughly a century.[15] On a clavichord, each key lifted a lever beneath the strings, striking one of the strings with a small metal tangent. One end of the string was dampened so that as long as the key remained depressed to hold the tangent in contact with the string, the string would vibrate between the tangent and its other end. The most complete diagram of an early clavichord comes from a manuscript by Arnaut, who made musical instruments in addition to those for scientific use, probably from the 1440s. It shows a detailed design for a clavichord with thirty-seven keys, including all five accidentals in each octave. Arnaut makes no reference to the colors of the keys, but he does draw them carefully to scale. One historian has estimated that they were 2.6 centimeters wide, slightly broader than modern piano keys which tend to be about 2.3 centimeters. In addition, the keys appear to have been much shorter than modern piano keys, extending only about 3 centimeters from the ends of the accidentals, because the player needed less leverage to operate the clavichord's simpler mechanism.[16] This results in a row of natural and accidental keys that are decidedly more square than on a

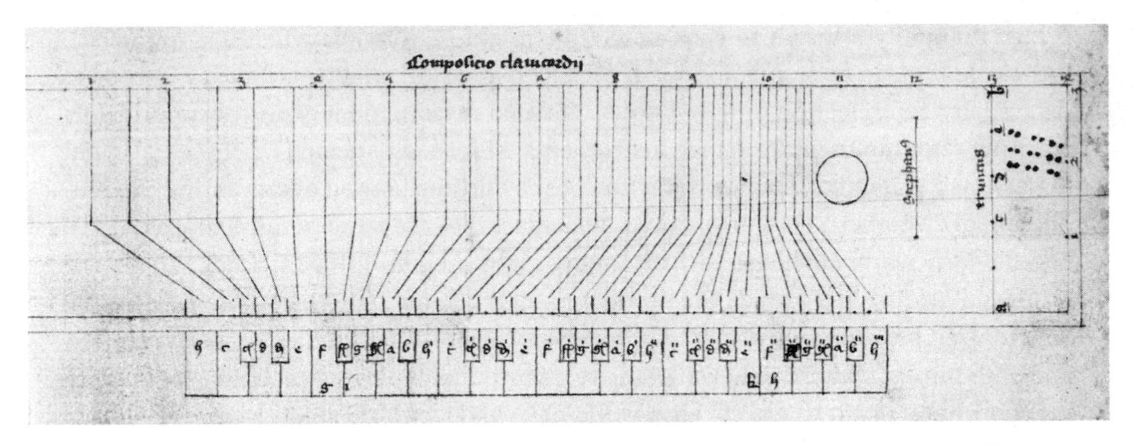

FIGURE 16. Illustration of the mechanism and keyboard of an early clavichord, from Henri Arnaut de Zwolle's *Compositio Clavichordi*, circa 1440. Bibliothèque nationale de France, MS Latin 7295, fol. 129r. Courtesy of the Bibliothèque nationale de France.

modern piano, which accords with even earlier but less technically precise representations of clavichords in other carvings and illustrations.[17]

The mid-fifteenth-century date of Arnaut's description and illustration of the clavichord roughly coincides with Leon Battista Alberti's first proposal to employ the grid in perspectival drawing around 1435. There is no direct causal connection between these events, nor even any indirect influence that I can see. Yet they apply a similar logic to the organization of analog continuities, whether of sliding pitch or curving shapes. In the realm of sound, the keys chop pitch into so many bits, evenly distributed and scaled to the fingers. They digitize sound in the process, but they also digitize the fingers, rendering the extremities of the body as numerical entities as well. Only once techniques digitize the fingers do they properly become digits, and thus become legible as the body's own distribution into multiplicity. But the musical keyboard does even more. It also enforces a rigorous equality on the fingers so extreme that keyboard players must train arduously to live up to it. Unlike identical keys, fingers come in different sizes, strengths, and dexterities. The gridding of sound thus entails, first, the gridded interface of the keyboard as a spatial expression of the mathematical intervals of pitch and, second, the alignment of the body to that grid through extensive training. No less than Dürer's draftsman, the pianist must become rigid to match such rational and rationalizing media. Pythagoras may have thought that the ratios were found in nature, but once the keyboard automated the storage and production of harmonic ratios in such fantastically efficient ways, it rationalized everything else it touched along similar lines.

Scientific Instruments

Like clocks, which developed in Europe around the same time, keyboard instruments attracted enormous technical energy and ingenuity, and their association with mathematics makes them hard to distinguish from other kinds of early scientific instruments. More than just aesthetic or ludic technologies, then, keyboards were part of the early modern science of sound.

However, unlike mechanical clocks that had appeared in sophisticated forms in China and the Middle East centuries before they appeared in Europe, the keyboard seems to have been peculiarly European from the start. And it was peculiarly complex compared to other world instruments. Nineteenth-century pianos could contain as many as twelve thousand parts, but even earlier and simpler versions required sophisticated engineering and mathematical understanding. Considered in the context of the music it was designed to play, the piano eventually involved an entire network of manufacturers and material supply chains, repairers, movers, tuners, and producers of sheet music, not to mention protocols of training and socialization. We could call it the "Steinway Galaxy," that constellation of mass media forms and functions that altered the consciousness of Europe through the mechanical production of music just as decidedly as Marshall McLuhan's "Gutenberg Galaxy" did through the mechanical production of writing. Only the musical cultures of Europe committed to mechanizing music on this scale.[18] Prominent or prestigious instruments of Arab and Asian musical cultures have tended to be smaller and more intimate, hand-held zithers or lutes, such as the Chinese Guqin, the Japanese Koto, the Middle Eastern and North African Oud, or the Indian Veena. In contrast, according to technology historian Hans-Joachim Braun, "a piano is the most appalling contrivance of levers this side of the steam engine."[19]

The first musical keyboards appeared in classical Greece, probably around the third century BCE. The Greek water organ, or hydraulis, used water to compress air that sounded the pipes. The Romans adopted the hydraulis as well, and later replaced the water-driven mechanism with manually operated bellows. Similar organs spread east as far as Constantinople.[20] A third-century Roman organ from the town of Aquincum in southeastern Europe has been recovered in remarkably complete form. Its remnants suggest lever-style keys that were spring-loaded, equally sized and spaced, scaled to the fingers, and responsive to a light touch, remarkably like the those of musical keyboards today.[21]

Little is known about the interface of the Greek hydraulis, but even less about the earliest medieval organs, not even fragments of which survive. There are brief references to three organs in the eighth and ninth centuries in the courts of Pepin the Short, his son Charlemagne, and Charlemagne's son Louis the Pious. They tell us little more than that the instruments came from Byzantium. The ninth-century Utrecht Psalter contains three illustrations of water organs, the technical accuracy of which is far from clear. The roughly contemporary Stuttgart Psalter shows two bellows organs as well. The most detailed description of an early medieval organ comes from Wulfstan of Winchester in tenth-century England, who describes a monstrous machine operated by seventy men working the bellows. Wulfstan may exaggerate, but even so, the machine must have been formidable, for he also says that "the melody of the pipes is heard everywhere in the city." About the interface he says much less, only that "the hand of organic skill . . . opens the closed and in turn closes the opened." He may be describing slides rather than keys, pushed and pulled when grasped by the whole hand, which later manuscript illustrations amply depict.[22]

Not until the twelfth century did organs begin to appear in western and northern Europe in larger numbers. Large slides remained, but something like pressable keys probably began appearing during this time too. The mid-thirteenth-century Rutland Psalter includes a depiction of King David and his musicians, reproduced as Plate 5. The illustration shows David operating a large positive organ while another man treads the bellows to power the instrument. The keys of the organ appear to be large and blocky, but also tightly spaced and arranged entirely on one level. Judging by the gesture of David's hands, they appear to be operable by the fingers rather than the fists, and other thirteenth-century illustrations similarly suggest interfaces designed for lighter and more rapid finger-based control.[23]

Unfortunately, no instruments from this period survive. The earliest surviving fragment comes from the Norrlanda organ built around 1380, and though it is exceedingly partial, it tells us a great deal. Its keys are levers protruding from slots in a vertical surface, like keys from a lock. They are large by modern standards, yet small and tightly spaced enough that they seem to have been designed to be played by the fingers. They closely resemble the keyboard of the Halberstadt organ from 1361, as depicted in Michael Praetorius's seventeenth-century *Syntagma Musicum*. Praetorius considered the Halberstadt organ the first fully chromatic keyboard. Like the slightly later Norrlanda organ, the Halberstadt organ had keys for all twelve notes of the chromatic scale, and both instruments organized the accidentals as a second row of levers above the naturals.[24] Surviving musical scores confirm that, even before 1350, at least some music was being written for fully chromatic keyboard instruments as well, so presumably other instruments with twelve keys per octave preceded even these.[25] The mid-fourteenth century thus marks the birth of the modern keyboard in northern Europe, with fully chromatic twelve-tone octaves, keys scaled to the fingers, and the separation of naturals from accidentals on two levels. When John Cage composed a work designed to last 639 years, he determined the duration by subtracting the year of the construction of the Halberstadt organ, 1361, from that of the looming millennium, in homage to the modern keyboard's supposed origin.[26]

Eventually, this orthodox arrangement would appear on a wide range of smaller and more secular keyboard instruments throughout the fourteenth and fifteenth centuries. These include the small portative organ, the clavichord, and the harpsichord, all of which standardized the keyboard considerably. Ironically, this period of tremendous innovation and experimentation in new keyboard instruments coincided with the standardization of the keyboard itself. Even as new methods of sound production proliferated through organs, clavichords, harpsichords, hurdy-gurdies, pianos, and more, the keyboard gradually settled into an orthodox form across all of them. Experimentation with sound production accompanied standardization of the interface. And both of these trends may have coincided at an early date in that obscure fourteenth-century keyboard instrument known as the *chekker.*

Termed "organology's Loch Ness Monster" by one music historian, the chekker first appears in the journal of expenses of King John the Good of France, who reports receiving *leschequier* as a gift while being held prisoner by King Edward III of England in the mid-fourteenth century.[27] No examples survive, but about three dozen references to the chekker have been found across Europe. Early historians imagined the chekker as a harpsichord or a hammered dulcimer with a keyboard, but Edwin Ripin suggests that it simply may have been a clavichord, and possibly a primitive one at that.[28] Others have noted that the chekker may have been a keyed version of the dulce melos, a portative organ, or even "a small, upright protoharpsichord."[29] Even in the fifteenth century, there may have been considerable confusion about just what counted as a chekker. As David Kinsela argues, later references to the chekker tend to be the most diverse, which suggests that, after about 1470, the term may have been applied generically to any keyboard instrument.[30]

Unfortunately, there are no fourteenth-century illustrations of a chekker. The earliest date from about 1420 and consist of illustrations accompanying references to the chekker in moral allegories by Jean Gerson.[31] All three illustrations depict a square cabinet with small keys on one side, which, according to accompanying text, are "struck by the fingers in meditation" and "sound the strings within."[32] Marginal comments claim, "Devotion makes the strings of the chekker sound within in accordance with diverse emotions."[33] All of this suggests that this instrument is some sort of protoclavichord. Still, caution about the details is warranted. Although the three illustrations appear to be by the same artist and of the same or similar instruments, they nonetheless depict different numbers of keys: eight in two illustrations, but seventeen in another. Both are plausible arrangements, but the fact that they appear side by side should warn us against taking any details too literally.

The cabinet itself is especially interesting. It features a checkerboard on top measuring eight squares by eight squares, a grid long in use for the game of chess by this time, and one that neatly aligns with the eight natural notes from the start of one octave to the first note of the next. In the illustration in Figure 17, what appears to be a musical instrument does double duty as an actual chess board in Gerson's moral allegory. On the left and right of the board, names of virtues and vices are arranged like chess pieces prepared to wage battle. Elsewhere, Gerson refers to what "we may call a *scacordum,* as a stringed chessboard."[34] Although historians of musical instruments are eager to pin down the chekker's method of sound production, I am more interested in the symbolic association between the name of this instrument, the grid that appears on its cabinet, and the presence of a keyboard. Even if the instrument that John the Good referred to as a chekker looked nothing like the illustrations in Gerson's text, it certainly went by the English name *chekker,* and by other continental equivalents, including the French *eschiquier,* the German *Schachtbret,* the Catalan *squaquer,* and the Latin *scacordum*—all of which refer either to chess or to the chess board's distinctive checkered pattern.[35]

If the name *chekker* really did refer to a wide range of instruments starting with primitive clavichords and later including a wider range of portative organs, keyed dulcimers, harpsichords, and more, perhaps its name referred not to the decorative cabinet but to the alternating black and white keys of the keyboard. Kinsela speculates that the chekker's alternating black and white keys may have recalled the alternating black and white squares of a chessboard.[36] After all, the instrument termed a *clavichord* seems to have been named after the *clavis,* the key or lever that distinguished its interface, so there was ready precedent for naming the instrument after the interface.[37] However, even though the fully chromatic keyboard with all five accidentals existed on clavichords by the early to mid-fifteenth century, given Arnaut's illustrations from the 1440s, such instruments may have been comparatively rare. On the continent, clavichords did not commonly have all five accidentals until the mid-fifteenth century.[38] Moreover, on the earlier mid-fourteenth-century northern European organs in Norrlanda and Halberstadt, most or all of the accidentals are present, but they consist of a separate row of levers arranged above the naturals, and there is no indication that they were colored differently.[39] So, a stringed instrument that appeared with a keyboard resembling those fully chromatic organs but with alternating black and white keys in the 1350s likely would have been a novelty, but hardly an impossibility. The name of the mysterious *chekker* thus may indicate not the apparatus of sound production, and not the form of the cabinet, but the appearance of its novel interface, a new kind of keyboard that distinguished naturals from accidentals with the suggestive alternation of black and white rectangular keys.

Ce sont les vertus de raison par dedans.	Ces pions sont preudes gaugnies selon les .v. sens.					Ces pions sont felons tricheurs selon les .v. sens.	Ce sont les vices de lame par dedans.
Le diligent	Ferme de pie.					Pie trebuchant	Le trop hatif.
Le couraigeux	Lactif de bras.					Main ravissant.	Le despiteux.
Le prudent.	La langue eloquent					Bouche mentant	Le mal engin.
Lumble de cuer.	Le cler oyant qui obeist.					La sourde oreille.	Le hault de cuer.
Domme amoureuse.	Le simple oeil o le tast chaste					Lubrique oeil et vil toucher	Fole amoureuse.
Saige desperit.	Le doulx fleurier et savourer.					Lort fleurer et savourer.	Saige charnel
Le victorieux.	Le fort de bras.					Lasche de bras.	Le vuyt doneur.
Le veillant.	Lysnel de pie.					Retif de pie.	Le sommeilleux

Roc offitier en temporel.
Chevalier en temporel.
Alphin conseiller en temporel.
Lesprit Joue.
Roy.
Royne.
Alphin conseiller en lespirituel.
Chevalier pour lesperituel.
Roc officier pour lespirituel.

Ce sont les clefs de leschiquier par dedans.

Devotion fait sonner les cordes de leschiquier par dedans selon diverses affections.

FIGURE 17. Jean Gerson's illustration of the chekker, which appears to be both a musical instrument with keys protruding from the bottom and a chess board organized as a moral allegory. Bibliothèque nationale de France, MS Latin 17487, fol. 228r. Courtesy of the Bibliothèque nationale de France.

Checkered History

The fact that the first reference to the chekker emerged from an English royal context further suggests that accounting may have been a primary connotation, not chess. Chess may have seemed like the natural association to a continental authority such as Gerson, searching for the meaning of such a strange name, and the fact that the eight squares of a chessboard align with the eight natural notes of an octave (counting the first note of the next octave) must have been suggestive, as two of the illustrations accompanying Gerson's text confirm. Even so, in the British Isles an equally important context might have been the *exchequer,* the king's chief financial officer responsible for keeping accounts. Ripin first suggested the possibility that the chekker was named after the exchequer, and though some like Nicolas Meeùs have entertained the idea since, no great evidence has been marshaled in its favor. Others have strongly dissented.[40] But the argument has more merit than most historians have thought, especially when we look more closely at the form of medieval and specifically English accounting techniques.

There are plausible, though far from proven, relationships between musical notation in general and the kinds of calculating practices previously discussed in chapter 5. Marin Mersenne's *Homonicorum* from 1636 regularly refers to the keyboard with the term *abacus,* as does Athanasius Kircher's *Musurgia Universalis* from 1650.[41] Given that the Roman abacus had disappeared from Europe by the Middle Ages, medieval and early modern accountants calculated with the counting board instead, with parallel lines to define place values and small, coin-sized *calculi* to record specific values and aid in calculation. According to Karl Menninger, medieval and early modern counting boards, especially secular ones, differed from their classical predecessors in two ways. First, probably in the thirteenth century, European counting boards rotated the lines 90 degrees, so that they ran horizontally, as shown in Figure 5 in chapter 5.[42] On those horizontal lines, the user formed numbers vertically and carried out computations horizontally from left to right. The other difference involved the replacement of the individual *calculi* with *apices,* a significant advance. *Apices* were counters marked with Arabic numbers, which could stand for a number greater than 1. A single *apex* marked with the symbol for 4 thus could replace four separate *calculi.*[43]

It is not clear why Europeans oriented the lines of the reckoning board horizontally, but Menninger suggests that it may have been related to parallel changes in musical notation. "In the 11th century Guido of Arezzo invented the horizontal lines which are still used for the writing of musical notes, and which are fully analogous with the parallel lines of the counting board," Menninger claims.[44] Around the same time, Gerbert of Aurillac (later Pope Sylvester II, one of the most learned men in Europe) reintroduced a modified form of the Roman abacus to the West. A mathematician who also introduced Arabic numbers to Europe, Gerbert was trained in music as well, and according to William of Malmesbury helped build both an astronomical clock and a powerful hydraulic organ at Reims.[45] There is no evidence that Guido and Gerbert met, or even shared any specific sources, though both did draw on learning from the East. Although the counting table and musical staff function differently, Menninger's point is that, at around the same time and in roughly the same place, two Europeans overhauled existing techniques with new and similar arrangements of place-value notation. The precise relationship between methods of European calculation and music notation must remain largely speculative, but if there were any influence, presumably Guido's horizontal staff would have contributed to the horizontal orientation of the European counting table, which came slightly later.

None of this is very conclusive, unfortunately. A slightly more promising association of accounting and musical techniques derives from similarities between calculating techniques and the forms of instruments themselves. Ripin first suggested that the clavichord may have consciously imitated the medieval counting board. With its long rectangular cabinet and strings arranged parallel to the length of the keyboard, the clavichord cabinet looked a little like the medieval counting table with its horizontal lines. Ripin even suggested that the chekker may have received its name from the similarity of its long, low rectangular cabinet to a medieval counting board.[46] The problem, however, is that the exchequer used in England did not resemble continental European counting boards in the least. In England, Ireland, and even Normandy, counting boards did not employ exclusively (or even primarily) either the vertical columns of the classical world or the horizontal rows of continental Europe. They employed both at the same time, for they took the form of checker boards.

English and Irish counting tables adopted the form of gridded checkerboards as early as the twelfth century, especially though not exclusively in royal contexts, as depicted in Plate 6.[47] That peculiar form gave the royal treasury in England its name, *exchequer,* which it retains to this day. The twelfth-century *Dialogue Concerning the Exchequer* describes these counting boards like this:

> The exchequer is a rectangular board, about ten feet long and five feet wide, which those sitting around it use like a table. It has a raised edge about four finger-widths high, so that nothing placed on it can fall off. Over this aforementioned exchequer is placed a cloth bought during the Easter term, not an ordinary cloth, but black, marked with lines a foot or a spread hand's width apart. Counters are placed in the spaces in a certain way.[48]

Counting practices and apparatuses varied between monks, merchants, and monarchs, but there is little doubt that, in the British Isles, by the thirteenth century, the checkerboard pattern suggested the reckoning techniques of the state.

The name of these English and Irish counting boards did derive from their resemblance to chessboards, for even the *Dialogue of the Exchequer* claims that, "just as, in chess, battle is joined between the kings, so at the exchequer there is basically a competition and struggle between two individuals, namely the treasurer and the sheriff."[49] Accordingly, a fourteenth-century English instrument called a *chekker* likely suggested a wide range of connotations, but it may be that Gerson's continental perspective obscured the instrument's association with England's distinctive calculating techniques. Perhaps the name of the chekker suggested ludic play too, as Moseley suggests, or even a contest between opponents, as the *Dialogue of the Exchequer* claims, but in the British Isles its primary association was likely the calculative rationality associated with period accounting techniques.

In slightly later works of music theory, the grid quite literally bleeds into the keyboard. The great sixteenth- and seventeenth-century mathematician and musical theorist Mersenne not only referred to the keyboard as an *abacus,* as noted above, but also imagined fanciful keyboards stacked with extra ranks of staggered accidentals. These "augmented keyboards," as Mersenne called them, seem to have been solely works of the imagination, because most would have been forbidding if not impossible to build. One example contains twenty-seven keys per octave in four staggered and overlapping ranks. Concerning playing such monstrosities, Mersenne insists, "It is of no importance that the difficulty of playing them is greater, inasmuch as it is not necessary to feel pity for the pains . . . that lead to perfection."[50] With practice, he continues, "they will be played as easily as the others when the hands become accustomed to them, because they follow the infallible rule of reason."[51] According to Mersenne, experienced organists should be able to master these new keyboards in eight hours or less. Here the justification for the keyboard's rationalizing techniques lies fully exposed: the rigid rectilinearity of the keys in Figure 18 materializes the otherwise abstract perfections of musical intervals, which the body naturally accommodates. To learn to play such a device is to align oneself with the fundamental truths of the universe. That same logic is implicit even in simpler, earlier, and more orthodox keyboards as well. To bend the body to such an interface is to partake of the perfect rationality of the machine, which is nothing other than the perfect rationality of God. Mersenne's extraordinary claim only makes explicit what had been implicit in cathedral keyboards all along, which is that the rationalizing rectilinearity of the keyboard and of the musical structures it controlled could also organize the rational capacities of people.

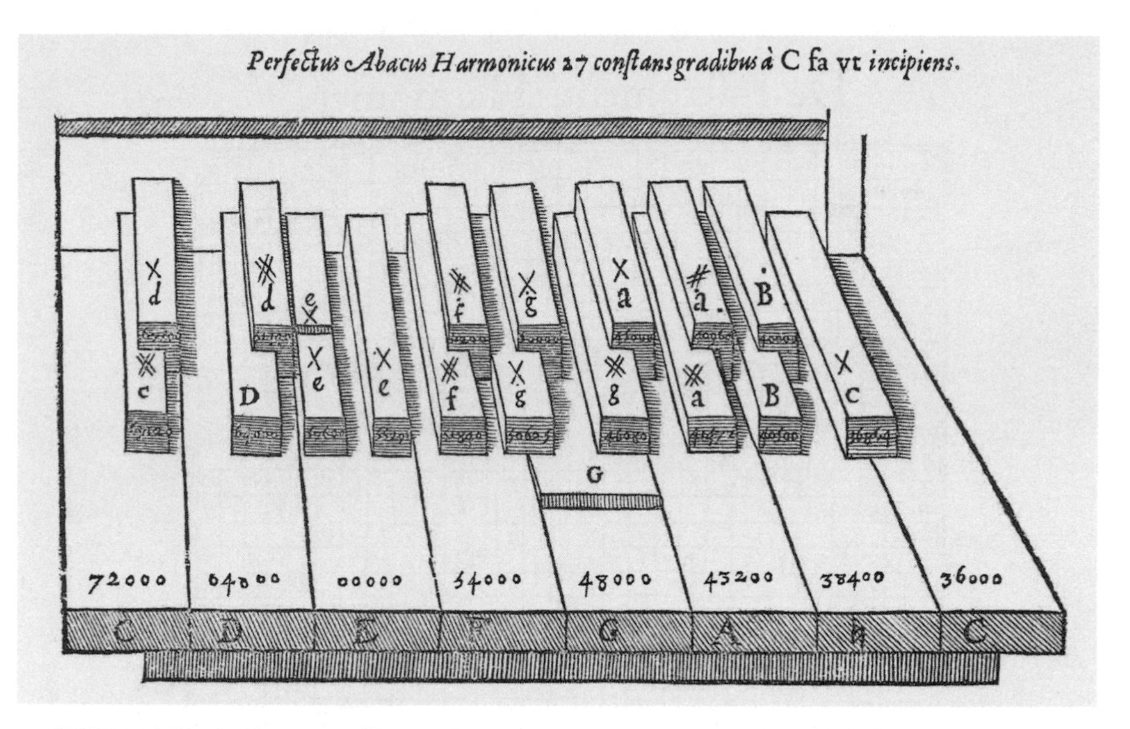

FIGURE 18. Marin Mersenne, illustration of his conception for an augmented keyboard with twenty-seven keys to the octave, from *Harmonicorum Libri XII* (Twelve books on harmonics), *"Instrumentorum Harmonicorum: Liber Tertius: De Organis, Campanis, Tympanis, ac Caeteris Instrumentis,"* 1648 edition, page 119.

But Mersenne goes even further, mapping this fantastic interface onto an actual grid in order to demonstrate his system's overall perfection. He maps the twenty-seven keys in an octave onto twenty-seven numbered rows and columns, as seen in Figure 19. One can trace any combination of two notes in the octave horizontally and vertically to a single cell to gauge its consonance or dissonance. The columns terminate at bottom in an extension of the grid into actual keys, each marked with the note. Mersenne renders the accidentals as shorter and the naturals as longer rectangles. Each key thus appears as an extension of the rationalizing grid itself, another cell in the table. One is reminded of the illustration of the chekker in Gerson's text, in which the eight keys align with the eight files of the chessboard, but here the link between the grid and the keys is far more explicit. And all of this has much broader import, because Mersenne aligns the seven consonant ratios (each of the eight natural keys paired with one of the seven remaining natural keys) with the seven planets of the classical solar system: Moon, Mercury, Venus, Sun, Mars, Jupiter, and Saturn. Trace any natural note to any

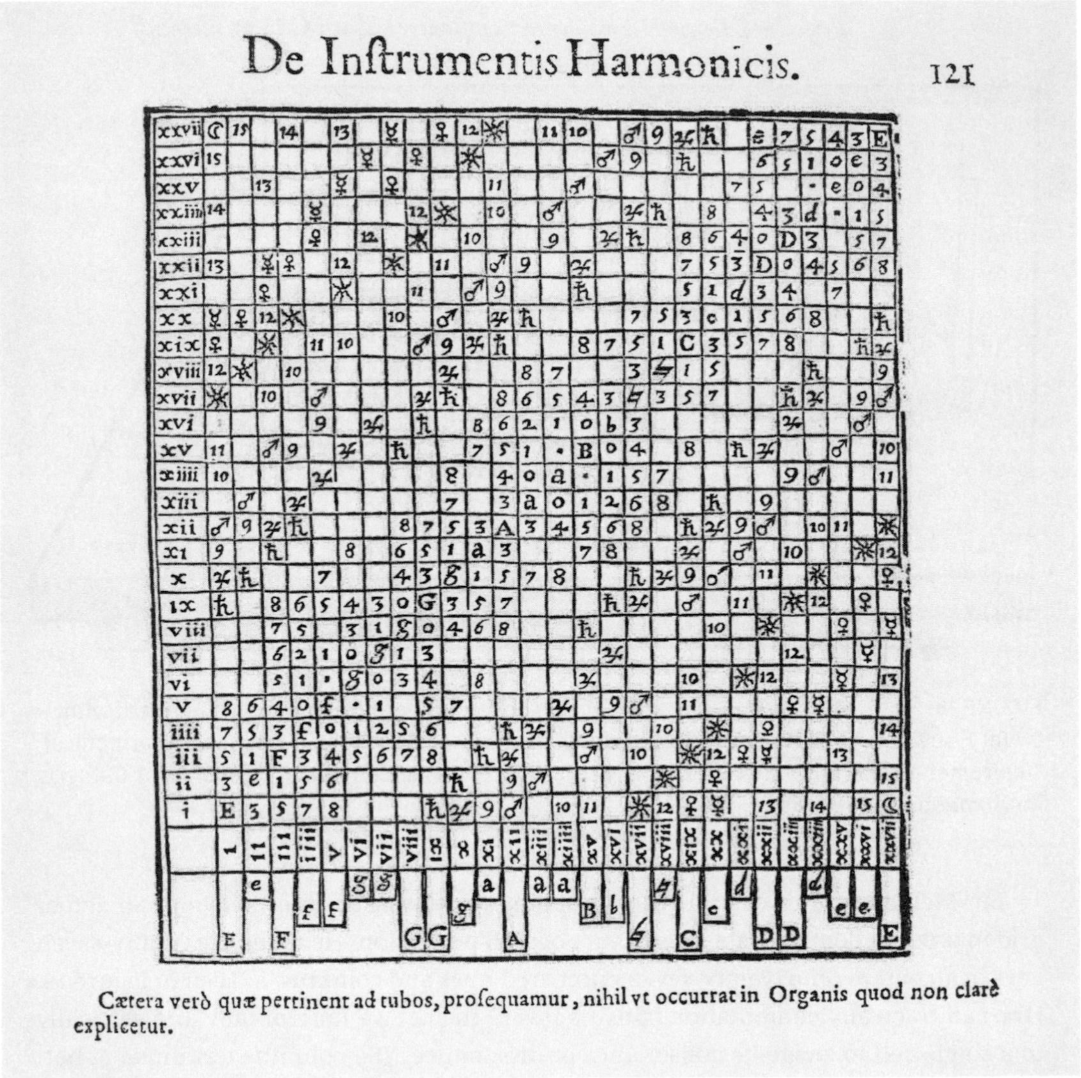

FIGURE 19. Marin Mersenne, gridded chart of the harmonic combinations possible with an augmented keyboard containing twenty-seven keys per octave, from *Harmonicorum Libri XII* (Twelve books on harmonics), *"Instrumentorum Harmonicorum: Liber Tertius: De Organis, Campanis, Tympanis, ac Caeteris Instrumentis,"* 1648 edition, page 121.

other, and they will intersect in the symbolic space of a celestial body. Other harmonies appear to have celestial relevance too. Mersenne thus devised a keyboard that not only renders the keys themselves expressly as extensions of grid logic, but further imagines the whole arrangement as having astronomical significance. Kepler had wondered which planet sings soprano and which bass, but Mersenne proposed that the keyboard itself can hail the heavenly spheres.

The fact that none of this gets expressed so directly before the seventeenth century does not mean that earlier keyboards were devoid of similar rationalizing techniques. Mersenne only confirms a tendency toward grid rationality present in keyboard design for centuries. He made manifest what had long been latent. He articulated practice as theory. In his imaginary instruments, he was able to show the values already concealed in material keyboards. Even the chekker did not truly inaugurate the grid rationality of the keyboard; it only made it more conspicuous, by name certainly and probably also in form. But the chekker likely did spread this grid rationality to a wider range of musical interfaces. First, grid rationality spread to different kinds of instruments with drastically different kinds of actions, all concealed behind a single orthodox kind of keyboard, culminating with the modern piano. Later, as we shall see, keyboards leaped to other kinds of instruments, including more recognizably scientific instruments including telegraphs, writing machines, and computers. As the keyboard spread to new technical contexts, it carried with it the calculative rationality that originated in Western music. It also carried with it the requisite configuration of the human as a rational being. Seated at my computer, my fingers on a staggered grid of keys, my eyes lifted to a grid of pixels, I tap out these words through grids that are the heirs of this musical, mathematical, and scientific legacy.

Still, this should not be taken to imply that such grid rationality was universally or uniformly distributed. As we trace the proliferation of keyboard culture beyond music and into other more recent contexts, we will see that that the keyboard's rationality could be selectively assigned to certain kinds of people deemed most deficient of it.

CHAPTER 8

Human Types

In the previous chapter we saw how the musical keyboard developed rapidly across a range of musical instruments beginning in the fourteenth century as a highly rationalizing interface. In this chapter, we turn to a much later stage in the history of the keyboard, as it escaped musical applications altogether. Starting with telegraphs and continuing through other kinds of writing machines, the musical keyboard gradually transformed into the interface most computer users write with today. However, the legacy of Greek rationality persists even in the keyboards of writing machines, not in the harmonic ratios described in the previous chapter but in their visual equivalent, the grid logic that discretizes writing into so many evenly sized and spaced keys. Letters that, in script, take different amounts of time and effort to produce become utterly equivalent on a keyboard. The keys resist equally too, so that the weakest fingers must work just as hard as the strongest. Everything succumbs to this relentless logic of equivalency that transforms the analog gradations of script into so many discrete digital units. In the process, the rationalizing interface of the keyboard has rationalized not just writing but writers too.

This chapter examines how keyboarding rationalized one kind of person, those deemed especially deficient in reason. We should not wonder that societies that attribute unequal rational capacities to adults and children, men and women, whites and nonwhites, would selectively distribute rationalizing remedies as well. Keyboarding, it turns out, is a technique not just for accomplishing certain practical ends, but also, and more importantly, for correcting inequalities imagined to be innate. In that sense, the keyboard is an assistive technology, a corrective for those whose reason was thought to be disabled, and who were thus seen as lacking the full capacity for rational choice and self-control. If the kinds of early musical keyboards we saw in the previous chapter belonged almost exclusively to adult white men, in time the opposite would be the case, as both musical and writing keyboards promised to elevate everyone else to humanity's highest and fullest condition.

The keyboard is so common, so unconsciously employed by so many people, that it can be hard to see with fresh eyes. When the senses become dulled by familiarity, sometimes art can productively estrange the status quo. Let us return to an example already discussed in a previous chapter, Roger Vadim's 1968 film *Barbarella,* which at this stage can help us perceive what it means to think of the keyboard as an assistive technology, a prosthetic. In the film, the title character played by Jane Fonda falls into the clutches of the villain Durand Durand, who proposes to execute her in his "Excessive Machine." Barbarella lies captive inside the machine, only her head visible at one end, as Durand controls the entire apparatus with something like a cross between a computer console and a musical keyboard. The machine deals death by orgasm, but it is also a musical instrument, consisting of a vocal woman who Durand plays from the keyboard using some sort of printed score. Musicologist Emily I. Dolan has read the scene as evidence of "a particular mode of instrumentality, namely one based on the idea of complete control."[1]

In the end, the scene is also about turning the tables on complete control, as we have already seen. Instead of being destroyed, Barbarella destroys the machine, which meets its own fiery climax. Accordingly, the male master at the keyboard ends up diminished too. For the Excessive Machine was evidently a sexual prosthesis all along, one that strongly suggests Durand's own impotence. Whatever mastery he enjoys derives from this supplemental tool, and the scene invites viewers to snicker accordingly. Barbarella herself might be little more than a sex machine, but Durand's virility also depends entirely on assistive technology. It turns out not to be very manly to play the keyboard after all, given that doing so signals some more significant incapacity. If playing the keyboard is both empowering and emasculating, both a method of domination and evidence of deficiency, the pages that follow will ask what it means that millions of Barbarellas in the late nineteenth century also turned the tables when they took over a different kind of keyboard, the one on the typewriter.

This story of women's mass entry into the workforce as typists has been told many times, and historians often have debated the reasons for it and the results of it. I will not rehearse that familiar history in detail, aside from a brief reminder of the astonishing transformation wrought by typewriting. In the early 1870s, after the first commercial typewriters appeared in the United States, only about 5 percent of typists and stenographers were women, and almost all secretaries were men. By 1930 the ratio was reversed: 95 percent of typists were women and only 5 percent were men.[2] Still, it has never been quite clear why women proved so successful in occupations that involved keyboards, such as typing, stenography, and bookkeeping, but somewhat less successful in other kinds of office work. By 1930 in the United States, they made up a much smaller percentage of the overall number of office clerks than they did of the overall number of typists.[3]

Friedrich Kittler has explained women's domination of the keyboard as the direct result of their practical proficiency operating musical keyboards, but this is only part of the story. After all, if managers preferred to hire women to operate the keyboards of typewriters, why did they not hire women to operate other keyboard machines, such as printing telegraphs? The answer, I will suggest, is that typewriters, unlike printing telegraphs, registered broadly as assistive technologies for the disabled, and thus seemed more suited to those whose full humanity and corresponding capacities already were in doubt. Because European and especially American societies understood typewriter keyboards as prostheses, they associated them not just with the exercise of power but also with the remedying of powerlessness. To conceive of the keyboard as prosthetic is to perceive it as simultaneously enabling and disabling, humanizing and dehumanizing. When women took over the first keyboards of the information age, the meaning was decidedly mixed. Although they staged a successful rebellion against certain forms of patriarchal advantage in the workplace, their very success probably confirmed assumptions about their innate inferiority.

Keyboard Culture

Women's piano practice was crucial to their later success as typists, but not for the reasons some critics have suggested. Kittler has argued that piano training allowed women to out-compete men in the new field of typing. As recorded sounds displaced pianos in middle-class homes, he suggests, women exported their manual dexterity to the office, where they proved more adaptable than men to new finger-based interfaces.[4] That may be part of the explanation, but difficult questions remain. For one thing, as noted at the outset, it does not explain why women should have taken over typewriting so rapidly even though they made far slower gains in fields like printing telegraphy. Moreover, if piano techniques were decisive, they should have dominated printing telegraphy even more rapidly than typewriting, given that all of the most common printing telegraphs in use in the 1870s employed actual piano keys, rather than the typewriter's array of staggered round buttons. We will learn more about printing telegraphs in the next two chapters, but a glance ahead to Figure 26, a photograph of London's Central Telegraph Office in the late 1890s, shows one problem with Kittler's argument: at a point in the 1890s when women already made up a majority of typists, the operators of these Hughes printing telegraphs were exclusively men, despite the fact that they sit before what appear to be dozens of toy pianos. More important, although it is pleasant to imagine that women's practical ability at the keyboard compelled the labor market to include them in droves, since when has practical merit alone led to such rapid gains for any excluded group? To believe that women took over typing simply because they were better at it implies that they were working in a gender-neutral meritocracy.

However, it is still accurate to say that women typists benefitted from an entire feminized keyboard culture initially organized around the piano, a culture that made some other kinds of keyboarding seem socially appropriate for them. Margery Davies has suggested that the typewriter was not "sex-typed" as masculine when it first appeared, which gave women an opportunity to claim it for themselves.[5] In truth, however, it probably was "sex-typed" as feminine, not by virtue of the QWERTY keyboard itself but rather because of a range of broader associations of typewriting with domesticity, disability, and femininity. Social suitability probably mattered more than practical proficiency. This is easier to perceive when we recall that earlier writing machines had long employed some imitation of the musical keyboard. Although Christopher Latham Sholes's Type-Writer of 1873 featured rounded buttons modeled on telegraph keys, earlier prototypes of the same device used piano keys. Sholes's patent illustration from 1868, reproduced in Plate 7, shows what appears to be a small clavichord or even a toy piano, but with letters arranged on the white and black keys.[6] Sholes seems to have borrowed this interface from other earlier writing machines rather than from the piano directly. For instance, Giuseppe Ravizza called his writing machine the Cembalo Scrivano (writing harpsichord). It featured a row of white keys below a row of black keys, both resembling stubby piano keys.[7] Samuel Francis created an influential writing machine in 1857 with more realistic piano keys.[8] And there were many others, including the printing telegraphs we will discuss in more detail in the next chapter. Until Sholes's later designs established the basic interface that typists today take for granted, writing machines with musical keyboards were common.

These machines entered a world where piano playing had become an important social refinement for middle- and upper-class women and girls, and piano lessons for many of them compulsory.[9] As men and women increasingly retreated to separate spheres in Victorian society, men found less time to master such a complex musical instrument than did their sisters, daughters, and wives, even though men continued to dominate the ranks of paid concert performers.[10] Women's facility with the piano enhanced men's social status too, because it connoted surplus leisure time, which reflected the prosperity of the household's male wage earner.[11] Partly because the piano bore so much social weight, it became one of the most popular media technologies of the nineteenth century. As mass production brought down the costs of piano manufacturing, ownership expanded dramatically and pianos became standard equipment in middle-class homes. By the end of the nineteenth century, one out of every 360 people in England purchased a piano each year, and by 1910 there were about three million pianos in England in all, roughly one for every ten people. Only the United States had a higher rate of piano purchasing around this time, where 1 in 260 people bought a piano every year.[12]

As nearly every history of the typewriter has observed, Sholes promoted one of his earliest typewriters with a photograph of his daughter Lillian seated before the keyboard. Reproduced as Figure 20, this photograph presaged the wave of women office workers to come, even though Lillian was not likely more expert at operating the machine than anybody else. Nobody was operating the machine very well in 1872, given that touch typing lay more than a decade in the future, and even Sholes was not sure who would use his new device. Lillian's presence instead reflects the symbolic currency of a woman seated before a keyboard, which decades of domestic piano practice had trained millions to approve.

The photograph repays closer attention. Lillian appears to be entirely at ease, her fingers resting on three separate keys in a position that few typists would have adopted, her right hand covering part of the left side of the keyboard. The visible fingers of her right hand rest on the space bar, the *B* key, and the *G* key (assuming this keyboard uses the standard QWERTY arrangement, as some early models did) while she adjusts the platen with her left.[13] A pianist might have to reach the right hand to the left side of a keyboard, but a typist probably would not. With her other arm extended, she looks like a young piano pupil keeping her starting position with the right hand while attending to the sheet music with the left. Perhaps it is no accident that her fingers rest on letter keys that correspond with the names of musical notes. Lillian's eyes also appear to be narrowed or even closed, an odd pose for someone typing and an impossible one for someone copying. However, it is more than a little reminiscent of a musician deeply absorbed in performance. This is not a photograph of a tireless drudge repetitively copying all day, but of a woman at leisure, entirely at ease before an interface that does not appear radically new after all.

Sholes was not the only one who associated the typewriter with women long before they had demonstrated any special proficiency with it. After he licensed manufacturing to Remington and Sons of Ilion, New York, in 1873, the rifle maker improved the machine and reissued it the next year, now emblazoned with elaborately painted flowers and curlicued gold filigree. Some machines have small pastoral scenes painted on larger panels. The whole apparatus is fastidiously pretty, designed to complement women's domestic spaces rather than the manly worlds of law or business. Remington even mounted some of its early models on a sewing machine stand, the treadle of which operated the carriage return. It is hard to think of an interface with which nineteenth-century middle-class men would have been less familiar, or an aesthetic presentation less calibrated to their tastes.[14]

FIGURE 20. Promotional photograph for Christopher Latham Sholes's "Type-Writer" produced around 1872, depicting his fifteen-year-old daughter Lillian seated before a late prototype of the final version of the machine. Courtesy of the Wisconsin Historical Society, WHI-5125.

All of this produced not just new kinds of musical instruments but also, and more importantly, new kinds of women and girls. Through the piano, women and girls gained admission to the digital techniques that would prove indispensable in the looming information age, even though nobody knew that at the time. And they entered this mechanical world not as exploited laborers, like factory girls yoked to steam looms, but as higher status workers whose conduct was entirely compatible with middle-class standards of Victorian propriety.[15] Indeed, musical keyboarding was a compulsory performance of middle-class femininity already. At the keyboards of nineteenth-century pianos, women established both practical credentials and social license to access later forms of machine-mediated writing, calculating, communicating, and coding. They followed the interface, and a society that required them to operate one kind of keyboard could not easily forbid access to others later.

But if typing became socially acceptable for women through the analogy of the piano, it remains to be seen why employers and even male secretaries and clerks either invited or failed to resist women's intrusion into the labor force. After all, women did not displace men as paid concert pianists, for women dominated the instrument only in the home, where their expertise was mostly confined to leisure, aside from giving lessons to other women and girls. Nor did women displace male telegraphers operating the small piano keyboards of early printing telegraphs, which date to the 1840s and the use of which expanded rapidly in the 1870s. If a feminized keyboard culture had been sufficient on its own to flood modern offices with women workers, women likely would have seen greater gains operating other machines that were even more reminiscent of pianos. But they did not. And the reason, I want to suggest, is that a second factor had to come into play. Unlike those printing telegraphs and unlike the piano, writing machines often had been understood as prosthetic remedies for people whose alleged disabilities could disqualify them from the ranks of the fully human.

Prosthetic People

By the *human,* I refer to something like full personhood, a status accorded unevenly to different kinds of people. There are many ways of according this status, such as constitutionally or through concepts of human rights, or other forms of legal recognition, and these have varied widely over time as Western liberal societies have gradually, if unevenly, accorded more rights to more kinds of people. Despite the gradual expansion of rights over the last several centuries, everyone recognizes that there are classes of people who are not accorded full personhood socially, legally, and politically, including, for these purposes, both women and the disabled. I will refer to the widest possible range of people simply as *people* and to that more privileged class that enjoys the fullest range of rights and responsibilities traditionally accorded to able-bodied adult white men as *human.*

Just as people become human when states accord them human rights, as Hannah Arendt argued, so too they become human when technologies help them perform certain qualifying actions. Technologies that aid mobility, cognition, hearing, communication, hygiene, or dexterity, to name only a few, help millions of people assume a fuller range of rights and responsibilities. For these purposes, one especially prominent but often overlooked assistive technology is the binary switch. The most obvious example might be the so-called "handicapped buttons" located at many public doorways in Europe and the United States, which facilitate access for those with limited mobility. Anyone can use the handicapped button to open doors, and plenty of able-bodied users apparently do. In doing so, however, even able-bodied users accept some implication of their own incapacity. At the absolute minimum, using the handicapped button confirms that mildest and most temporary of disabilities, fatigue.

Then again, able-bodied users of handicapped buttons probably do not conceive of themselves as disabled in the least, but as having augmented or extended their capacities. In that way, prostheses often appear to be transparent or even additive to their users, not remedial. When prosthetics resist, cause pain, inhibit other actions, or require additional care, they resume their obdurate form as objects. Observing this possibility, Sigmund Freud once noted, "Man has, as it were, become a kind of prosthetic god. When he puts on all his auxiliary organs he is truly magnificent; but those organs have not grown on to him and they still give him much trouble at times."[16] No organ feels magnificent when it chafes. Prosthetic technologies thus can be more or less transparent, which is to say, they can register both to users and to observers as more or less of a remedy for some prior deficiency.

As Vivian Sobchack, Sarah Jain, and Steven Kurzman have noted, there has long been something far too totalizing and pejorative about the metaphor of the prosthesis in media studies.[17] Writing from the position of someone who uses a prosthetic limb, Sobchack has pointed out that the experience of someone using a prosthetic is often drastically different from that of a someone witnessing its use. This is probably true even of eyeglasses, which I sometimes forget I am wearing for long periods, but which signal clearly to others that my eyesight is limited. According to Sobchack, the prosthesis signals that body parts are "missing or limited," even though the person using it might "successfully incorporate and subjectively live the prosthetic and sense themselves neither as lacking something nor as walking around with some 'thing' that is added on to their bodies."[18] Indeed, allowing one to forget the fact of disability may be the prosthetic's highest function of all.

Sobchack's point is that the outsider's perspective has dominated prosthetic theory in the field of media studies, which has tended to be deeply suspicious of any medium that becomes transparent or naturalizing. Marshall McLuhan sometimes tried to summon people to a higher awareness that media both causes and remedies some corresponding injury.[19] For McLuhan, "the wheel as an extension of the foot" is also "the immediate occasion of the extension or 'amputation' of this function from our bodies."[20] To use the wheel is to lose the foot. To extend the body is really to deplete it. Accordingly, while prosthetic theory may seem like an analysis of technology, it also betrays a sweeping disposition toward people. To conceive of technology prosthetically from the outsider's position is really to conceive of people as a collection of technically remediable deficiencies, as beings short of wholeness and self-sufficiency. By most standards I would count as an able-bodied adult, but a quick inventory of my prostheses suggests otherwise, heavily dependent as I am on corrective lenses, dental fillings, and the titanium screws holding together one knee. Reading, eating, and walking remain easy for me, but only because these assistive technologies unobtrusively help me approximate an ideal of human wholeness. And yet I rarely think of them that way. I rarely think of them at all. Only when I do think about them prosthetically do the deficiencies they remedy return to consciousness too. Decades of media studies have insisted that there is something politically valuable in this consciousness, in that it helps debunk the alleged naturalness of the human. That may be true for all sorts of good reasons, but it also pays to recall Sobchack's reminder that, for those enduring some kinds of disability, there can be something politically valuable about sustaining an illusion of the naturalness of the human too.

Prosthetic theory thus cuts two ways. If it sometimes defines all people as damaged and incomplete, at other times it fondly conjures the ideal of organic humanity. Both tendencies are usually present in some degree. For instance, Henry Ford once measured industrial production as equivalent to the disabled bodies that his factory could restore to a full level of economic productivity. Ford determined that fewer than half the jobs in a factory required "strong, able-bodied, and practically physically perfect men." The rest "could be performed by the slightest, weakest sort of men," and even by "women or older children." In fact, Ford goes on, "670 could be filled by legless men, 2,637 by one-legged men, two by armless men, 715 by one-armed men and ten by blind men."[21] Here we see a sliding scale of humanity fully articulated, as it descends from "perfect men" through the "weakest sort of men," through women and children, and finally to amputees and the blind. Only those "perfect men" at the top of the hierarchy can enjoy a relationship to technology that is not prosthetic, at least in Ford's account, because they are naturally whole already. For everyone else, technology remedies defects. Humanizing workers through industrialization requires, at a prior stage, dehumanizing them as damaged goods.

Not everyone is nostalgic for natural and organic humanity. For many prosthetic theorists in media studies, everything is prosthetic, which means that there can be no pure, whole, or complete human prior to techniques of hominization.[22] Bernard Stiegler retains the term *prosthesis* precisely to indicate that technical prosthetics both create and destroy traditional conceptions of the human.[23] Stiegler argues that tools did not just adapt to the capacities of humans; the capacities of humans also adapted to tools. So which came first, the hand or the knife? They co-evolved, Stiegler says, such that the human "body and brain are defined by the existence of the tool."[24] And not just defined by, but physically shaped by the tool. After all, it would be strange if early hominids suddenly began walking around on their hind legs before they had anything they needed to do with their especially dexterous hands. It would be even stranger if they evolved especially dexterous hands before they had anything they needed to do with them. Flint and fingers must have mutually modified each other, which is to say that the human is itself partly an effect of manual techniques. An opposable thumb is itself a tool-mark, etched by obsidian into our very genes.

For Stiegler, the technical origins of the human raise an important question: what "if we already were no longer human?"[25] That is, what if that which passes as human has always derived only from its alleged inferior, mere technics? We may never have been purely, organically human in some prelapsarian moment before Adam had to take up the plow. A further point is equally important. Even though the technical cuts people off from traditional conceptions of their own humanity, it allows them to claim the status of full humanity too.

One might note a universalizing tendency of some posthumanist versions of prosthetic theory that sometimes smuggles a universal subject back into the conversation, and in doing so, distracts attention from categories such as race, class, or gender.[26] As we have already noted, the human is a status distributed unequally among people. As a result, it pays to consider not just how the human might be constituted technologically in the most general way, but also how some people end up being humanized more than others. Prosthetic theories always are social interventions, for better or for worse. The question, then, is not just whether the typewriter keyboard was prosthetic; in rather obvious ways, it surely was. The real question is who these prosthetics served, especially in that period before keyboards had become practically universal, and thus mostly transparent to everyone who depends on them today.

Disabling Gender

In machine writing, the logic of the prosthesis was close to the surface from the start, so initially there was not much of an illusion of transparency to be dispelled. Many inventors designed early devices specifically to remedy disability, especially visual disability and disabilities affecting manual dexterity.[27] These include Pellegrino Turri in 1808,[28]

Claude Lechet and Franz Huber around the same time,[29] Johan Knie in 1838, Charles Thurber in 1845, Pierre Foucauld and a Dr. Saintard in the same year,[30] and Alfred Ely Beach in 1856.[31] Giovanni Ravizza made several prototypes for the Milan Blind Institute around 1860, and Danish minister Hans Johan Rasmus Malling-Hansen, while working as the principal at the Royal Deaf and Dumb Institute in Copenhagen in the late 1860s, created his successful *skrivekugle,* or *writing ball,*[32] the most famous user of which was Friedrich Nietzsche, who experimented with it enthusiastically as a remedy for his own rapidly failing vision.[33]

However, the inventor of the first writing machine to enjoy widespread commercial success, and the model for most writing machines that followed, paid little or no attention to disability. Before Sholes and his collaborators licensed their designs for their Type-Writer to Remington in early 1873, they had much larger markets in mind, such as telegraphers who might use the new machines to transcribe incoming messages. But so associated was machine writing with disability that, even after Remington started mass producing typewriters for sale in 1874, observers continued to describe them in prosthetic terms, often specifically invoking disability. "Deaf and dumb people take naturally to the typewriter," one journalist claimed in 1888. "There is one boy, Thomas M. Caton, who uses the Remington with rapidity and accuracy, who is not only deaf and dumb but also blind."[34] Similar appeals invoke different kinds of disabled people, such as "people who . . . are nervous, and liable to make a wretched scrawl of any manuscript." And they frequently invoke that paragon of limited human capacity, the child: "The machine is so simple that a child can readily learn to use it," one journalist enthused in 1880.[35] Although Sholes did not market to the disabled, blind users found the typewriter extremely useful. By 1890 the New York Institute for the Blind provided fourteen machines to its students.[36] Even Helen Keller bought a Remington typewriter in the 1890s and for the rest of her life rarely travelled without similar portable models.[37]

These explicit references to disability from the 1880s and early 1890s betray deep and probably only semiconscious anxieties about what was being lost: not limbs or eyes themselves, but a more general sense of self-sufficiency. For that reason, early typing implied a more common kind of disability than blindness; it confirmed that whole complex of disabling conditions that the late nineteenth century associated with being a woman. The commentator quoted above who recommended typewriters for "nervous" people "liable to make a wretched scrawl" was, as we shall see, likely thinking specifically about women. Although a great deal of recent work has focused on the intersection of gender and disability, rather less has attended to gender as disability in historical terms. Yet, as Sarah Jain has argued, "certain bodies—raced, aged, gendered, classed—are often dubbed not fully whole."[38] Similarly, Rosemarie Garland-Thomson has emphasized that, for centuries, Western societies have routinely treated women as damaged men. Aristotle called women "mutilated males," and Freud saw them as having endured a different kind of amputation, symbolic castration.[39] As a result, until late in the nineteenth century, most women were seen as constitutionally unsuited even for clerical work. Deemed nervous to the point of hysteria (an affliction sourced from the uterus) and sensitive to all manner of shocks, distempers, and infirmities, women endured medical restrictions justified by their alleged incapacities. However, in these dynamics, cause and effect were liable to be confused: the remedies did not necessarily follow the disability; they sometimes caused it. As philosopher Iris Young has argued in an essay titled "Throwing Like a Girl," even subtle forms of sexism can have serious effects on overall bodily comportment, effects that then get glossed as evidence of innate disability. Young concludes, "Women in sexist society are physically handicapped."[40]

The disabilities that the typewriter remedied were of this culturally inflicted variety. All disabilities have a physical component, but they count as disabilities only within the context of social and cultural expectations. It is true, as Kittler claimed, that typewriting disrupted what he called the "sexually closed feedback loop" of male-dominated script, which is a slightly different matter from how script became a disabling for women in the first place.[41] As Tamara Plakins Thornton has shown, women were forced to learn inferior versions of Spencerian script known as "ladies' hands" for many decades. According to Thornton, "Ladies' hands were considered easy both to learn and to execute, and so symbolized female physical delicacy, intellectual inferiority, and constitutional flightiness."[42] Later, Victorian education gendered handwriting even more strictly. Both genders learned basic penmanship, but only men received further training in script that connoted the manly self-control thought proper to professional life.[43] Having been habituated to an inferior kind of handwriting, women found themselves bound to demonstrate their allegedly natural incapacity. But the typewriter obliterated this difference, suddenly and with no need for laborious modification of ingrained habits of penmanship. As an assistive technology, it made their writing entirely equal to that of men, indeed indistinguishable from it.

Ladies' hand was a disability deliberately inflicted on women from an early age, like foot binding. Promoted as a social grace, it ended up debilitating them instead. No wonder so many women experienced typewriting as a welcome escape from this trap. In 1880, less than a decade after the appearance of the first commercially successful typewriters, women already made up 40 percent of all typists. Just a decade after that, in 1890, they amounted to more than 60 percent of all typists. They continued to claim roughly 10 percent more of the labor force each decade, until by 1930 they amounted to more than 95 percent of all typists.[44] They made other gains in other industries, but not as rapidly. By 1920 they amounted to only about half of all bookkeepers, using keyboard machines like those discussed in chapter 5. They made even fewer inroads working as telegraphers, though they are represented as early as the 1850s in Switzerland and the United States, and the 1860s in France, Germany, and Russia.[45] By 1890, there were more than eight thousand in the United States, only about 15 percent of the total.[46] In contrast, women typists by that point were already a decisive majority.

Such rapid gains involved more than just changes to the culture of penmanship. Margery Davies has argued that "capitalist expansion, not the typewriter," brought women into the office in droves.[47] Women's high rates of literacy helped as well, as did their willingness to work for significantly lower wages than men.[48] As a result, she concludes, women's domination of secretarial work would have happened even if they had continued using paper and pen. Perhaps so, but that does not quite explain why telegraphers did not hire women at similar rates, or why women did not dominate other clerical jobs such as bookkeeping. Although Davies worries that crediting the typewriter for women's gains "can encourage the conclusion that a technological invention causes women's employment because said invention is particularly appropriate for women," that seems to have been precisely the case.[49] The typewriter promised not only to remedy women's penmanship, but also to regulate the rest of their allegedly nervous and unreasonable natures so they could approach a male standard of productivity. Even early terminology persistently conflated women with mechanical typewriters. Until the 1890s, the term *typewriter* referred ambiguously both to the machine and to its operator, which betrays a deeper conflation of woman-*at*-machine with woman-*as*-machine. That conflation persisted even after the term *typist* replaced *typewriter* as the name for human operators. A photograph on a 1912 cover of the company trade journal *Remington Notes* depicts Russian novelist Leo Tolstoy with two women secretaries and a Remington typewriter. Its caption reads: "Count Tolstoy and his three assistants."[50]

Women's domination of these keyboards thus contains both revolutionary and retrograde potentials. It was revolutionary for all the reasons historians have documented for decades. Women typists found safer, higher paying, and more fulfilling work in offices. They thereby gained more economic mobility, increased social status, and greater independence. But their domination of keyboards was retrograde in that women could claim those advantages only by acquiescing to their description as less endowed with a capacity for creativity and judgment, unable to write properly, better suited to repetitive work, and more mechanical by nature. Deficient of the fullest measure of humanity, they could produce at a male standard only when remedied by machines. Women's reason thus seemed to be an inferior kind, recruited from without as a prosthetic supplement for something lacking, something like the natural and organic Cartesian mind. We saw in a previous chapter that women's bodies were commonly deemed excessively mechanical, a condition that interfered with their capacity for self-discipline, constancy of purpose, emotional and moral self-regulation, and precision.[51] All those qualities might be restored by machines. The cure was homeopathic. An excessively mechanical body demanded extensively mechanical remedies.

Accordingly, it is not just that the typewriter leveled the playing field for women clerks and secretaries by eliminating the disabling system of script, but also that men likely interpreted typing as a confirmation of the users' incapacity. Women flocked to the typewriter, but men also probably fled from it. There are other complexities too subtle to address fully here, such as prevalent ideas that women were so infirm that they could not endure even the mechanical remedy of typewriting. Those arguments clearly did not prevail. And obviously the suitability of this mechanical remedy intersected with issues of class and race in ways that deserve much more attention. Still, the fact that millions of mostly white, mostly middle-class women willingly accepted this bargain suggests a deeper renegotiation of gender itself. In her account of "technologies of the gendered body," Anne Balsamo argues that gender is "a social consequence of technological deployments."[52] She has in mind invasive reconfigurations of the body such as pacemakers, cosmetic surgery, and bodybuilding, but much more mundane activities such as keyboarding end up incorporated as gender as well. Gender is made not just out of technologies, but also out of techniques, reciprocal relations between human practices and technical artifacts. And these techniques rehearse and impose the full range of gender's social meaning, evincing in this case, and at the same time, both a claim for women's full and equal humanity and a counterclaim for their alleged inferiority. As a result, we should take care not to overstate how uniformly allied the binary switch is to power and mastery. As with Durand Durand, the situation is often contradictory and mixed, for the instrument of power can also affirm a deeper and more pervasive powerlessness.

Over a much longer period of time, the typewriter keyboard may have shifted the seat of reason from the mind toward the hands. As reason ceased to be only something that people *have* and also became something that people *do,* it appeared less as a fact of nature (innate mental property) and more as a feature of culture (acquired manual ability). If reason is as reason does, what difference does it make whether rationality is mental or manual? Male office workers still could insist on their superior reason, but even they probably sensed that the boundaries of the human were shifting around them. Once reason lodged in hands and not just in minds, many people, including employers, recognized that the most conspicuously rational people might be the ones most willing and best able to keep their fingers on the keys.

In that sense, many women became digital before men, for better and for worse. They externalized digital rationality first, largely because men excused themselves from having to do so. In the process, women also externalized reason itself, which they mapped onto the extremities of their hands. Men, in contrast, continued emphasizing their own superior mental status, deployed through the analog media of script and speech, and through managerial roles that deemphasized direct manual techniques. Over time, however, the distinction between women's manual reason and men's mental reason proved hard to sustain. Gradually, men also began demonstrating their rationality by communicating and calculating using those grids of binary switches that women mastered first. Today, the techniques of male middle managers look more like those of nineteenth century women typists than ever. As a result, their capacity for reason has been similarly organized around digital interfaces, and their own claims to natural superiority correspondingly reduced, though certainly not eliminated. Upper managers still seem to delegate many manual functions to others, such as email, which affirms that they remain creatures of the mind in the most traditional ways. But for the rest, digital rationality and the keyboarding it entails have become compulsory. Not so long ago, it was common to meet older male academics who could not or would not type, and whose wives had typed their publications from manuscript for decades. But they are rare today. Word processing has automated the amanuensis (a term tellingly built on the Latin for *hand*), so that now almost everyone must take the mechanical remedy of the keyboard for themselves.

Accordingly, the most revolutionary moment in the history of typewriting might not have been when women took over the field, but decades later with the advent of modern computing, when millions of men had to start typing for themselves. At that point, the keyboard brought most men down to a traditionally female level of disability, forcing them to compete not on the tilted terrain of fanciful biology, but on the more level ground of now standardized and nearly universal techniques.

CHAPTER 9

Chording and Coding

Douglas Engelbart's 1968 demonstration of interactive computing technologies at the joint conference of the Association of Computing Machinery and the Institute of Electrical and Electronics Engineers, a demonstration known ever since as the "mother of all demos," is well known for introducing many of the innovations that computer users take for granted today. These included the use of windows, hypertext, file linking and revision control, copying and pasting, and even the computer mouse, which Engelbart and his team both invented and named. Engelbart was a pioneer in early interactive computing, determined to harness computers to address what he saw as the increasing complexity of the problems facing the world. But computers, Engelbart thought, should be more than just high-speed calculators for complex mathematical problems; they should be integrated more fully in the operation of individual and especially collective human thinking. His Augmentation Research Center at the Stanford Research Institute had the goal of "augmenting human intellect," as Engelbart had put it in a 1962 conceptual framework, which entailed nothing less than "finding solutions to problems that before seemed insoluble."[1]

Thierry Bardini has insightfully described the mother of all demos reverberating through the field of computer science, but my concerns are much narrower here, limited to the manual interfaces Engelbart used to operate the system.[2] For the duration of his ninety-minute demonstration, Engelbart sat before a futuristic console for his NLS ("oN-Line System") designed by the Herman Miller furniture company, pictured in Figure 21, containing at its center a decidedly antiquated QWERTY keyboard devised almost a century before.[3] The keyboard was flanked by two other input devices. On one side, Engelbart operated the first computer mouse, which looked remarkably like those still in use today. On the other side, Engelbart operated a third input device that has fared less well, a small chorded keyset consisting of five long keys, sized and shaped roughly like piano keys. By pressing these five keys separately and together in their thirty-one different useable combinations, Engelbart could enter text with one hand without removing the other hand from the mouse, an enviable arrangement. Together,

FIGURE 21. Workstation of Douglas Engelbart's NLS system, used in his 1968 demonstration of interactive computing technology. Courtesy of the Division of Medicine and Science, Smithsonian National Museum of American History, catalog #2015.3073.11.

the mouse and the chorded keyset could have made the keyboard redundant, if only subsequent computer users had embraced the gadget on Engelbart's left as eagerly as they embraced the one on his right.

But they did not. Instead, the vast majority of users continue to operate desktop computers by combining the mouse with an interface cribbed from a Victorian writing machine, the typewriter keyboard that even Engelbart could not quite do without. The typewriter keyboard had been part of information processing from an early stage, including early computing. IBM made electromechanical card-punch machines as early as 1933 that employed QWERTY keyboards, and early computers such as ENIAC (Electronic Numerical Integrator and Computer) read input data from such punch cards.[4] By 1951, the UNIVAC I (Universal Automatic Computer) had several QWERTY keyboards, one at the center of its massive console, a Remington electric typewriter for output, and a separate QWERTY-based Unityper to prepare punch tape offline. By the mid-1960s, the ubiquitous Teletype Model 33 and other similar teleprinters often served as computer terminals as well. It is not surprising that mechanical writing techniques would have transferred to analogous functions in computing, but it is more surprising that QWERTY has persisted for so long, even as manual typewriters became electric, as electric typewriters became electronic computers, and as electronic computers began using graphical user interfaces, virtual touch screens, and voice commands. Historians sometimes assume that the keyboard remains in its current form because it would take too much effort to change the settled practices of so many institutions and individuals. Changing from QWERTY to something like Engelbart's chorded keyset would be like changing from the imperial to the metric measuring systems, or like requiring automobiles to drive on the opposite side of the road. Once established, large systems compounded of extensive material infrastructure, ingrained training protocols, and generations of human habit require massive investments to dislodge.

Bardini explains the failure of the chorded keyset largely in these terms, but there are other factors in play.[5] After all, the history of technological change has as many examples of the rapid overthrow of existing technical systems as it does of their stubborn persistence. QWERTY itself overthrew a stubbornly entrenched competitor to become the dominant mechanical writing interface when Christopher Latham Sholes's stacked array of round buttons replaced the musical keyboards that had been used on earlier writing machines, printing telegraphs, and stenographs for decades. A musical keyboard familiar to millions suddenly gave way to something quite different. If QWERTY could displace musical keyboards, why have chorded keysets failed to displace QWERTY? The answer, I will suggest, is not that the chorded keyset was too radically new, but that it was too old, too laden with the explicitly mathematical rationality of the musical keyboard, which QWERTY had partially concealed. That mathematical legacy is most evident in musical chording itself, the simultaneous pressing of multiple keys to produce consonant harmonies. Although Sholes designed many of his prototypes with musical keyboards, neither his prototypes nor any of the other early writing machines with musical keyboards permitted chorded inputs. Keys had to be pressed one at a time, and if two keys were pressed simultaneously, machines were likely to jam. There is thus something deeply counterintuitive about using musical keyboards for writing on such machines, because the form seems to invite chorded input, but the function absolutely prohibits it. In contrast, the chorded keyset requires the simultaneous combination of key inputs, and not only were those inputs functionally like the musical keyboard's production of harmony, but they usually were defined mathematically too, not as ratios but as binary sequences of ones and zeroes mapped onto keys, and thus also onto the fingers of the hand.

My contention in the pages that follow is that the QWERTY keyboard has persisted for the same reason it initially displaced musical keyboards on writing machines: it enforces a rigorous seriality in the production of written language, and unlike a musical keyboard, its form invites no attempt at chording functions. If harmony is the musical register of simultaneity, melody is the register of seriality, one note after another. A piano excels at both, but to the extent that a writing machine is like a piano, it is a piano for melody alone. That largely accords with widespread expectations about writing, which, in alphabetical languages, proceeds one graphic symbol at a time. Writing is a serial operation, while computing permits and even requires simultaneous registry. However, even though a computer does many different things at the same time—processing binary digits in batches of eight, sixteen, or thirty-two—its Victorian keyboard ensures that, at the interface, its human user can only do one thing after another.

Thomas Mullaney has recently noted that "code consciousness" is peculiarly absent from Western typewriting, in contrast to its conspicuousness both in telegraphy and in most modern versions of Asian machine writing.[6] But, Mullaney says, "we would do well to remember that the keys upon which the symbols 'Q,' 'W,' 'E,' 'R,' 'T,' and 'Y' appear are simply *actuators*—they are *switches* not unlike those we use to ring doorbells and turn on light fixtures. That the closing of those particular switches should, in most of the world, result in the near-instantaneous appearance of those same symbols on one's screen is the result of a complex cascade of mediations that only a few could readily explain."[7] The code is still there, Mullaney is saying, but the modern computer conceals it within the machine. We might say that the QWERTY keyboard is a machine for producing code *unconsciousness,* for insulating language from apparent mediation by additional layers of code, and thus for creating the illusion of the immediacy of writing. Accordingly, Engelbart's chorded keyset violates deeply embedded assumptions about the seriality of writing, and by extension, deeply embedded fantasies about the way people should relate to language. Code unconsciousness makes language appear to be much more than just one code among many. It helps secure its status as the foundational code, part and parcel of the human condition, and innate to one particular kind of primate that has come to think of language as its exclusive and prized possession.

Against Utility

Sophisticated versions of Engelbart's chorded keyset are available today, but despite impressive efficiencies, none have succeeded in attracting more than a small number of users. Working versions have been on the market since the late 1970s, rightly extolled as fast, small, and mobile. The most sophisticated model at present is the Twidder3, which sells for less than $200 and combines the functions of both keyboard and mouse in a device roughly the size and shape of a small television remote control. Designed for one-handed use, Twiddler3 could conceivably be used to write while walking, driving, or lying in bed in the dark.[8]

If one-handed chorded keysets make machine writing more mobile and versatile, two-handed versions permit incredible gains in speed. According to researchers at the Max Planck Institute for Informatics, while a professional typist typically produces about 50 to 80 words per minute, a professional pianist playing "Flight of the Bumblebee" strikes 17 notes per second, which, if translated into text input, would amount to 204 words per minute.[9] Elite pianists can go even faster, they note, striking up to 30 notes per second. The researchers wondered what would happen if touch typing were retooled for the interface of the piano, and in a way that could exploit its chording principles. They produced a device that looks much like a twenty-first-century version of nineteenth-century writing machines with musical keyboards. Called PianoText, their system consisted of a modified Yamaha MIDI-capable (Musical Instrument Digital Interface) electronic piano programmed to produce text on a screen, which they used to train pianists and nonpianists in this very different writing technique.

They did not expect anyone to produce 204 words per minute. Pianists achieve such speed by practicing the same pattern of notes repeatedly, while typists have no such luxury. Besides, piano keys are larger and more spread out, and they travel much farther, which introduces other delays. But the advantages are still significant. Because a standard keyboard has more keys than letters in the Roman alphabet, common combinations can be mapped onto multiple locations, creating "'islands' of statistically co-occurring letters on different parts of the piano."[10] Common letter sequences also can be mapped onto familiar musical structures, so that musical habits reinforce typing efficiency. And because certain letter combinations and even entire words recur regularly in any language, an electronic musical keyboard can associate whole strings of letters with simple chord structures, and those too might be input at various octaves along the keyboard. The results will be sobering for anyone who imagines that their computer keyboard is the best of all possible interfaces. Novice typists who trained for 200 hours on a QWERTY keyboard reliably achieved typing rates of 50 words per minute. Those who trained on PianoText for just 140 hours achieved rates of more than 80 words per minute. With 30 percent less training, PianoText typists gained 60 percent more speed. If the goal is to type faster and more accurately, and to learn to do so more quickly, we should still be writing on piano keyboards today.

In the Italian Senate, some people still do. Since 1880, the stenographic system used to transcribe that body's proceedings has employed a device known as the Macchina Michela, pictured in Plate 8, the first true stenographic writing machine. Named for its inventor, Antonio Michela Zucco, who devised his first model in 1862, the Macchina Michela appeared a full decade before Remington brought the first commercially successful typewriter to market.[11] Most stenography machines outside Italy little resemble piano keyboards, but they work on roughly the same principle, using chorded key combinations to input entire phonemes or orthographic combinations of letters, parts of words, entire words, or even whole phrases. Zucco was a pioneer in phonetics, and his "phonographic palette," which was one of the earliest phonetic alphabets, linked spoken phonemes to an alphanumeric code. The machine he built printed this phonetic code onto a paper tape as operators pressed chords on the machine's musical keyboard, which could be translated back into conventional written Italian later. The Macchina Michela features two groups of ten keys each (six white and four black), one group for each hand. A gap separates the two group of ten keys to make space for the printer and paper roll. Today, computerized versions have an almost identical interface, but computers automatically translate the code back into written Italian, which can be displayed in real time on a screen. The speed of stenography comes at a price. The effort required to learn stenographic code is forbidding, and the fact that stenography machines record in code rather than natural language is not very useful for anything other than transcription. For transcription, the overriding requirement of speed justified overlaying a condensed code onto natural language, the addition of a subsequent step to translate that code back into natural language, and the creation of an entire profession to master the necessary techniques. In the nineteenth century, it probably would not have been possible to build a machine that would receive chorded inputs and output standard alphabetical writing. For that reason, many early nonstenographic writing machines employed musical keyboards even though they forbade chorded inputs. Miles Berry's 1836 English device used a piano keyboard, as did Antoine Dujardin's 1838 Tachygraphe.[12] Guiseppe Ravizza's *cembalo scrivano,* or "writing harpsichord," also featured miniaturized musical keys starting around 1847.[13] Some writing machines made by English telegraph pioneer Charles Wheatstone used piano keyboards in the 1850s, pictured in Figure 22.[14] So did Samuel Francis's sophisticated American machine from 1857, Adolphe Charles Guillemot's French one from 1859, and the first and second models of John Pratt's successful machines of the mid-1860s.[15] There are many others, including Sholes's prototypes of the late 1860s.[16] Yet none of these employed chorded inputs, even if the form of the musical keyboard seems to invite it. They turned an interface designed for harmony into one capable only of melody; they used an interface designed for simultaneity exclusively for seriality. Nor did later machines revert back to musical keyboards once electrification and computerization made chorded input more technically feasible, except stenography machines like the Macchina Michela and experimental efforts such as PianoText.

FIGURE 22. Charles Wheatstone's third and final design of a writing machine, circa 1855–60. Copyright by the Science Museum Group.

But the example of PianoText suggests that, even without reverting to musical interfaces, a computerized QWERTY keyboard easily could be programmed with chording functions. For instance, simple key combinations might produce common words such as *that* or *the* if, for instance, *T* and *H* were depressed simultaneously in any order. Among the 26 letter keys there are 325 possible simultaneous two-key combinations, and surely at least a quarter of those co-occur in convenient positions.[17] The fifty most commonly used words in any language might be linked to some of these two-key combinations, and another twenty-five common prefixes, suffixes and letter combinations to others. As it is, the only readily used combinations are those involving shift keys for capitals and control keys for certain special characters and accent marks, but even these must be pressed in serial order, with shift and control always pressed and held first. The other main exception involves those fifty-nine two- and three-key shortcut sequences designated for various administrative functions on my desktop computer, some of which can be pressed in any order. Surely it matters that computers restrict this largest number of most complex combinations to administrative functions such as saving and printing, which do not contaminate the seriality of language with chorded simultaneity.[18]

There were very few chorded writing machines in the nineteenth century that output alphabetical writing. The most sophisticated, Benjamin Livermore's extraordinary Permutation Typograph or Pocket Printing Machine from 1863, is a small hand-sized device with six buttons on one end, each of which prints a segment of a single letter.[19] By pressing between one and four of the keys simultaneously with the fingers of one hand, the machine formed an approximation of a single Roman letter. Livermore's six-segment orthography consisted of the four sides of a rectangle, each of which could be printed together or separately, depending on which buttons were pressed, as well as two additional segments in the middle, each of which formed the top and bottom halves of an *X*. When all segments were depressed, the device printed a box with an *X* inside. The printed letters resemble those of the Roman alphabet, but not always recognizably. Although some letters such as *I, L, M, O, U, P,* and *W* appear in forms close to standard orthography, others are approximations, such as *K,* which appears as an *X* with a vertical line down the left side. Others are counterintuitive, such as *E,* rendered in a form closer to *C.*[20] The results look oddly familiar and strange, reminiscent of the Roman alphabet but different enough that words do not quite swim into view: ⩞ᒪ⊏ ᑭ⊏ᐳMᑌ⩞∧⩞I ☐⊓ ⩞⊻ᑭ☐ᐃᐳ∧ᑭᒪ.[21] By defamiliarizing the alphabet, the Permutation Typograph makes written language appear to be a new kind of code, but a reader can grow accustomed to these modified letter forms quite quickly and read them with relative ease. The phrase reproduced above in Livermore's orthography is simply the name of his machine, The Permutation Typograph. But until these modified letter forms became second nature, a reader would have to translate from this new code back into familiar Roman orthography, conscious at every step of a second layer of symbolic mediation.

For the writer holding this device, the harder task would be to learn which buttons produced which letter segments, and to develop the dexterity to press the correct ones simultaneously. These patterns of digital input amount to a more complex gestural code, roughly equivalent to the inputs of a stenographer, though much simpler because far fewer in number. Because each gesture corresponds to a single letter, not to phonemes or strings of letters, only twenty six of the sixty-four possible key combinations needed to be memorized, rather than many hundreds in traditional stenography. With enough practice, writing with Livermore's Permutation Typograph probably felt as intuitive and as unconscious as touch typing on a QWERTY keyboard. Livermore himself apparently achieved this. According to his grandniece, he kept a device under his pillow in order to record stray thoughts while lying bed.[22]

N. Katherine Hayles has called automatic activities like these "incorporating practices," which are actions "that are encoded into bodily memory by repeated performances" until they become habitual.[23] Many manual techniques become incorporated in this way, including touch typing, as Hayles notes. If I were asked to label a blank drawing of a QWERTY keyboard with its proper letters, I would have to put my hands on the page and pretend to type to do so. The memory lives in my fingers. Or rather, an intervening code of gestures-to-letters resides in my embodied habits, in practical techniques rather than in visual memories of symbols. As a result, this gestural code does not feel much like code at all. It feels like *me.* Having been incorporated into my body as gesture, the code recedes from consciousness almost entirely, and I am little more aware of how I move my fingers while typing than I am of how I move my tongue while speaking. Both can be so thoroughly internalized, so fully naturalized as bodily comportment, that people can overcome the traditional Western denigration of writing as secondary or derivative and speech as primary and natural.[24] Just as I never pause to think that my vocal cords cannot make sounds without manipulating that object known as *air,* so, when absorbed in typing, I generally forget that my fingers cannot make letters without manipulating that object known as a *keyboard.* As a result, I can more readily experience the action of typing as being just as unmediated as speaking, which is to say that my fingers seem to be immediately in contact with the effects I seek.

It is humbling to consider how much media is devoted to fostering the illusion of immediacy, as ever more layers of media obscure the fact of mediation. How much machinery it takes to make me feel autonomous and self-sufficient! Such incorporating practices are not just things that people achieve with their tools, a comfortable collapse of boundaries between people and things, a reassuring Heideggerian *ready-to-hand,* but also something that technology does to people, as technology naturalizes itself as human body. These gestures were all learned socially and enforced technically, but they are now so deeply incorporated that they seem part of me, like the naturalized prostheses discussed in the previous chapter. Any second layer of code might be incorporated this way; once incorporated, it largely ceases to register as code at all. So it is not as if the simultaneous inputs of chorded writing are impossible to incorporate, just that it takes much more practice and deliberate effort to do so. After all, every writer has had to master a complex code already, writing itself, which equates letter forms (in alphabetical writing) with phonemes of speech. That is rarely a simple one-to-one operation, especially in a language as irregular as English, and yet, over time, writers become mostly unconscious that they are doing any such thing, at which point they experience an illusion of naturalness, a sense that writing and speaking perfectly coincide.

Whenever another layer of code is added on top of writing, it brings code to consciousness again. This effect is evident even with Zucco's phonetic code or Livermore's gestural code, but it may be most pronounced when the additional layer of code is explicitly mathematical. This is apparent in the most influential chorded writing machine of all, one that shaped the form and function of computers ever since: the telegraph transmitter created by French engineer Émile Baudot.

Baudot was the source of two great innovations in telegraphy. The first, time-multiplexing, allowed multiple messages to be transmitted along a single wire at the same time. This drastically improved volume efficiency and cut down on the amount of infrastructure required. According to one period source, the Baudot system could transmit three hundred messages of twenty words each in one hour over a single line, which "very much surpasses in rapidity" a single Morse machine.[25] Because he pioneered such immense improvements in transmission rates, his surname has been used ever since to refer to the rate of information transfer in a communications channel, the *baud* rate. Second, and related to the first innovation, Baudot patented a telegraphic system in 1874 with a small five-key transmitter that resembled a toy piano. The operator depressed these keys in chorded sequences to transmit binary codes to a receiving device on the other end, which translated the codes back into Roman letters and inked them on a strip of paper. Baudot code is a 5-bit code, like Engelbart's, consisting of five digits of ones and zeroes, which permits thirty-one useable combinations, enough for the Roman alphabet and various control codes.[26]

Although Baudot notated his code using plus and minus signs, with plus indicating a depressed key, it could just as easily be represented with ones and zeroes. Those strings of ones and zeroes can then be translated into other number bases for easier reference, such as decimals. For instance, pressing the three keys to the left and leaving the two at the right untouched would correspond to the binary number 11100. This would transmit the letter U if the machine were set to transmit letters rather than numbers. That unwieldy binary might be made more comprehensible to people if rendered as the familiar decimal number 28. Operators, however, had no such luxury, for they had to speak the machine's binary language, and instead of pressing a key with a *U* on it, as on period typewriters, or instead of inputting the decimal number 28, they had to input a five-bit string of binary numbers by simultaneously pressing only the three keys on the left of the keyboard.

It must have been a punishing machine to use. Even though its limited five-bit code was not too difficult to memorize, operators had to enter text rhythmically in strict

FIGURE 23. Émile Baudot's five-key keyboard telegraph transmitter. This model was produced for the British Post Office at some point between 1897 and 1926. Copyright by the Science Museum Group.

synchronization with the cycles of the machine, each of which lasted roughly one third of a second.[27] A metronome ticked off fractions of a second to help operators synchronize their input with this cycle. No assembly line could be as brutally inflexible as this, as its split-second timing overran all ebbs and flows of human attention and imprecision. Because there would be no time for operators to translate binary numbers into alphabetical equivalents, the binary sequences had to be entirely incorporated into gestures of the fingers. As a result, probably for the first time ever, embodied digitality consisted of both finger-based numeracy and expressly binary code. The Baudot terminal converted the fingers from a base-10 to a base-2 numerical reference system and submitted them to the ruthless exigencies of the machine. In the case of Baudot's transmitter, chording is coding, nothing other than coding, and binary coding at that.

So the nineteenth-century chorded writing machines of Michela Zucco, Livermore, and Baudot all confirm that chorded writing has long been possible and even advantageous in some circumstances. But precisely because of that fact, they also reveal that the obstacles to wider adoption were not just technical but also experiential. The simultaneity required by chorded writing machines increases the human user's code consciousness, and thus interferes with the illusion of immediacy. Because chording is coding, and often expressly mathematical coding, it serves as a stubborn reminder that so-called "natural language" is not so natural after all.

Codes of Conduct

According to Vilém Flusser, digital computation was supposed to liberate people from earlier and more traditional ways of thinking and being, including their stubborn adherence to linear or serial procedures. For Flusser in the late 1980s, modern computation would transform tedious linearity into an exhilarating simultaneity. This would stem largely from the computer's use of "mathematical codes" rather than "letter codes," the first of which correspond, in Flusser's view, to a swarm of particles, a chaos of interacting factors, a welter of simultaneous signifiers that escape language's tedious linearity.[28] Put differently, Flusser says the computer inaugurates a shift from Heraclitus's flux, imagined as a river's linear flow, to Democritus's atomism, imagined as something like rain. From lines to particles, from the bounded to the sprawling, from one-thing-after-another to everything-all-at-once, digital computation might invest people with a similar aptitude for simultaneity. Flusser's vision borders on the utopian, but extends even to humble computer users tapping away at the keys. "Behind the keyboard whose keys they press is a swarm of particles, and this swarm is a field of possibilities to be realized," he says.[29] Perhaps so, but in Flusser's account, it is not entirely clear just where or how those swarming particles ever do get realized, or what difference they make to individual people, human societies, or anything else. Certainly the swarm of particles is not released through the keyboard, a relic of nineteenth-century technology that forbids most kinds of simultaneous input. At its interface, at the level of embodied technique, the computer remains doggedly committed to the serial input of strings of characters, one after the next.

Friedrich Kittler similarly imagined that modern computing restores a distinctly mathematical sensibility to literacy. Because the Greeks used letters of the alphabet both as numbers and as the names of musical notes, Kittler posited a golden age in which language, math, and music were all united by the same set of shared symbols. And this unity was not just abstractly symbolic, but organized through technical media such as musical instruments, especially the lyre and its bow. The lyre, Kittler says, "is not only a musical instrument such as exists in any culture, but a magical thing that connects mathematics to the domain of the senses."[30] But if Aristotle started to resist this unity, according to Kittler, "the European Middle Ages demolished this trinity of writing, tone and number for good," at least until modern computing began to restore it again.[31] As computers fed numbers back into the human, Kittler says, numbers came to seem less like neutral tools for people who think in language, and more like the substance of thought itself. As a result, in Kittler's view, modern computing reinjects the nonhuman into the humanities, and vice versa, recovering through its technical media that earlier condition in which language and math were united in a combined mode of reference.

I am far from convinced that computers accomplish any such thing. Perhaps some richer conflation of language and math happens among computer scientists who may have incorporated coding practices much more deeply, and who may in fact be committed to the priority of mathematics over language, but Kittler's argument would be more persuasive if more people wrote on chorded keysets like those of Baudot or Engelbart. Doing so would literalize Flusser's *swarming particles* as so many simultaneous ones and zeroes and incorporate them into the fingers. Even PianoText unites mathematics with writing as surely as Pythagoras had united mathematics with music, in that it aligns those codes with the traditional ratios of Western harmony. But the QWERTY keyboard does no such thing. In fact, early versions even underwent minor modifications to separate number and letter keys more completely. On early typewriters, the letters *I* and *O* did double duty as the numerals 1 and 0, but at some cost of efficiency, later designs mapped these potentially interchangeable figures onto separate keys. Afterward, the mathematical made almost no intrusion at all into the linguistic.

To understand why writing machines resist the intrusion of the mathematical, we need to distinguish two different kinds of rationality present to varying degrees in different kinds of keyboards. The first kind is the specifically mathematical rationality that characterizes only some switches and only some keyboards. At a minimum, mathematical rationality involves basic quantification and enumeration, but it usually also includes more sophisticated mathematical relations, such as the harmonic structures of Western music, the arithmetical functions of late-nineteenth-century mechanical calculators, or the binary code of Baudot transmitters. We can trace the mathematical rationality of switches back to the rudimentary mathematics of finger counting, as discussed in chapter 5, but machine mediation made far more complex operations possible. In that chapter, I also noted that the first widely adopted key-based calculators, the Comptometer and the Burrough's Arithmometer, also used chorded inputs; operators could input a five-digit decimal number by pressing five separate buttons at the same time. Wherever we find the chording principle, we typically find not just code, but often expressly mathematical code too.

The second relevant kind of rationality consists of the causal rationality of inputs and outputs, a feature of all switches. Causal rationality involves regularity of operation and the necessity of predictable results. Through machine logics, causal rationality ensures that certain things happen invariably for a reason, the reason being the operation of the switch. But what is the reason for the operation of the switch? Usually it amounts to some primitive gesture by a human being. The causal rationality of switches sorts actions into causes and effects and aligns humans decisively with the former. We have already traced the causal rationality of switching back to the indexical gesture and associated it with the imperious agency of early childhood. This kind of rationality is closely connected to the social and political values of freedom and autonomy at a profoundly individual level, for causal rationality often expands individuals' radius of action and magnifies their force.

The key point is that the interfaces of the most prevalent writing machines today have largely dispensed with mathematical rationality in favor of causal rationality, at least at the surface. Coupled with the computer's visual interface, the QWERTY keyboard creates the illusion of a noncoded system, which is also to say, a realm of strictly causal rationality in which gestures such as pointing and clicking directly compel invariable results. In a causally rational world, everything happens for a reason, and the causal rationality of switching ensures that those reasons emanate from people. In fact, the function of a switch is not just to compel action when touched, but to prevent action in all other circumstances. Absent human contact, the world remains static and inert. The more action ends up governed by switches, the more humans appear to be action's exclusive source.

Obviously, a modern computer keyboard does not exactly replace mathematical rationality with causal rationality, but rather conceals the mathematics as much as possible. As Mullaney noted at the outset of this chapter, the illusion of immediacy in word processing depends on a "complex cascade of mediations that none but a very few could readily explain," most of which are mathematical, and few of which users can even perceive. This was already happening in a limited way in Baudot's machine. On the transmitting end, operators had to translate Roman letters into binary code, incorporating it gesturally into the body, but on the receiving end, the machine automatically translated binary code back into Roman letters. Baudot's system concealed the math for reading but exposed it for writing. The next step was to automate the translation of alphabetical letters into code on the transmitting end too, which Donald Murray accomplished in 1901 with his Telegraphic Typewriter, a second-generation printing telegraph and precursor of the teletype. It replaced those keyboards derived from musical interfaces long standard on printing telegraphs made by Baudot, Royal Earl House, and David Hughes with a standard QWERTY keyboard.[32] Murray ensured that the mathematics of binary code no longer had to be incorporated into the body after all, by automating the translation of letters into and out of mathematical equivalents. Today's computers record and transmit text with descendants of the 7-bit American Standard Code for Information Interchange (ASCII), introduced in the mid-1960s.[33]

And yet, even as the mathematical codes themselves change, the interface of the keyboard remains the same, arranged as it was in 1873 and requiring roughly the same input techniques. When attached to a modern computer, the keyboard simulates the direct causal linkages of the mechanical typewriter, in which information appears not as a swarm of millions of mathematical operations per second, but more quaintly, as the kinds of analog and directly causal operations proper to a nineteenth-century clerk.

For instance, my word processor presents me with dozens of simulated analog objects that stand in for hidden algorithms. These include images of folders, scissors, clipboards, a waste basket, paint brushes, rubber erasers, simulated paper pages, and a magnifying glass. Each one conjures the illusion that a mathematical operation is really a causal operation involving familiar material things. But my computer's thorough simulation of analog office work includes one other Victorian implement just as anachronistic as scissors and rubber erasers, and the only one that does not appear on the screen: the keyboard itself.[34] Matthew Kirschenbaum's recent history of word processing rightly emphasizes the enormous differences between typing and word processing, as do many of his period sources who were enthralled by word processing's new possibilities.[35] But, precisely because word processing was so revolutionary, it seems all the more surprising that the QWERTY keyboard has persisted with so few changes. At the level of embodied human experience, the causal reason embedded in the keyboard is at least as meaningful as the mathematical reason that Flusser and Kittler associated with computation more generally. When coupled to the computer, the QWERTY keyboard is a machine for concealing mathematical reason largely by translating it into the illusion of direct one-to-one causation.

But there is an interface even more traditional than the Victorian typewriter, one that both recalls the keyboard's musical lineage and more openly acknowledges its mathematical investments: the chorded keyset. The QWERTY keyboard is one hundred and fifty years old, but through its relation to musical keyboards, the chorded keyset is more than two thousand years old. To return to Engelbart, we can now see that the three different input devices on his NLS console represent three different historical stages of input techniques. The mouse is the most innovative interface, a truly novel reconfiguration of existing technologies, and a means of simulating the indexical gesture so users can point to what they want. The QWERTY keyboard is much more traditional, invoking the techniques and values of nineteenth-century machine writing after it distanced itself from the mathematical implications of the musical keyboard. But the chorded keyset, which might seem like the most innovative interface, is actually the most traditional, drawing as it does on musical keyboards and chording techniques that strongly imply mathematical mediation. Moving from one end of Engelbart's console to the other, one travels further back in time, and thus closer to that classical period where math and language were not so clearly distinguished after all.

The QWERTY keyboard is thus an antisimultaneity screening device, a machine for enforcing seriality at all costs. As such, its effects have been rather different from those observed in the earlier chapter on calculating machines, which digitized the fingers by enrolling them in calculative techniques. In machine writing, the interface strove to purify itself of math, to render the digital as causal rather than coded, and thereby to foster a more powerful illusion of immediacy. In the last part of this book, we will look more closely at some of the social and political consequences of causal rationality when we examine the relationship between switching and liberal political values, especially in relation to one kind of switch that is the most causally rational of all, the trigger of a gun. But before that there is still a little more to say about the QWERTY keyboard. One thing at a time.

CHAPTER 10

The Archeology of QWERTY

One popular mythology of the QWERTY keyboard holds that the letters had to be scrambled to slow down nineteenth-century typists in order to prevent the type arms from jamming. Because input had to be serial, never simultaneous, as we saw in the previous chapter, the scrambled arrangement of letters would prevent typists from accidentally pressing two keys at once. The absurdity of this explanation defies even the lowest expectations, but it nonetheless betrays a widespread desire to understand technology in purely utilitarian terms. Any complex technical object might have been shaped exclusively by utilitarian design principles consciously formulated and intentionally applied from the start, but usually the matter is more complicated, as generations of scholars in science and technology studies have amply shown. We already have begun to excavate the cultural history of the keyboard in a media archeology that has exposed some of its eclectic sources. At the bottom layer, we uncovered the residue of the Greek hydraulis and the European chekker, and the legacy of mathematical rationality associated with both. Higher up, we encountered the feminization of the piano, and with it, bodily comportments and gendered expectations that transferred readily to machine writing. Higher still, we found a decisive break with the form of the musical keyboard, as machine writing rejected chorded simultaneity in favor of the seriality associated with natural language.

When we sift the material history of the keyboard even more finely, other surprising historical details appear, but still we find no overriding standard of utility that explains the result. The keyboard's form and function derived from many different intersecting goals and values, some that conflict with one another, others unconsciously applied, still others borrowed opportunistically but not very purposefully, and many that survived beyond their period of usefulness. The keyboard is a heterogenous, multifarious, contradictory, messy amalgamation of temporary work-arounds, short-term solutions, guesses about future uses, shameless appropriations, and hasty analogies. This is true of the material form of the keyboard, as we already have begun to see, but it no less true of the one feature that gave QWERTY its name, and which we have not yet addressed in any detail: the arrangement of letters on the keys.

The most common explanation for QWERTY's stubborn persistence over a century and a half involves what economist Paul David has influentially called "path dependency," a controversial term in economics that David developed specifically from the example of QWERTY. For David, path dependency explains how certain settled economic arrangements get locked in early and then remain resistant to change even once they are no longer optimally efficient. Or even if they were never all that efficient to begin with. According to David, path dependency explains how "important influences upon the eventual outcome can be exerted by temporally remote events, including happenings dominated by chance elements rather than systematic forces."[1] David's argument has been controversial among economists, mainly because path dependency implies historical processes that are anathema to the reigning neoclassical economic orthodoxy. To neoclassical economists, economic laws are no more historical and changeable than the laws of thermodynamics.[2] And one of the highest and most sacred economic laws of all is that individual economic agents in a functional free market will relentlessly maximize their utility. In such a system, market processes will always yield the best available allocation of resources.

The problem is that QWERTY seems to be far from optimally efficient for reasons that could be easily remedied, and so seems to exemplify a poor allocation of resources globally. Chorded keysets offer decided advantages, as we already have seen, and the Dvorak layout (patented by August Dvorak) and others for serial input using similar equipment promise similar gains but without the overhead of extra specialized training. Given the sheer volume of information that passes through QWERTY keyboards today, even tiny gains in speed, accuracy, or safety could yield significant economic benefits. Yet the liabilities of QWERTY are easy to see. For instance, in a world dominated by right-handed people, QWERTY is left-hand dominant: 56 percent of keystrokes in English come from the left hand, while the right controls less frequently used marks of punctuation.[3] The smallest and weakest finger on the left hand, the pinkie, is also severely overworked, controlling three letters, one of which is very common, along with the 1 key and the shift key. Low-frequency letters such as *K* and *V* are located centrally for stronger and more agile fingers, while high-frequency letters such as *A* and *O* are located on the periphery.

If QWERTY really is as inefficient as critics have long alleged, for these reasons and others, neoclassical economics would have to opt for one of two explanations: either a "market failure" occurred, involving the nearly universal adoption of an inefficient option for a century and a half, or the alleged inefficiency is economically negligible, and hence not really an economic matter after all. Given that there clearly is no "market failure" such as monopoly conditions or a government mandate to use QWERTY, any other kind of market failure operative since 1873 would amount to a significant challenge to economic orthodoxy. That is an intolerable conclusion for most neoclassical economists. As a result, most of David's critics have opted for the second explanation, concluding either that QWERTY is perfectly efficient for its purposes or that its inefficiencies have been overstated.[4]

We need not try resolve the debate among economists here; I mention it only to show how QWERTY sits at the center of a great question about whether the way things are has more to do with economic laws tantamount to laws of physics, and hence with utility, or with a highly contingent history in which things might have happened in some drastically different way. I side with the latter position, and it must be acknowledged that, to most historians, David's argument might seem a little self-evident. Unlike his audience of contemporary economists, historians take for granted that the way things are has been profoundly shaped by how they previously unfolded, even in economic arrangements. For instance, people may believe that they drink red wine from distinctive stemware with a certain shaped bowl because of its utility for bouquet or temperature, but the real reason they do so is that their parents drank wine that way, and all their parents' friends, and their parents' parents too, and that the stores are stocked with glasses of that shape, and mass media full of representations of them. Wine glasses are a path-dependent technology too, which is just to say that they got started in a certain form for reasons that are obscure to most, and possibly insignificant from the standpoint of utility, but ever since, they have been hard to change. "Path dependency" is one name for cultural history, the very thing that neoclassical economics has generally been determined to exclude from its mathematical description of economy.

But path dependency implies something more specific about cultural history, especially as it relates to technology. A great deal of energy has gone into trying to explain how technologies change, but path dependency addresses the equally interesting question of why they stay the same. It assumes that any technical arrangement is not just an effect of some prior determination, but also a cause of its own continuation. Technical forms have inertia, to hazard a physics metaphor of my own, which can make them self-perpetuating. One might attribute this technological inertia solely to people who stubbornly persist with their own poor choices, but looking at it that way overstates the amount of choice that people exercise. It makes more sense to say that this inertia lodges in both people and machines, which is to say, in techniques. When we set aside the fetish of utility, we can see that the form of a technical object and the techniques associated with it can persist as stubbornly as any other cultural form. Nobody ever asks whether a dress style or a wedding ritual is optimally efficient. They take for granted that utility or efficiency are not their primary guiding values, and for that reason they are not typically deemed economic concerns. But the forms of machines often are attributed to utilitarian standards, as if the shape of our keyboards must somehow be fundamentally different from the shape of our clothes.[5]

As noted in an earlier chapter on design, mid-twentieth-century design tended to take a holistic view of the entire design and production process. Not only are design principles consciously articulated in advance, but they typically serve some larger goal related to human experience, identity, sociality, or morality. Better design builds better people, and better societies for them to live in. But no such comprehensive design led to the QWERTY keyboard, even though there certainly were many individual design decisions along the way, and even though some of those apparently were based on local assessments of utility. The design of the QWERTY keyboard seems to have been precisely the kind of unplanned, unguided, undertheorized process of haphazard experimentation that mid-twentieth-century design rebelled against. As such, typing techniques today remain a monstrous assemblage of competing goals and values devised in the century before last, a clumsy Victorian muddle that eventually came to seem inevitable. Somehow, this ungainly monstrosity never was consigned in the past. It is the Frankenstein monster up from the grave, a mummy from the tomb, a vampire from the crypt. It should be dead and buried with countless other nineteenth-century designs, and yet it continues to walk among us. As a result, the machine-writing techniques that billions of people practice every day around the globe are hauntingly anachronistic amalgamations of what previous generations of musicians, typists, and telegraphers used to do. Typing is an old ghost that haunts the present. QWERTY is its vault.

Fossil Utility

It is plausible that technical constraints could have determined the layout of letters on a QWERTY keyboard, but even if that were so, the persistence of that arrangement would be useless at this point, like carriage wheels on a Ferrari. Fossil utility is the persistence of obsolete forms that once were useful, relics shaped by environments that no longer exist. The most obvious fossil utility in the QWERTY keyboard involves the precise spatial arrangement of the keys themselves, rather than the arrangement of the letters on them. The keys in the middle row of letters beginning with *A* are staggered exactly halfway between the keys in the bottom row of letters beginning with *Z*. This is also true of the numbers in relation to the top row of letters beginning with *Q*. But the position of the top row of letters beginning with *Q* in relation to the row immediately below it is different. Those rows are staggered from each other by a lesser amount, by a quarter of a key rather than by a half. The result of this curious staggered arrangement is that the keys are not arranged in straight diagonal lines, but in a more irregular pattern of varying offsets, such that no key is aligned directly above another.[6] Chances are, the keyboard on your own computer is still arranged with these irregular offsets today.

The principle of utility behind this form has nothing to do with making typing easier or more efficient for people. It has everything to do with accommodating the mechanism connected to each key—or rather the mechanism that used to be connected to each key. On a mechanical typewriter with four rows of letters and numbers, each key had to connect to the corresponding type arm with a metal rod, which meant that no key could be directly above another. The pattern of irregularly staggered keys ensured that each rod would have a clear path under the keyboard to the basket of type arms inside the machine. And even now, on electric typewriters and computer keyboards that have no such physical constraints, the keys typically remain arranged in this irregular way, as if the ghostly remnants of mechanical rods still might interfere. Human typists who have struck millions of keys have habituated themselves to nineteenth-century mechanical constraints that no longer exist. Or perhaps it is better to say that the constraints do survive, but not in the requirements of metal and mechanism. They survive in the far more durable material of cultural techniques.

Given that the fossil utility of key spacing persists, it would seem at least possible that the layout of letters could similarly reflect other overriding but obsolete technical constraints. And yet this seems not to be the case, at least not in any all-encompassing way that would explain the QWERTY layout as a whole. As I noted at the outset, the most prevalent and least convincing argument has held that typing too rapidly on early machines caused type arms to jam, which drove Christopher Latham Sholes and his collaborators to scramble the letters to slow the rate of input.[7] An early version of that claim appears in a 1925 reminiscence by Henry Roby, one of Sholes's early business partners, who said, "At first the keys were set in alphabetical order and we found certain keys colliding and jostling with one another." As a result, Roby said, Sholes determined to reorganize them by incidence of use in English, "like a printer's font."[8] Aside from the fact that the keys are in no way arranged like letters in a printer's font, the main problem with this theory is that it assumes that typists will not memorize any new arrangement in short order. It further ignores the evidence of the keyboard itself, which does position several uncommon letters centrally, but which also consigns several rarely used letters such as *Q* and *Z* to the edges. Further, it locates common letters such as *T, E, H,* and *R* more centrally, which surely facilitates faster typing, especially given that those letters often appear in English in close proximity.

Given that touch typing developed long after the QWERTY keyboard, Sholes would not have designed it for ten-finger typing, but given the precedent of the piano, surely he anticipated the use of multiple fingers on each hand in at least an improvisational way. Six- and eight-finger typing methods developed shortly after the typewriter's commercial introduction.[9] In any version of two-handed typing, the keyboard's left-handed bias might be seen as an attempt to slow entry, given that the left hand controls the three most common letters in English, *E, T,* and *A*. But if Sholes wanted to slow down entry by making the keyboard left-hand dominant, he likely would have separated common letters more widely. Why make it hard to type common letters with the left hand, but easy to find common combinations by grouping them closely together? Moreover, some of the most common letter combinations in English are also on the left hand, but instead of being widely spaced, they are tightly grouped, including *ER, ED, ES,* and *RT.* Even worse, the common combination of *T* and *H* involves nearly adjacent keys easily operable by different hands, which permits the most rapid entry of all.

Other arrangements suggest that Sholes may have grouped similar kinds of characters for reasons that seem to have little to do with rate of input. For instance, except for *A,* all vowels appear in the top row of letters, including three of the most commonly used letters in English: *E, I,* and *O.* Whatever the reason for this, surely it is not a matter of slowing entry. Along with the vowels in the top row are two other commonly used letters in English, *T* and *R.* Five of the eight most commonly used letters in English are in QWERTY's top row, hardly evidence of randomized distribution, let alone of deliberate separation of commonly used keys.[10]

The arrangement of the vowels in that row may derive from other decisions about the placement of *I* and *O.* Because early typewriters used the letter *I* for the number 1 and the letter *O* for the number 0, the top row of numbers started at 2 and stopped at 9. *I* and *O* are thus conveniently grouped just below 9, where they can do double duty as both numbers and letters. But why not position *I* in the place of *Q,* at the start of the numbers, just below 2, rather than at the end? Perhaps this is because one of the most commonly typed numbers would have been the year at the top of a piece of correspondence. At the date when the Sholes invented his first machine in the early 1870s, positioning *I* at the end of the top row of letters conveniently groups 1, 8, and 7 so they can be typed in rapid succession. So the evidence that the keyboard scrambled letters to slow down typing not only finds little support, but seems to be contradicted by evidence that for some purposes, the keyboard may have been designed to facilitate speed.

There is a much more sophisticated version of the argument about the utility of slower entry that deserves more serious consideration. On Sholes's machines and on the many imitators it inspired, each key is linked by a rod beneath the keyboard to a circular basket of type arms that hang vertically beneath the rubber platen, or roller. As on later typewriters that struck the front of the platen rather than the bottom, these type arms could jam if activated simultaneously or in too rapid succession, and the arms may have been more likely to jam the more closely they were located to one

PLATE 1. Permutating Keyboard of the Cooke and Wheatstone telegraph, similar to the keyboard depicted in the 1837 English patent. The rest of this apparatus has been lost. Copyright by the Science Museum Group.

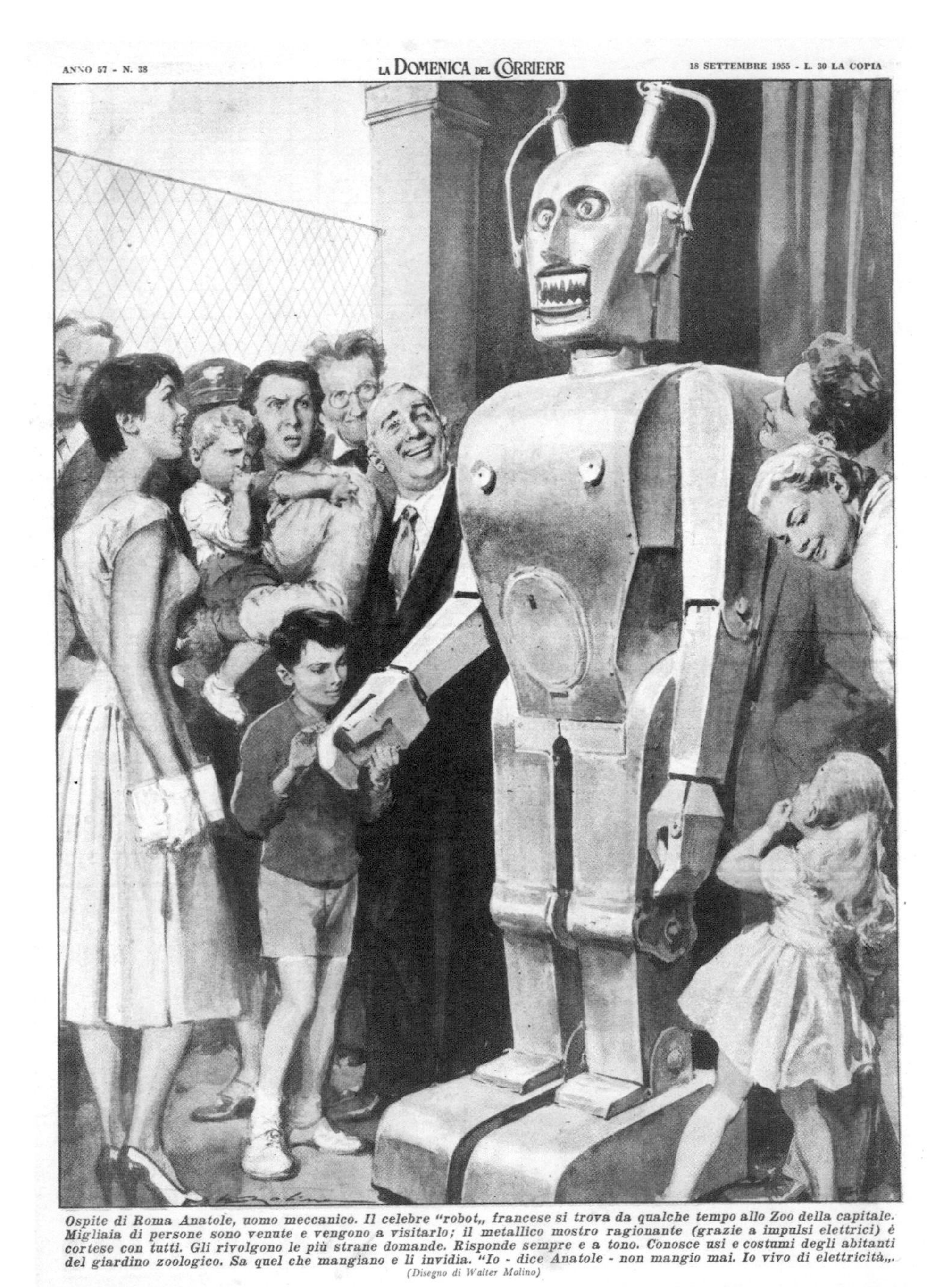

ANNO 57 - N. 38 LA DOMENICA DEL CORRIERE 18 SETTEMBRE 1955 - L. 30 LA COPIA

Ospite di Roma Anatole, uomo meccanico. Il celebre "robot„ francese si trova da qualche tempo allo Zoo della capitale. Migliaia di persone sono venute e vengono a visitarlo; il metallico mostro ragionante (grazie a impulsi elettrici) è cortese con tutti. Gli rivolgono le più strane domande. Risponde sempre e a tono. Conosce usi e costumi degli abitanti del giardino zoologico. Sa quel che mangiano e li invidia. "Io - dice Anatole - non mangio mai. Io vivo di elettricità„.

(Disegno di Walter Molino)

PLATE 2. Walter Molino's illustration of the robot Anatole from the back cover of the Italian illustrated weekly *La Domenica del Corriere*, September 18, 1955. French inventor Jean Dusailly developed Anatole from an earlier robot named Gustave created about a decade earlier by a different inventor. The robot's nipples seem to be nonfunctional, but a window allows viewers to peer into the mechanism within the lower abdomen.

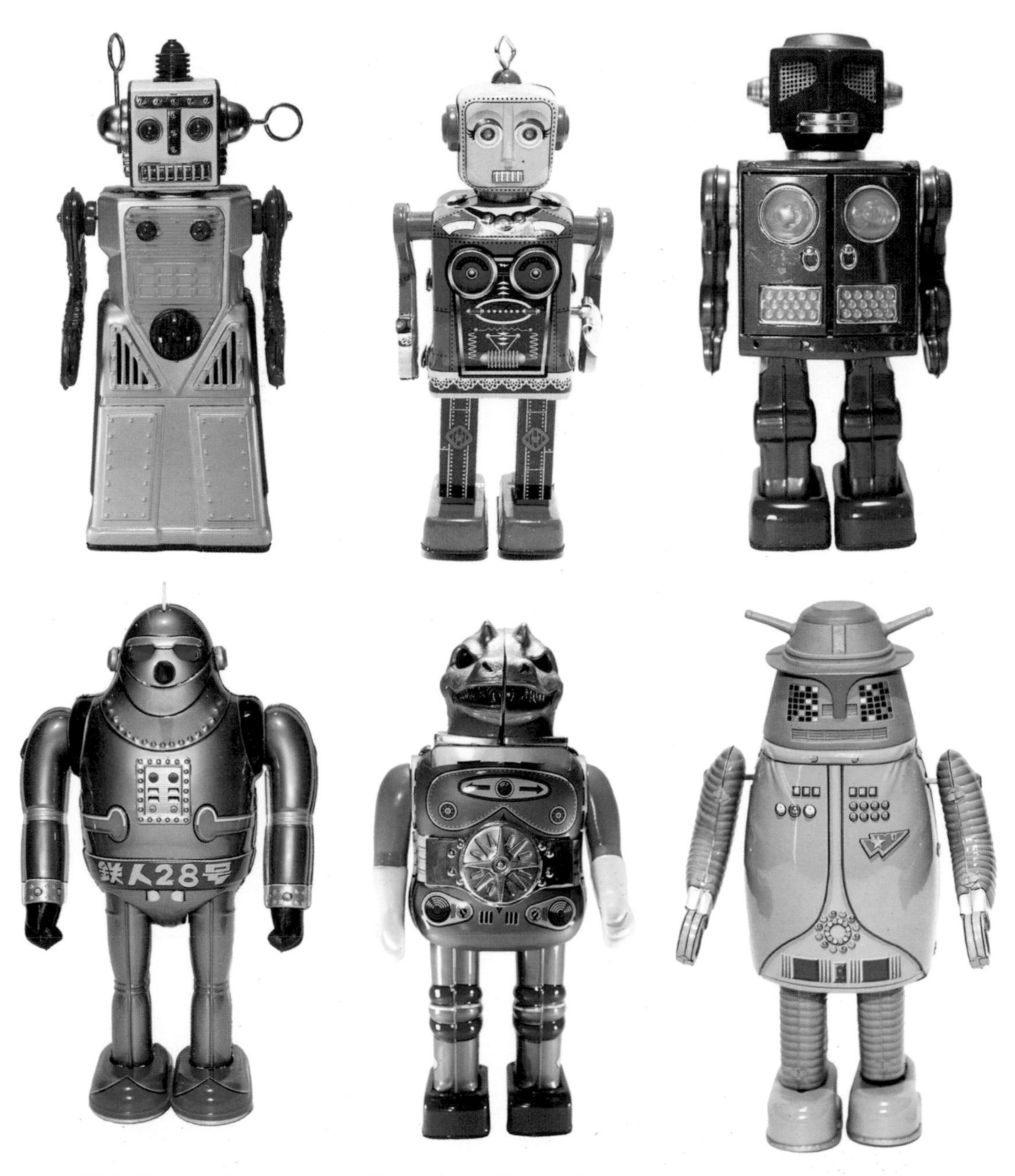

PLATE 3. Six toy robots produced from the mid-twentieth century to the present, all displaying some configuration of mechanical breasts or nipples. Clockwise from top left: (1) Radical Robot, manufactured in Japan and marketed by the American brand Cragstan in the 1950s and 1960s; (2) Roberta Robot, by the American toy company Schylling, circa 2013; (3) Attacking Martian, by the Japanese toy company Horikawa, 1962; (4) Blue Robot Hack, a wind-up robot by the Japanese tin toy company Billiken, 1997; (5) Changeman Robot, a toy prototype by the Japanese toy company Horikawa, 1972; (6) Tetsujin 28 #5, by the Osaka Toy Institute, 1990s. All images courtesy of tin-robot.com.

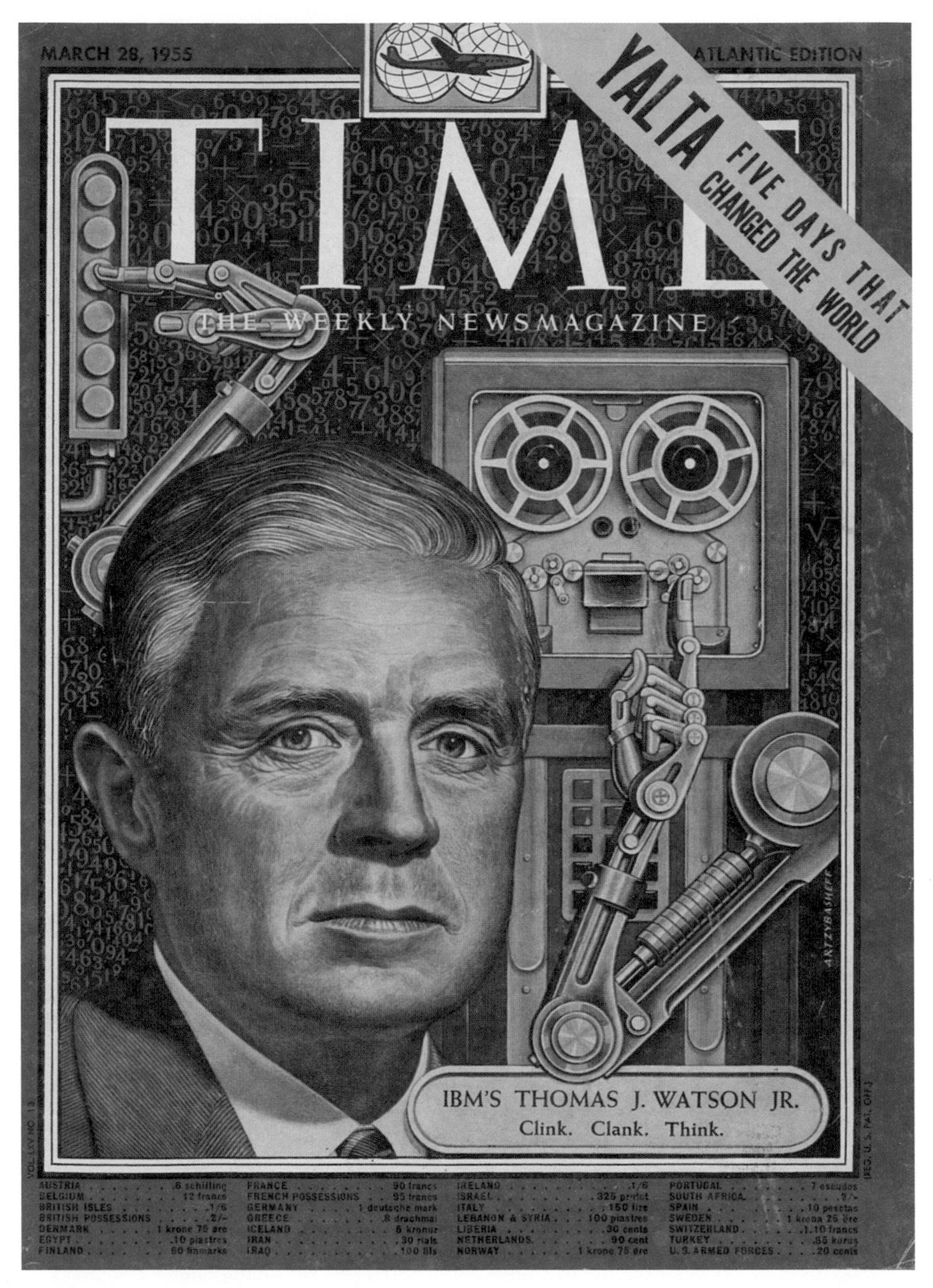

PLATE 4. Boris Artzybasheff's illustration of IBM founder Thomas J. Watson Jr., accompanied by a devious robot, cover of *Time* magazine, March 28, 1955.

PLATE 5. Manuscript illumination of King David playing an organ while another musician plays a hurdy gurdy, from the Rutland Psalter, circa 1260. British Library Add. MS 62925, fol. 97v. Courtesy of the British Library Board.

PLATE 6. Illustration of the Exchequer of Ireland, from the *Red Book of the Exchequer,* a thirteenth-century manuscript containing records and remembrances of the English Exchequer. Reproduced in *Facsimiles of National Manuscripts of Ireland,* 1884, Part 3, Plate 37. Courtesy of the Department of Special Collections, Memorial Library, University of Wisconsin–Madison.

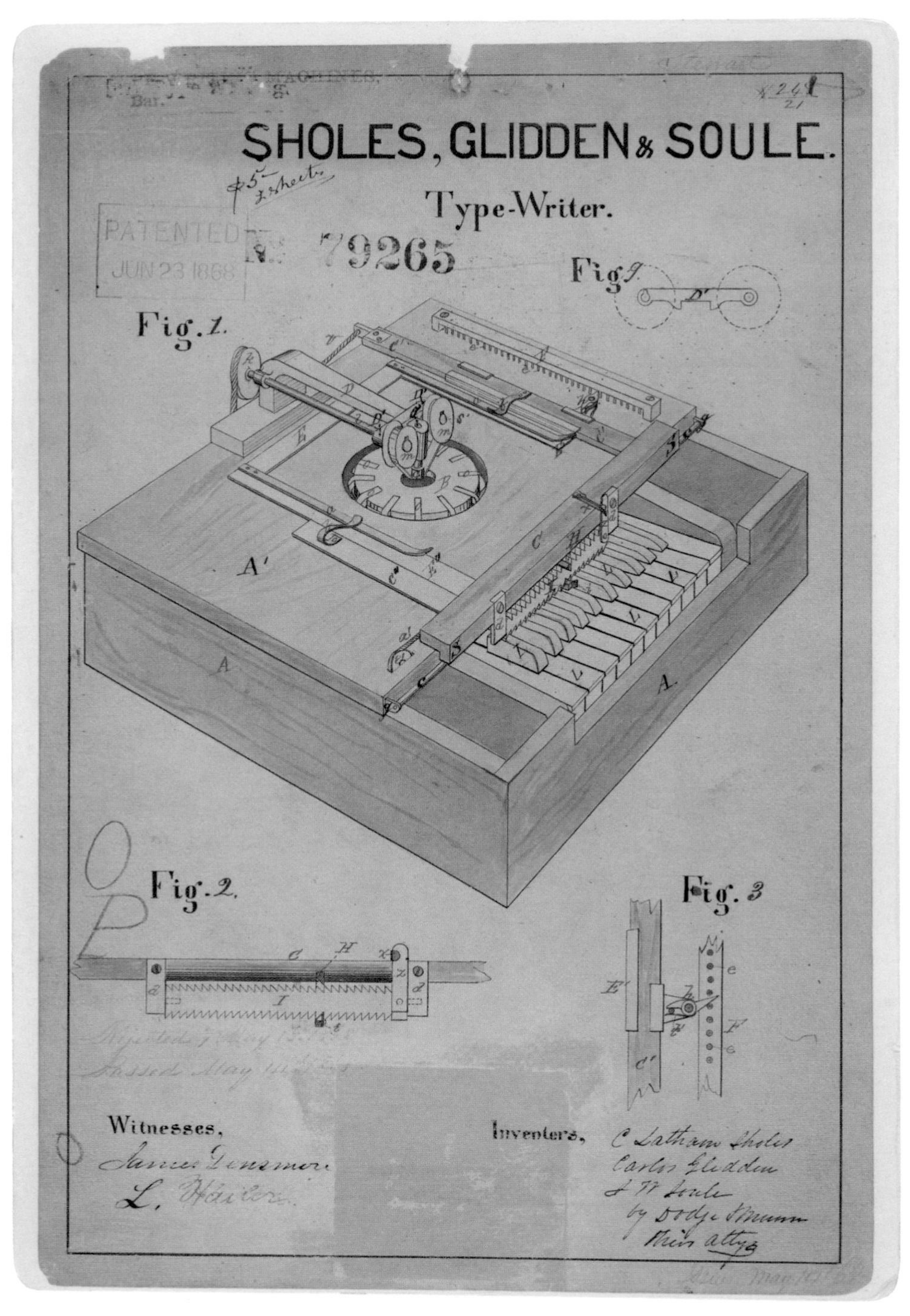

PLATE 7. Christopher Latham Sholes's patent illustration for the Sholes, Glidden, and Soule Type-Writer, U.S. Patent 79,265, Drawing for a Typewriter, issued January 23, 1868; Series: Utility Patent Drawings, 1837–1911; Record group 241: Records of the Patent and Trademark Office, 1836–1978; National Archives at College Park, Maryland.

PLATE 8. Antonio Michela Zucco's stenography machine, invented in 1862 but not patented until 1878 in Italy and 1879 in the United States. Later known as the Macchina Michela, and first used in 1880 in the Italian Senate, where versions with a similar interface remain in use today. Copyright 2022 by Auction Team Breker, Cologne, Germany (breker.com).

PLATE 9. Elmer Fudd pressing "the red one" on the control panel of his home automation system, from "Designs for Leaving," *Looney Toons,* Burbank, Calif.: Warner Brothers, 1954.

PLATE 10. Nintendo Entertainment System (NES) game controller, first introduced in 1986 but based on the Nintendo Famicon controller introduced three years earlier. Photograph by Patrick Manning.

PLATE 11. Thomas Skaife's Pistolgraph camera from around 1858. The camera is somewhat smaller than the image suggests: from the front of the lens to the back of the box that holds the plates, the entire device is just a little more than three inches long. Copyright by Science Museum Group.

PLATE 12. Still from *Star Trek,* episode 21 of season 1, "Return of the Archons," directed by Joseph Pevney, written by Boris Sobelman, aired February 9, 1967, on NBC.

another in the basket. In Michael Adler's reliable and detailed history of typewriting, he argues that, "in order to prevent the sluggish type-bars from jamming at the printing point when the speed of the operator outstripped that of the machine, the immediate solution was to place the most frequently used type bars as far away from each other as possible in the type basket (not on the keyboard)."[11] This theory thus holds that, instead of scrambling the letters on the keyboard to slow down the operator, Sholes separated common letter combinations in the basket of type arms, which compelled a mostly random reordering of the keys.

And yet, this theory does not correspond very well to Sholes's designs, though the matter is complicated by the fact that Sholes's layouts kept changing. His partner and principle financial backer, James Densmore, later said of their prototypes around 1870, "We never made two alike, and never put the same key-board on two successive experiments."[12] An 1872 article in *Scientific American* depicts a machine close to its final form, but with a QWE.TY keyboard that swapped the positions of the period and the letter *R*.[13] That arrangement may have sped up entry by moving *R* to the right hand while leaving the letters it most commonly combined with on the left, including *E* and *A*, but it also might have helped separate the type arms to opposite sides of the circular basket. Perhaps that was an ideal arrangement: maximum speed by using different hands, with widely spaced type arms for minimal risk of jamming. Other frequent letter combinations such as *T* and *H* are separated on opposite sides of the basket as well, as Adler notes, and they too are easily reachable by different hands.[14]

However, as Koichi Yasuoka and Motoko Yasuoka recently have shown, there are reasons for doubting that Sholes arranged type arms systematically to prevent jamming. They helpfully tabulate the frequency of the most common letter combinations in English and compare those against the degrees of distance in the type basket of the original Sholes and Glidden Type-Writer that reached market in 1874. They find that, while *T* and *H* are 180 degrees separate (on opposite sides of the basket), *E* and *R* in their final positions are only 16 degrees separate. Other comparatively uncommon combinations are spaced surprisingly widely, such as *T* and *N* (172 degrees) and *V* and *E* (123 degrees), which further complicates this theory. In fact, of the fifteen most common letter combinations in English, only three are separated by more than 145 degrees. Yet, in the next, lower group of fifteen most common letter combinations, six are separated by more than 145 degrees.[15] Less common combinations tend to be more widely separated overall.

It would be nice to have a single reason to explain why the keyboard developed as it did. That would confirm that technologies are rational, and by extension that people are too. It would also affirm not just that technologies develop reasonably, but that they develop for *a reason,* and that everything can be boiled down to a single guiding principle that structures all the results. But the keyboard developed for many different reasons, some of which sped up typing and others of which slowed it down. Some changes seem conducive to ease of use; others clearly inhibit ease of use. Overall logics are easy to imagine but hard to discern. The QWERTY layout seems to have resulted from an ongoing process of trial and error, experimental in its details and unguided by any single unifying principle. And, as we shall see in a moment, much of the QWERTY layout was not compelled by utilitarian concerns related specifically to typing at all, but borrowed wholesale from other, earlier technologies devised for different purposes. Much of what we covered so far involves the top row of letters. We now turn to the bottom two rows, where the keys are positioned largely as they would have been on a mid-1870s printing telegraph.

Zombie Utility

Fossil utility does not always stay put, but can leap by analogy to new and different contexts. Perhaps we should call this *zombie* utility: fossil utility that gets up and walks. In Plate 7, I reproduced Sholes's elegant patent illustration for an 1868 prototype, featuring a small musical keyboard. Piano techniques undoubtedly influenced typewriter design and machine-writing practices, but they probably did not do so directly. Even before Sholes replaced the musical keys of his 1868 prototype with round buttons derived from the Morse-Vail telegraph key, he seems to have been drawing directly from other telegraphic interfaces. The most important of these were far more technically complex than the Morse-Vail system, for they were the precursors of later teletype machines, in that they received inputs of alphabetical characters on one end and output printed alphabetical text on the other.

As early as 1844—the same year that Samuel Morse sent his first long-distance message from Washington, D.C., to Baltimore, Maryland—a Vermont inventor named Royal Earl House demonstrated a marvelous new device, a printing telegraph with a piano keyboard that transmitted individual Roman letters to a printed paper tape at the receiving end. Like an organ, the House printing telegraph was powered by compressed air, and so required a second operator to turn a crank while the first operator input the message at the keyboard. It allegedly could transmit forty words per minute, but it was temperamental, expensive, and difficult to manufacture and maintain. Still, it was convenient and familiar too: convenient because operators spelled out their messages letter-by-letter in any language they liked and with no need for special code, and familiar because they selected those letters while seated before what appeared to be a miniature piano.

FIGURE 24. Patent model of Royal Earl House's Printing Telegraph, U.S. Patent 9,505, issued December 28, 1852. Courtesy Division of Work and Industry, Smithsonian National Museum of American History.

After just a few decades, printing telegraphs like House's were common in the United States and Europe, though not as numerous as simpler and cheaper Morse-Vail models. The masthead of *The Telegrapher,* a technical and trade journal founded in 1865 in the United States, represented printing telegraphs with musical keyboards from the start, and with increasing prominence over the decades. By 1873, the year Remington licensed the rights to manufacture Sholes's Type-Writer, the Morse-Vail transmitter does not even appear in the masthead, which featured two different printing telegraphs instead.[16] Largely forgotten by popular culture today and rarely represented in films depicting the nineteenth century, printing telegraphs like these were not only common but also the most glamorous, high-tech manifestations of the burgeoning information age.

The machines depicted in the masthead of *The Telegrapher* were much more capable and reliable than the early House models. Faced with resistance from the Morse syndicate, House sold his company to the growing American Telegraph Company in 1859.[17] He also faced competition from newer and better printing telegraphs. In the

FIGURE 25. Masthead of the American trade journal *The Telegrapher* from 1873, showing two Hughes machines with musical keyboards, but no Morse-Vail system. Courtesy of Google Books.

early 1850s, a Welsh immigrant named David Hughes became enamored with telegraphy and began working on his own version of a printing telegraph. Born in London, Hughes immigrated to Virginia with his family in 1840 at ten years of age and showed great aptitude for both music and mechanics. His father had exhibited the young pianist as a child prodigy in England, and he may have performed for the royal family at Windsor Palace. As an adult, Hughes eventually made his living teaching music but continued to experiment with printing telegraphy. He replaced the pneumatic power on the original House machine with much simpler cable-drive weights like those used to power grandfather clocks. He also increased the speed significantly by devising a rotating printing wheel that operated so rapidly that it did not have to stop spinning while printing a letter. After investors purchased the rights to Hughes's machine, they formed the American Telegraph Company, which later bought House's company.[18]

Printing telegraphs were thus not experimental devices, but the first working writing machines in widespread use. Ironically, people commonly typed messages that would be printed hundreds of miles away long before they could type messages that would be printed just a few inches away on their own desks. The American Telegraph Company installed Hughes printers in New York, Philadelphia, and Boston by 1858, but Hughes eventually proved more successful selling them in Europe, where he moved with his family. The French national telegraph administration installed Hughes printers on all the main lines by 1862.[19] By 1863, Hughes printers were on the line from London to Birmingham, and by 1865 he was experimenting with a long-distance line from St. Petersburg to Paris, an impressive feat for any system at the time.[20] In the years that followed he contracted to install machines in Prussia, Austria, Holland, Switzerland,

FIGURE 26. Exclusively male operators of Hughes's printing telegraphs with their small musical keyboards in London's Central Telegraph Office in the late 1890s. From *The Queen's Empire* (London: Cassell, 1897–99), 7.

and Turkey, among others. By the end of the nineteenth century, there were about three thousand Hughes machines operating in Europe, some of which remained in service in France until the 1940s and in the Netherlands until 1950.[21]

Unlike the Baudot terminals discussed in the previous chapter, these machines did not transmit code, and they were no more able to receive chorded inputs than were typewriters.[22] Accordingly, Hughes's first prototypes dispensed entirely with the musical keyboard. As an accomplished pianist, he may have been attracted to the piano interface, but he also may have perceived the discrepancy between the seriality of writing and simultaneity of chording.[23] His 1856 improvement to House's design employed an array of small round touch interfaces in two rows, fourteen on top and thirteen staggered below. In his patent, he referred to these buttons as *keys,* the residual language of the musical keyboard, but in fact they were repurposed from actual clothing fasteners, the buttons from a vest.[24] It may even be that Hughes's buttons literalized those metaphors already circulating among journalists attempting to describe the "ivory button" of the Morse-Vail transmitter.[25]

In the long run, however, Hughes reverted back to the familiar interface of the musical keyboard, which ensured that it would remain the norm throughout the mid-nineteenth century. And at this point, some features of Hughes's keyboard migrated to QWERTY. As we already have seen in a previous chapter, Sholes imitated the piano keyboards of printing telegraphs on some of his prototypes in the 1860s, including his 1868 patent model with its truncated keyboard and unlabeled keys, as depicted in Plate 7. Another model shown in Figure 27 from around the same time survives, and though it is incomplete and badly damaged, its full keyboard is mostly intact. That keyboard consists of fifteen white keys and fourteen black keys, a common arrangement on Hughes printing telegraphs. Unlike Sholes's miniature 1868 patent model, this larger prototype also contained a third row of raised buttons above the black keys for numbers and marks of punctuation. And although the black keys are damaged and labels for them no longer visible, the lower white keys are labeled with letters, which confirms that Sholes was borrowing the precise layout of letters from printing telegraphs. From right to left on the white keys, the letters proceed from *O* to *Z,* suggesting that the black keys would have proceeded from left to right from *A* to *N,* exactly as they did on Hughes's machines.[26]

This arrangement persists in QWERTY, as Sholes developed it incrementally from this starting point. The middle row of letters begins with A, just as the upper black keys did on Hughes telegraphs, and presumably on Sholes's prototype. Aside from the intrusion of S after A, the middle row of a QWERTY keyboard progresses in perfect alphabetical order from left to right: between D and L, only the vowels are missing. They had moved to the top row with other commonly used letters. The bottom row of letter keys wraps around in alphabetical order from right to left, just as they did on the white keys of House and Hughes printing telegraphs. Only B and C break the pattern, having been relocated from the row above, but in their new location they also are arranged in reverse alphabetical order, like the other letters from the second half of the alphabet. Accordingly, on most modern QWERTY keyboards, just as on Hughes's telegraph keyboard, *A* stands just above *Z* at the left, and the middle of the alphabet wraps around at the right end of the middle and bottom rows of letters.[27] Exactly half of the Roman alphabet, thirteen letters in all, remain positioned on the QWERTY keyboard as Hughes had determined, on a different sort of machine, for a different purpose, with a different interface, and under different technical constraints. When people type at modern computer keyboards today, they are in a sense possessed by nineteenth-century telegraphic techniques.

FIGURE 27. The lower portion, including the keyboard, of a prototype of Christopher Latham Sholes's Type-Writer, probably from 1867 or 1868. From the Collection of the Milwaukee Public Museum, #E4175.

American Morse

This monster has other telegraphic parts stitched into its awkward frame. When Sholes and his partners Carlos Glidden, Samuel Soule, and James Densmore began shipping their first trial models to preliminary customers around 1868, an early customer was Edward Payson Porter, the principal of Porter's Telegraph College in Chicago.[28] One of Porter's 1868 advertisement claims, "By touching keys like a Piano this machine produces letters faster than the most rapid penman."[29] In an accompanying illustration, a man sits at a large table-sized console with what appears to be an interface of musical keys. On top of it sits a Morse-Vail transmitter and receiver. Given that Sholes had not switched entirely to round buttons by 1868, Porter's advertisement probably depicts a device much like the prototype in Figure 27. That fragmentary device may have been one of these trial models, as its decorative carving around the base suggests some degree of public exposure.

Porter's college trained students in five departments, including "Penmanship" and "Type Writing." These were complementary abilities, because a Morse operator would have to write down the message by hand clearly and legibly, without errors or ambiguities, and very rapidly in order to keep up with transmission. Because Porter recognized that the typewriter might aid both speed and legibility, he alleges in the same advertisement, "Its use in this College enables Students to become expert Telegraphers without regard to their penmanship."[30] It is possible that such an appeal might have been calibrated to women, given that they were taught what were deemed to be inferior forms of script, but judging by the telegrapher shown in Porter's advertisement, he was primarily marketing his courses to men.

Historians have amply attended to the fact that, after a certain point in the late 1860s, Sholes began experimenting with replacing piano keys with Morse-Vail telegraph keys, which he eventually employed in his final versions, and which remained recognizable in typewriter keyboards for decades after.[31] But other less obvious evidence of telegraphic connections remained as well. By 1873, some versions of the Sholes and Glidden Type-Writer contained a key to the left of A with four dots arranged in a vertical line, a key that also appears in the 1878 patent illustration, which was filed earlier. As Yasuoka and Yasuoka point out, the key is not a mark of punctuation but an indicator of a paragraph break in the peculiarly American version of Morse code: "–——–". With this key, an operator transcribing Morse code with a typewriter could strike a single key to signal a paragraph break, rather than laboriously inserting an actual paragraph break by advancing the platen and returning the carriage. As late as 1873, then, Sholes was anticipating the demands of telegraphic transcription and suiting the interface to those familiar with Morse equipment, conventions, and code.

As Yasuoka and Yasuoka also show, telegraph operators had to contend with some ambiguities of American Morse code. By 1865, Europe had adopted an improved version of American Morse first known as Continental Morse and later as International Morse. International Morse consists of only dots, equal length dashes, and equal length gaps between dots and dashes, but the original American Morse code that continued to be used in the United States consisted of dashes of variable lengths, pauses between dots and dashes of variable length, and more dots than International Morse employed. Because it used more dots than dashes, American Morse was faster and more efficient, but it was also more prone to error. Five letters were particularly troublesome because they contained both dots and a pause inserted between two of the dots. These letters were *C, O, R Y,* and *Z.* The internal pause within the code for each of these letters could make it hard to determine whether a single letter or two different letters had been transmitted. For instance, the code for the letter *S* was three dots (. . .), while the code for the letter *E* a single dot (.). The problem was that American Morse code represented the letter *Z* with three dots followed by a brief pause followed by another dot (. . . .). In practice, this meant that an operator might not be sure whether an initial sequence ". . . ." referred to the letter combination *SE* or the single letter *Z* until subsequent letters eventually clarified things. Yasuoka and Yasuoka reason that *S* may have been moved to its location on the QWERTY keyboard in order to position it below E and above Z, so that a Morse receiver pausing to disambiguate *Z* from *SE* could have his finger poised near all three at once, ready to strike and move on.

We can finally perceive the full historical complexity of QWERTY, which developed as it did for many different reasons, rather than for one overriding reason. The best we can say is that Sholes borrowed about half of the keyboard from Hughes telegraphs, relocated most vowels to the top row, moved *I* and *O* to the end of the top row because of their double duty as numbers, and inconsistently managed the placement of common letter combinations, but whether all this was done in order to increase or decrease speed of input is unclear. He certainly did worry about jamming type arms, but the final arrangement of the arms in the basket does not suggest that he implemented a general solution to the problem. And finally, anticipating use by Morse telegraphers, he probably did give some thought to the ambiguities of American Morse code and arrange at least some keys accordingly. The result is a mishmash of practical solutions to small-scale problems, a collection of tweaks implemented continuously but haphazardly, a record of determinations made improvisationally, on minimal evidence, and with mixed results. Some keys probably ended up where they are for no good reason at all. Like some of the more idiosyncratic items on my bookshelves, the *Q* and *B* keys finally just had to go *somewhere.*

I began this part of chapters on keyboards by associating them with the mathematical rationality of the musical keyboard, and though that remains part of the legacy of the keyboard, even reason gives out at a certain point. Many commentators have noted that the top row of letters on the QWERTY keyboard contains not just all the vowels, but all the letters needed to spell the word *typewriter.* That is also the longest English word that one can spell on a single row of the QWERTY keyboard, tied with *proprietor* and *repertoire.* Some have tried to explain even this fact as utilitarian, a clever gimmick to allow salespeople to spell out the machine's name in the showroom, but I find no historical evidence in support of that view. Even in the history of technology, there can be a place for whimsey too. We simply do not know whether it was intentional or accidental, and if it *was* intentional, we have no insights as to the reasons for it. Perhaps it's even a kind of joke, the negation of utility in favor of playfulness, and thus a demonstration that people can act with a kind of spontaneity and purposelessness that ever after continues to survive in their techniques.

Part IV

»» »» Objects of Play

CHAPTER 11

The Toys of Dionysus

When Zeus's son Dionysus was born to his human mother Semele, Hera became jealous, as she sometimes did, and sent the Titans to kill the child. They distracted him with toys—spinning tops, dolls, golden apples, a mirror, and knuckle bones—before devouring all but his heart. Zeus rushed in, dispatched the Titans and sewed Dionysus's heart into his own thigh, from which the god of wine and ecstatic pleasure later was reborn. Dionysus became the twice-born god, having issued first from his mother and later from his father, while dying in between. The story is well known, but the common children's toys at its heart rarely receive much notice. Toys rarely appear in classical literature, and almost never in criticism about it. Yet these toys are instrumental to Dionysus's death and rebirth, and thus presage his later commitment to the chaotic inutility, destructiveness, and irrationality of play.[1] Seemingly trivial, grossly material, and entirely ubiquitous, the toys of Dionysus remind us that toys are serious business after all, for they mediate between the worlds of childhood and adulthood, humanity and divinity, even life and death.

The Dionysian mania, usually equated with drunkenness or sexual ecstasy, is an adult version of childhood play. This is why the devotees of Dionysus kept toys in their homes as ritual objects, according to Clement of Alexandria, symbols of temporary rebellion against reason and responsibility.[2] The role of toys in the cult of Dionysus also confirms that child's play has more in common with sex and drinking than many today might like to think. Sex, wine, and toys all functioned as playful escapes from Apollonian reason and responsibility. But this involved much more than just the suspension of certain kinds of conduct expected from adults; it also involved the dissolution of adult subjectivity itself, an escape from that firmly bounded self into some other mode of being that toys temporarily facilitate.

This chapter takes up the question of what happens when people play with one certain kind of toy, those with binary switches or buttons. As we have already seen, keyboard culture has been profoundly rationalizing, its millions of ubiquitous switches organizing subjectivity according to gridded and binary logics, causal logics, and sometimes expressly mathematical logics. So what happens when such rationalizing techniques end up enrolled in the activity of play? Can Apollonian reason and order coexist with Dionysian play and mania? And in more concrete and historical terms: what happens to modern subjectivity once children's toys with switched interfaces begin shaping it so pervasively? After all, toys with switches appeared only about a century ago. Before the late nineteenth century, children grew up in a world of analog playthings like the toys of Dionysus, such as sticks, balls, hoops, wheels, dolls, and more. Such toys are well attested in the archeological record, from a clay dog some eight thousand years old found in present-day Israel, to mechanical string toys in Egyptian tombs three thousand years old, to wooden animal figures from two thousand years ago bearing child-sized fingerprints unearthed in the American Southwest, and continuing through Roman and medieval dolls, early modern rattles, and Enlightenment pinwheels and tops. Only around the start of the twentieth century did toys with binary switches appear, first in imitations of adult mechanisms such as pianos and typewriters, and later in electrical toys festooned with switches and buttons. If toys really are involved in the making and unmaking of the self, as the story of Dionysus supposes, this chapter will consider what happened when the children of digital modernity suddenly began toying with binary switches.

This is not the first attempt to chart the relationship between switches and play. In his analysis of *ludomusical* practices, Roger Moseley has traced the playfulness of music into keyboard culture more broadly.[3] The keyboard articulates what Moseley calls *digital analogies,* fluid relations between the digital and the analog, never reducible to simple opposition. The analog complexity and expressive potential of keyboard music relieves the digital of some of its rigid binarism by introducing playful elements hard to reconcile with the strict on–off logic. Part of Moseley's point is that even the array of binary switches on which I type these words may be sedimented with a ludic potential passed down from its musical ancestors. Children's toys certainly derive something from this musical heritage, as evidenced by the many infant playthings that imitate pianos, but toying with buttons goes beyond ludomusicality, because it often involves playthings that have no real relationship to music or musical keyboards, including makeshift playthings such as light switches and elevator buttons. My goal is thus not to trace the ludic properties of music into other kinds of activities, as Moseley has done, but rather to trace the rational and rationalizing functions of binary switches into one kind of activity that would seem to be most antithetical to it: play.

In this chapter, I set aside for the moment more complex forms of button-based play such as pinball or video games and attempt to isolate the simplest possible playful interactions with the simplest possible switches. Even toying with something as simple as a light switch turns out to be complicated, more complicated than either playing with analog toys or operating switches for practical reasons. Because of its mixed nature, the practice of toying with switches both rationalizes and derationalizes the subject simultaneously. In the process, it consolidates and dissolves the subject too, rendering the person at the switch both a masterful adult and an absorbed child. Although toying with switches probably helps solidify adult forms of subjectivity over time, it can also, at least temporarily, dissolve adults back into children again. Unlike the analog toys of Dionysus, then, the binarized toys of the twentieth century have a profoundly double effect. They consolidate stable subjectivities by establishing boundaries between self and other while also blurring those boundaries so that self and other, subject and object, bleed together. Further, unlike the toys of Dionysus, which foster a settled state of full absorption in them, toy switches foster a more dynamic oscillation between subjectivity and objectivity, between gaining and losing the self. This is not just another name for *experience,* which is too normative, comfortable, and permanent to explain this condition. Nor is it the radical collapse of subject-object differentiation that Julia Kristeva called *abjection,* which is too threatening and horrifying. It is rather a dynamic process that I will refer to as *ad-jection,* an unstable but also pleasurable wavering of subject-object differentiation.

Enlightening Childhood

The disruptive potential of the toys of Dionysus may seem surprising to modern readers who have learned to think of toys as practical and pedagogical tools, as theorists as different as Plato, John Locke, and Sigmund Freud all in different ways supposed. Plato says at the beginning of his *Laws,* "The man who intends to be a good farmer must play at farming, and the man who is to be a good builder must spend his playtime building toy houses."[4] The conception of play as an adult-supervised mode of pleasurable pedagogy solidified during the seventeenth century in Europe. In *Some Thoughts Concerning Education* from 1693, Locke says, "playthings I think children should have, and of divers sorts," though he emphasizes that children will learn more by making their own toys than by receiving "toys from the shops."[5] Locke himself seems to have first recommended one of the most familiar children's toys, toy blocks with letters of the alphabet on them. Through the pleasure of play, he says, "children may be cozened into a knowledge of letters, be *taught to read* without perceiving it to be anything but sport, and play themselves into that which others are whipped for."[6]

Once playing became a component of education, Europe witnessed the rapid proliferation of a wider range of toys, including board games, jigsaw puzzles, toy animals and vehicles, kites, tops, dolls and puppets, and much more. As the middle class expanded in the nineteenth century, more families had more time to indulge children in play, and more resources to organize that play through toys. Southern Germany had been an important site of toy production dating back to the Middle Ages, but industrial mass production allowed Nuremberg's toy industry to expand dramatically, supplying playthings to much of Europe and America. Ready supply satisfied growing demand.[7] By the later nineteenth century, European and North American countries had established as a central rite of childhood that modern Bacchanal of toy consumption known as Christmas. Thus did Christianity turn its most pagan-inflected holiday into a recapitulation of Greek mysteries, marking the miraculous birth of a half-human, half-god with a similar orgy of toys.

If toys really are educational, as Locke thought, they have been especially powerful instruments for inculcating social and political values, such as gender socialization through dolls. For example, Amy Ogata has shown how mid-twentieth-century American toys narrowly promoted the romantic value of creativity.[8] And Brian Sutton-Smith has argued that "the rise of the toy as a child's gift in modern society can also be seen as the rise of an instrument that would accustom children to . . . solitary preoccupation and solitary striving for achievement." He continues, "With the toy, we habituate children to solitary, impersonal activity; and this is a forecast of their years to come as solitary professionals."[9] By this view, toys smuggle social and political values into childhood, as if inside so many tiny Trojan horses.[10]

At this stage, we arrive at the difficult question of why so many children seem so enthralled with switches and buttons, when those interfaces are so often vanishingly small, minimally moveable, dully repetitive, and only sometimes rewarding in terms of sensory experience. The toys of Dionysus were all analog, and so admitted a much wider range of options. With his toy mirror, Dionysus could study reflections, redirect light, inspect hidden recesses, create visual effects, and more. A doll admits an even greater range of imaginative role playing. With analog toys, children explore a rich array of material, sensory, and social possibilities. In contrast, playing with a switch would seem to be the most impoverishing experience, a drearily repetitive operation of just two mutually exclusive alternatives and the automated effects they unfailingly induce.

And yet, many young children evidently are enthralled by even the most pedestrian switches and buttons, including my own children. I know of no empirical research on children and button-based play, but I doubt I am the only parent who has observed that many young children are more than usually interested in doorbells, remote controls, calculators, and light switches. My own were obsessed with elevator buttons and the switches on the car's dashboard. My oldest could capably operate the television remote control before he could speak. Other children evidently share this enthusiasm. In my hometown of Milwaukee, Wisconsin, children who visit the city's renowned Public

Museum line up beside one of its most popular and sensational exhibits, the world's largest three-sided diorama, depicting a Native American buffalo hunt. But the children appear to take little interest in the dramatic display itself. They are there to press the "rattlesnake button," a simple push button concealed in some fiberglass rocks off to one side that makes a small automated snake inside the display shake its tail. In a museum full of interactive and tactile exhibits geared to children, nothing is more alluring than this simple switch. All around it, the fiberglass rocks have been rubbed smooth and shiny by generations of avid little hands.

Commercial toys readily exploit but probably also inculcate this interest, even to children as young as six months old. The VTech Little Apps tablet contains a computer keyboard and a piano keyboard, a clear elaboration of Moseley's ludomusicality scripted for infancy. Similarly, the Busy Learners Activity Cube has four fist-sized buttons that light up to "keep toddlers 6 months to 3 years old engaged and entertained."[11] These contemporary toys have many earlier precedents. In 1961, Louis Marx and Company introduced Buttons the "Push-Button Pup," a battery-powered toy dog controlled by eight electrical switches, each of which triggered actions such as "paw point" or "tail wag." How long any of these toys would engage children is hard to know, but one suspects that as soon as children familiarized themselves with any given switch's unvarying effects, there would be very little left to play with. After all, the yoking of cause to effect in these toys is so rigid, so unvarying, that it is difficult to imagine what the reward of playing with such a toy could be after some period of initial uncertainty had waned. Only as long as the results are surprising can the activity register as play. The "rattlesnake button" might have enthralled millions of children, but I doubt it enthralled any single child for long. As Roger Caillois has claimed about play: "An outcome known in advance, with no possibility of error or surprise, clearly leading to an inescapable result, is incompatible with the nature of play."[12] By extension, once a switch becomes predictable, it ceases to be a toy.

This helps clarify that the rewards of playing with switches derive from two different and opposite sources. The first reward is simply the experience of power and mastery, as switches confirm one's own instantaneous control over complex process, formidable power, or distant action. This satisfaction, however, is not primarily playful, but rather involves the rapid rationalizing of action through a demonstration that the world works in highly regular ways, and that people can compel regular results. This is the pleasure of individual agency and control. The second reward is different but more important for these purposes. It consists of the pleasurable uncertainty about the effects of any switch. A child confronting a new switch likely realizes that the switch must do something, but exactly what will remain unclear until revealed by playful experimentation. Over time children learn that switches positioned on a wall near a doorway likely control a light, because experimentation with one switch rapidly generalizes expectations about similar switches. Usually these two kinds of rewards reinforce each other. The pleasure of playful uncertainty gradually acclimates the player to the pleasure of agency, as a child playing with a light switch transforms confusion into certainty, experimentation into mastery, and thus one kind of reward into the other.

Because adults have demystified so many switches, they have little access to the reward of uncertainty, and for that reason uncertainty must be deliberately scripted into the effects as randomness or complex interactivity, as in the slot machine. Typically operated by push buttons today, the slot machine randomizes outcomes in order to limit the achievement of mastery. But even on slot machines both pleasures are present. Players can be absolutely sure that pressing a button will spin the reel or change the display on a video screen, so they are in complete control of starting and stopping the operation. However, what will be displayed on the reel or screen is beyond their control. The slot machine thus restores the pleasurable uncertainty of outcome that adults have generally lost in other mechanical interactions. It reenchants the switch with mystery. But most switches do not work this way for adults, because over time they have transformed the playful uncertainty of childhood into the predictable rationality of adulthood. In the process, they also have transformed childhood subjectivity into adult subjectivity. There will be more to say about that in the next part, but for that reason, switches may be the most powerful pedagogical toys of all, devices that transform ignorance into knowledge, uncertainty into certainty, mystery into mastery. Locke's blocks might have cozened children into learning their letters, but the Busy Learner's Activity Cube teaches an equally profound lesson, which is that all action is rational and that people are its exclusive source.

Dionysian Rites

The condition of play frequently registers as deep absorption in the object, a loss of self in something that seems larger, such as sexual passion, intoxication, video stimulation, or language. But what does it mean to be absorbed in some *object* of play? Dionysus, we might recall, fell victim to the Titans once he became absorbed in the toys they gave him. In Latin, the verb *absorbeo* refers specifically to an act of consumption. Without the prefix, the verb *sorbeo,* meaning *suck up,* gave English (via Dutch) the verb *slurp.* But with the preposition *ab* prefixed, the verb refers more explicitly to devouring. Thus did the Titans give Dionysus absorbing toys, then gobbled up all but his heart. When we say that a person is absorbed in a toy, a craft, or a book, then, we note this loss of subjectivity to objectivity, of self to other. However pleasurable it might seem, to be absorbed is also to be eaten alive.

Moreover, as we have already begun to see, the toy always functions doubly, gradually inculcating the most orthodox modes of adult subjectivity while also affirming that the person playing is partly exempt from adult standards of rationality and practicality. Playing with toys might be childish, but doing so over time forms the child into an adult. Crucially, however, the toy also can topple the adult back into the child's more distributed subjectivity, a condition tantamount to the transport of Dionysian ecstasy. In Plato's *Laws,* the Athenian, advocating for the benefits of wine, asserts that a man drinking "reverts to the mental state he was in as a young child."[13] As with drinking, so with playing with toys: the altered consciousness amounts to a collapse of stable, bounded, adult subjectivity. Neither self nor other, neither subject nor object, the toy seems to dissolve the self in ways that are not necessarily as traumatic as some theorists have supposed.

Recent popular mythologies register these subtleties as surely as classical precedents. A 1954 *Looney Toons* cartoon episode titled "Designs for Leaving" (a superficial parody of the 1933 film *Designs for Living*) features Daffy Duck as a fast-talking door-to-door salesman from the Acme Future-Antic Push-Button Home of Tomorrow Appliance Company, Inc.[14] As a Bacchic lord of misrule sanitized for children, Daffy intrudes on Elmer Fudd's suburban home with a free trial of a new home automation system, and forcibly sends Elmer on a bus trip to Duluth while he installs a "beautiful push-button kitchen," a push-button "hat and coat remover," and a massive panel of wall-mounted buttons to operate the rest of the house. The results are predictably anarchic: a massage chair malfunctions, a necktie machine strings Elmer in a noose, and an automatic wall cleaner scrubs the clothes from figures in a painting. By the end, the house is all but destroyed. Elmer eventually concludes that bad things happen when Daffy presses the buttons, so determines to press one himself, but as he extends his finger toward a large, unlabeled red button, as depicted in Plate 9, Daffy yells, "No, no, no, no, no! Not the red one! Don't ever push the red one!" In 1954, the implication would have been clear: the "red one" is the "big red button," or just "the button," a synecdoche for a nuclear trigger that could annihilate the world.[15]

The cartoon exaggerates both potentials of the binary switch. Nominally, the switches are harnessed to labor saving devices, and so they automate domestic drudgery, reduce human effort, and contribute to an experience of mastery. In that form, the buttons in the cartoon resemble those used for home automation today, especially in idealizing advertisements for them. However, this "control panel" affords anything but control, as the allegedly rational system quickly dissolves into ludic anarchy. And not just innocent anarchy, but rather the Dionysian kind in which the misrule is tantamount to violence, culminating even in the threat of nuclear annihilation. The deeper insinuation is that children had it right after all, not just at home but in the wider world where not all of the buttons make sense yet. If Daffy is the riotous architect of this mad world, Elmer is the child trying to navigate it, complete with a child's mispronunciations, hesitations, and confusions. Squat and pink, he reflects back to the children watching a version of themselves, fascinated by buttons they do not understand, curious about their effects, but perhaps not quite fearful enough of their potential.

Then again, most adults do not have much more detailed knowledge than Elmer about how their domestic switches work, but they operate everything reliably nonetheless. Sometimes I will accidentally turn on the in-sink disposal instead of the kitchen light, but that is about as anarchic as it gets. Knowledge of switches is empirical, born of repeated experiences that confirm that certain regular actions cause certain regular results. Causal rationality is empirical in the sense that it is not logically necessary, does not derive from knowledge of a priori principles inherent in actions themselves. Skeptical philosophers since David Hume have insisted that causation is nothing more than a reliable expectation that one thing will follow another. "It is certain, that the most ignorant and stupid peasants—nay infants, nay even brute beasts—improve by experience, and learn the qualities of natural objects, by observing the effects which result from them," Hume said.[16] My cat does not reason that the back door must necessarily open after I lift the kitchen trash bag from its bin. He knows only that one usually follows the other, and so bolts for the door as soon as he hears me gather the bag, hoping to escape. Hume's point is that people's causal reasoning is a lot like my cat's, not a matter of discerning logical necessity hardwired into the nature of things, but just the result of the repeated experience that certain kinds of actions generally follow others.

Toys are a major source of this sort of empirical causal knowledge for children still learning the properties of the material world and of their own bodies. For that reason, it is rather surprising that psychologists have not taken more interest in toys, even though toys appear in some of the most quoted passages of some of the most foundational works of the twentieth century. In *Beyond the Pleasure Principle,* for instance, Freud describes an eighteen-month-old boy who played a game of concealment and revelation devised, Freud argues, to process his separation anxiety from his mother. Freud describes a child playing with "a wooden reel with a piece of string tied round it." The child would "hold the reel by the string and very skillfully throw it over the edge of his curtained cot, so that it disappeared into it, at the same time uttering his expressive 'o-o-o-o.' He then pulled the reel out of the cot again by the string and hailed its reappearance with a joyful *da [there]*."[17] According to Freud, by throwing and retrieving the spool, the child replayed in symbolic form the periodic disappearance and reappearance of his mother, and in the process worked through his feelings of anger, fear, or loss. Even though the toy is absolutely central to the development of a mature psychology in this example, it matters only symbolically to Freud, as a surrogate for something more important, the absent mother.

Jacques Lacan's account of the mirror stage is similarly indifferent to the mirror itself, even though mirrors have been popular children's toys for centuries, including for Dionysus. According to Lacan, the subject forms at a distance from itself, a subject-as-object perceived in the mirror. This reflected subject is far more cohesive than infants' fragmentary perceptions of their own bodies, so it appears to have the bounded, cohesive, integrated qualities more familiar in objects. The *I* thus amounts to "the armour of an alienating identity," the self as idealized other, integrated only when perceived from a distance.[18] But just as Freud has no toy in the *fort–da* game, Lacan has no mirror in the mirror stage. What he has instead is an abstract reflection, a perceived image dematerialized and utterly unmoored from the mirror itself.

One might multiply similar examples in which the toy appears in psychology only to disappear into an all-consuming humanism, including Derek Winnicott's "transitional objects," or the toys used in Melanie Klein's influential "play technique."[19] They register largely as after-effects of absent people. But perhaps children are more engaged with the toy itself than some psychologists give them credit for.[20] Balls and dolls might be used imaginatively to process separation from a mother, but they also might be used to process separation from balls and dolls. Even without the overlay of family allegory, playing with toys is already a complex negotiation of subjectivity and objectivity, a working out of the boundaries between people and things, selves and others, proximity and distance, presence and absence. Interactions between people and toys can be complex in themselves, complex precisely because the activity of playing with toys both erects and erodes the boundary that separates subjects and objects.

Jean Piaget was one of the most insightful theorists of the relationship between children and the material world, and one of the few early- and mid-twentieth-century psychologists to take toys seriously. His interest stems from his staged theories of developmental psychology, which presume that children's play progresses from symbolic play (such as imaginative make believe) to more realistic or mechanistic interactions with other people and things (such as sports). There are reasons to doubt that progression, but in any event Piaget's faith in a linear development from the symbolic to the realistic led him away from symbolic readings of toys and compelled him to take their material affordances more seriously.[21] In *The Child's Conception of Physical Causality* from 1930, Piaget wonders what kind of causal logic children have before they adopt what he saw as the mechanical mentality characteristic of most adults most of the time. Piaget was no Humean skeptic, for he denies that "intelligence is entirely moulded by its environment." Instead, causation is, he says, "assimilated by means of schemas that are drawn from internal experience."[22] It is far from clear to this reader what "internal experience" includes, but it apparently does not amount to Kantian phenomenalism that involves a priori categories of understanding or universal structures of thought. Against those who "make causality into an a priori form, fixed once and for all in the structure of the mind," Piaget objects that there are at least seventeen different kinds of causal thinking in children alone, which wax and wane as they grow. In truth, he argues, causality exists somewhere between empiricism and apriorism, as both individual experience and Kantian mental structures somehow combine. In the process, objects modify organisms experientially, even as organisms assimilate objects into their own governing structures.

Piaget was thus interested in observing how children understand causality in the everyday events all around them, such as the moving of air, floating of objects, or casting of shadows. But he was especially interested in the ways in which machinery makes drastically different experiences available. Those are especially important because "the spectacle of machinery has been imposed upon children by quite a recent civilization."[23] As a result, he also wondered what effects the mechanization of childhood has had on children's assumptions about causation and agency. By asking children to explain how a bicycle works, Piaget tried to separate out a wide range of causal understandings, all of which precede the kind of fully mechanical explanations that would be expected from adults. When asked what makes a bicycle go, some young children imagined that the light on the handlebars somehow propels it. One thought the pump that inflates the tires adds a "force" or "current" for movement. Some simply synthesized the whole without being able to break down the causal connections. One child said, "You pedal, you hold the handle-bar and that's all."[24]

These explanations are characteristic of various causal logics common in childhood, many of which, Piaget thought, amount to an excess of subjectivity, a sense that the self magically compels action around it. The world is suffused with ego, which is to say that subjects and objects are imperfectly differentiated. When children imagine that everything is saturated with self—that the moon follows them around, that the flowers want to be picked—things have not yet solidified as objects, but remain enrolled in children's sense of themselves. For Piaget, the child's world is replete with these *adherences,* these "fragments of internal experience which still cling to the external world," such as a belief in animism.[25] For Piaget, then, the split between subject and object is not achieved early and completely, suddenly and traumatically, but very gradually and through long experimentation, and much of it pleasurable, such that some scant residues of subjects and objects stick to each other for years, perhaps forever.[26] The process of growth and development involves learning to experience the world without these adherences of subjects to objects, but rather with the clear boundaries between the two that adults are generally expected to observe. Although there are exceptions, to be an adult is generally to recognize the separateness and independence of the object world, and to think otherwise is to risk discredit as childish, superstitious, or insane.

Crucially, Piaget denies that any sort of stable subject precedes an engagement with these not-yet-stable objects. "Our earliest experiences are not referred to a central 'I,' but float about in an undifferentiated absolute. The self would be the result of a gradual and progressive dissociation, and not of a primitive intuition," he says.[27] Most early-century psychoanalysts would agree. However, he continues, "In order to build up the idea of objects and to distinguish the different physical objects from one another, the mind proceeds by a series of the most laborious experiments, of which the essential task is to become conscious of the innumerable resistances set up by the external world." The self is thus not generated from an Oedipal drama, but produced continually through pragmatic experiments with all sorts of material things, none of which are necessarily symbolic. Objects can resist in their own right. Or rather, to put it more accurately, material resistances solidify into what come to be recognized as objects. The "idea of object is undifferentiated from the idea of resistance," according to Piaget, for anything that does not resist becomes incorporated into a conception of the self.[28] This is an important insight for explaining how the minimally resistant switches of the modern world allow everyone (not just children) to experience switching effects as extensions of an arrogated self.

None of this requires a symbolic reading of toys. Piaget might reply to Freud that a child playing the *fort–da* game is learning to differentiate himself from a spool on a string, no trivial matter. However, Piaget also says that the "atmosphere of mechanicism and scientific explanation" so prevalent in the modern world hastens the decline of children's "primitive mythological states," their ego extensibility.[29] "It is not that adults bring any pressure to bear upon children's minds, but simply that nowadays to walk down the street imposes a whole conception of the world," he says.[30] If that is the case, what conception of the world is being imposed by a world festooned with binary switches, especially as they have proliferated throughout domestic spaces and specifically on children's toys? Piaget does not address the question, but the effects of switching are likely the same: the direct-causal couplings effected by switches probably accelerate the transition to putatively adult mechanistic thinking, and in the process, also accelerate the transition to stricter forms of subject–object differentiation.

Even so, the causal logic of the switch differs from that of the bicycle in at least one important way: the linkages of its causal chains are almost entirely occult. Even as the switch affirms a strict causal logic of inputs and outputs, it also spawns new primitive mythologies from the recesses of its black box. Piaget concluded that, by age eight, most children could accurately account for the mechanical operation of a bicycle, but how old or highly trained would a person have to be to accurately account for the mechanical operation of a radio or a remote control? If Piaget had asked an adult what made the radio turn on, what would most of them answer except, “You press the button, and that’s all”? Between inputs and outputs, mysteries flourish. The highly rationalized world of switching is also a new world of occult forces and invisible connections, all designed to make it appear that everything really does respond to the slightest touch. Children shed the artificialism of childhood by committing to a mechanical view of the world, and switching no doubt accelerates this. But at the same time, switches reconstitute that primitive mythological state in technical processes. Having dispensed with artificialism in the psyche, adults inhabit a world that is really artificial, having reproduced in technical forms those primitive mythologies too painful to live without.

Adsorption and Adjection

Absorption has been long been considered a key component of play, perhaps even a defining feature. For Hans-Georg Gadamer, “play fulfills its purpose only if the player loses himself in his play.”[31] According to Johan Huizinga, “this intensity, this absorption, this power of maddening” is “the very essence, the primordial quality of play.”[32] Obviously, not all absorption is playful. Many kinds of work can be absorbing, but often the worker is not absorbed specifically in the object at hand. When I wash dishes, I am not exactly absorbed in the water and plates, but perform an activity automatically while my mind strays elsewhere. This is characteristic of the way in which a tool recedes from consciousness during certain kinds of work as part of that condition that Martin Heidegger terms *ready-to-hand*.[33] When a skilled carpenter is engaged in hammering, Heidegger argues, the hammer gets subsumed into his world. It becomes transparent, such that subject and object collapse together. Only when the tool malfunctions—Piaget would say when it *resists*—does it suddenly tumble back into the status of an object, at which point firmer subject and object distinctions reemerge.

But this absorption *with* a tool is somewhat different from absorption *in* a toy. To be absorbed in a toy is not to practice something so adeptly that the object gets subsumed by the subject, like a hammer that seems to become an extension of the arm. The toy is always resistant, otherwise it would not be a toy at all. So, although the carpenter might absorb the hammer, the toy absorbs the person playing with it. In work, I can forget the tool and absorb it as part of myself. In play, in contrast, I can forget myself and be absorbed by the toy. With tools, subjects absorb objects; with toys, objects absorb subjects. This is because of, not despite, the toy’s tendency to resist, for its resistance must be that pleasurable kind that solicits deeper identification.

We have few terms available to anatomize the full variety of states that might fall between strict subject-object differentiation, on one hand, and the total collapse of subject-object relations, on the other. I have been relying on the term *absorption* so far, but this has limitations. By implying the total consumption of subjects by object, the term names only one idealized extreme, the full intoxicating mania of play, the Dionysian ecstasy. That does not quite capture the complementary and simultaneous experience of mastery that playing with switches also affords. Anyone toying with switches likely experiences a mix of playful abandon and utilitarian mastery as subject-object differentiation waxes and wanes. In this unstable process, subjects and objects oscillate and waver, collapsing together and drawing apart again. From moment to moment, the switched toy both absorbs the subject and is absorbed by the subject. Subject sometimes yields to object; object sometimes to subject. This probably happens to some extent with all mechanical playthings, even all toys, but it must be especially pronounced for children playing with switches, who are simultaneously hastening their transit to adult models of mechanical causation and, at the same time, indulging in the absorbing qualities of all play.

Recalling Piaget's reference to the *adherences* of subjects to objects, a better name for this oscillation might be *ad-sorption,* a condition in which they temporarily join together, neither one devouring the other completely, but both chewing away at the boundary between. In physics, the term *adsorption* refers to the adhesion of atoms or ions from a liquid or gas onto a solid surface, a slightly better metaphor for these purposes than full devouring. In adsorption, it is not so much that objects consume subjects, or vice versa, but that they fuse at a boundary that both separates and conjoins. If *absorption* names an extreme condition of full immersion, *adsorption* acknowledges the persistence of boundaries as fertile contact zones that facilitate the partial and reversible mingling of subjects and objects in ways that are not quite so all-consuming.

Perhaps we also need a slightly different word to name this boundary condition, something different from either *subjectivity* or *objectivity,* even if it is made up of nothing other than their unstable and dynamic relation. Kristeva coined the term *abjection* as one especially horrifying mingling of subjects and objects. It names an encounter of self and other that has not been stabilized as strict subjectivity and objectivity, and that is associated, in her telling metaphor, with actions like involuntary gagging at the smell of spoiled milk. In such moments, the self expels not the separate object of the milk—for no milk was consumed—but *itself,* a purging without object that twists self into other.[34] For Kristeva, abjection is something like a raw encounter with presymbolic reality. But abjection is too horrifying to capture the more dynamic and even pleasurable process that I am associating with adsorption. On the model of adsorption, then, we might refer to the formation that congeals at the boundary of the switch as an *adject,* that which is thrown together, rather than the *abject,* that which is thrown away. Unlike a subject or an object, an adject is bonded to the other but has not been swallowed up by it. They are mouth-to-mouth, not devouring, but perhaps kissing. For this is pleasurable play, mutual and reciprocal, and equal in the sense that neither one contains the other. *Adjection* names a certain experience of intermittently but never completely losing oneself, more pleasurable and less horrifying than abjection. Crucially, adjection is also reversible. Adjects can quickly turn back into subjects or objects again once they retreat to safer quarters, and this can happen quickly and repeatedly.

It must be acknowledged that the adject finally does sustain a more traditional, humanist version of subjectivity by preserving some experiences of mastery, if only intermittently. The adject oscillates between absorption into the other and mastery over the other, between conflation and separation, between tool relations and toy relations, and thus also between adult and child identities. There are pleasures and perils at both extremes, hence the oscillation. Absorption in toys promises an ecstatic distribution beyond the narrow confines of self, but it does so as a form of devouring tantamount to loss. Conversely, mastery over the tool affirms human supremacy over inert materials, but it becomes tedious and repetitive. Playing with switches makes both experiences available in rapid alternation, as each alleviates the burden of the other. Adjection might even be seen as an acceleration and prolongation of Dionysus's transformation from absorbed child to god. Through childish absorption, he was reborn to superhuman mastery, a transformation that adjection makes continuous by rapidly alternating loss of self with magnification of self.

Up to this point, we have been trying to define this condition rather abstractly and exemplifying it only through the simplest kinds of switches, but we can now turn to more complex situations. In the chapters ahead, we will consider the role that button-based game pads play in video gaming and the ways in which that interface might afford satisfactions that real or simulated analog input might not. But before we get to video games, we need to look more closely at how binary switches came to mediate games in the first place, especially in an important precursor to video games that is also a close cousin to the slot machines discussed above: the pinball machine.

CHAPTER 12

Pinball Wizards

Adjection is not just child's play. The previous chapter examined some of the ways in which toying with switches both indulges children's presubjective or semi-subjective states and hastens the onset of normative adult subjectivity and the mechanistic models of causation associated with it. At this stage, we turn instead to an opposite dynamic, to the consequences for adults of toying with switches. If toying with switches hastens the development of normative adult subjectivity for children, in adults it temporarily and partially restores childish forms of presubjectivity. And yet, as with children, adult adjection amounts neither to a sustained demonstration of mastery over objects nor to a total dissolution into objects, but rather oscillates between the two. Neither all absorbed nor all absorbing, the adult adject simultaneously enjoys the exhilaration of mastery and the pleasure of its alleviation.

We thus turn to certain kinds of machines designed to foster adjection for adolescents and adults, games that expressly yoke switching techniques to forms of play: computer games and their antecedents, especially pinball. There can be no doubt that computer games have had enormous influence around the globe today, or that they have become one of the most important forms of imaginative, aesthetic, or even literary experience for many millions of people. In 2018, the computer-game industry in the United States alone earned $43.4 billion, more than the entire movie industry, which earned just over $11 billion. Globally, computer games earned almost three times as much as movies around the same time, more than $120 billion total.[1] Scholars of computer games often have focused on interactions that are predominantly visual and auditory, but they have paid significantly less attention to what players do with their fingers. Even the term *video game* overassociates computer games with visual media, defining them fundamentally as a medium for the eyes. Accordingly, scholarship on gaming has long struggled to escape the influence of film and television studies, forms of visual media that have sometimes seemed to be video gaming's closest ancestors. Although more recent scholarship has strongly resisted the tendency to treat games as interactive films, ocularcentrism still persists, which can make it hard to appreciate the complexities of touch in computerized play.[2] When we start with the controller rather than the screen, the hands rather than eyes, feeling rather than seeing, we can better appreciate that something profound is happening at that point where people get in touch with the binary inputs of gaming machines.

None of this began with computer games. We need to recover a longer history of binary input in games and play to understand the haptic dimension of computer games. Starting in the nineteenth century, games with analog inputs like billiards developed slowly but steadily into games with much more binarized inputs, such as pinball—before computer games, undoubtedly the most popular button-based game. Pinball taught generations of people to take pleasure in radically restricting their inputs to the timing of button presses to control the flippers, a crudely limiting method for managing the chaotic, high-speed dynamism on the other side of the glass. By tracing the gradual development of an analog game like billiards into a much more binarized game like pinball, we can begin to discern some of the stakes of playing with buttons for adolescents and adults.

Much of what follows amounts to a prehistory of computer gaming, and specifically of the computer game's touch interfaces. That is a much narrower focus than in most scholarship on the relation of pinball to computer games, which has tended to focus on the social space of the arcade. For instance, Michael Newman has recently detailed the transition from pinball to computer games in American arcades, which entailed significant shifts from urban and working-class to suburban and middle-class players.[3] Around the same time, Newman shows, home computer-game consoles made gaming more private and domestic than pinball ever had been. And yet these key differences can obscure important similarities, especially those related to methods of manual control. The goal will be to understand how the haptic binarism of switches and buttons made adjectification much more available, identified it specifically with adolescent and adult forms of play, and sanctioned it for those who might have outgrown other kinds of toys. For them, the switch confirms that a master is in command, and thus affirms the power and autonomy of players who deploy the will with the twitch of a finger. At the same time, however, the game as a whole persistently disrupts that pretense of mastery, and in doing so absorbs the subject in resistant objects never securely controlled after all. Adjectifying play with and through binary switches sets subjects and objects in motion relative to one another, sometimes separating and sometimes conflating them, throwing them together in chaotic and continual reconfigurations, without ever entirely dispelling the underlying structure that defines them both. Adjectification may be thought of as something that happens not just to subjects, but rather to subjectivity. Especially for adolescents and adults with mostly or fully formed selves, adjectifying play may even be a way of playing *fort–da* with subjectivity itself.

Binarizing Bagatelle

Despite its association with twentieth-century arcade culture among youth, pinball derived from European public house amusements as early as the sixteenth century. Initially, English patrons of Elizabethan public houses brought a version of nine-pin bowling inside, which they first played on the floor and then later miniaturized on table-top boards. They eventually replaced the nine bowling pins with holes in the board, and the game came to be called "nine hole."[4] The French adapted some version of this game to the use of the *billiart,* or little stick, and eventually named it Bagatelle. By the eighteenth century, bagatelle had recrossed the channel to heavily influence British amusements once again. In Charles Dickens's 1836 *Pickwick Papers,* the patrons of the Peacock Tavern "beguiled their time" with a "bagatelle-board on the first floor."[5] By the late nineteenth century these early antecedents of pinball had thoroughly colonized drinking establishments on both sides of the Atlantic, at which point many of the formal properties of pinball already were in place.[6]

The first great leap in automating bagatelle occurred in 1871, when Montague Redgrave, an English immigrant working in Cincinnati, Ohio, replaced the billiard cues with a version of the spring-loaded plunger found in classic pinball machines. Instead of shooting a ball directly at cups or holes with a cue, Redgrave's table-top bagatelle, pictured in Figure 28, required the player to pull back the spring-loaded plunger to launch the ball, which then rolled back toward the player on a slanted table toward the available cups. "This is a great advantage over the ordinary cue," Redgrave claimed. "With it all the mathematical science may be possessed without the firmness of nerve necessary to execute, and without the ability to impart the exactitude of force which is the foundation of success."[7] Redgrave's game reduced the player's activity to a single radically restrictive input, not quite binary but getting closer. The plunger automated how much force to apply to the ball so thoroughly that dexterity hardly entered the question at all, which brought the operation nearer to an all-or-nothing scenario.

Games of all sorts frequently make binary distinctions between certain significant actions, such as a ball crossing a line. But analog gradations quickly intrude, requiring still more distinctions about what constitutes *crossing,* such as the specification that any or all of the ball must cross the front or the back of the line. Even then, these apparently binary distinctions often require human judgment to ascertain, and not always to the satisfaction of everyone. In other games, mechanical methods help disambiguate such situations, such as the pocket of a billiards table. The pockets are mechanical disambiguators that generate a strictly binary alternative: either the ball falls into the pocket or it does not. Many games involve both kinds of distinctions, such as tennis, in which the net is a special kind of mechanized three-dimensional line, one that eliminates all ambiguity about whether the ball crossed it or not. These mechanical

FIGURE 28. Patent model for Montague Redgrave's "Improvements in Bagatelles," U.S. Patent 115,357, issued May 30, 1871. Note the scoring bells in addition to the scoring cups, which further associate his game with the sensory stimulation of pinball. Courtesy of the Division of Cultural and Community Life, Smithsonian National Museum of American History.

affordances function as rules of the game, parameters within which play must always occur, but enforced by physics rather than referees. A game with weak forms of mechanical disambiguation, such as wrestling, will necessarily enforce more of its rules through active human judgment.

Bagatelle games like Redgrave's were not wildly successful, but by the late 1920s electrification made them more stimulating and their popularity began to grow. Electrification also allowed pinball to be binarized even further. In 1928, James Nicholas put in place virtually all the components of pinball except for the flippers, such as rubber bumpers and an automatic ball trap and return, and his patent established terminology used in pinball ever since.[8] What his machine still lacked, however, was a clear way to differentiate pinball from games of pure chance, such as roulette. Players did have some ways to control the ball after it was launched by the plunger, which proponents claimed distinguished the game from slot machines or roulette wheels. By shifting or shoving the entire cabinet, players could redirect the ball away from hazards or toward goals, an intervention referred to as "nudging" or "gunching," a form of input that tilt sensors later limited but did not eliminate. In 1947, Chicago pinball manufacturer Bally automated the technique in a game they named Nudgy. Its promotional copy explained, "Merely by pressing a button on the side of the cabinet, a player could cause the entire playfield to jerk toward the rear of the cabinet."[9] The game turned out to be a failure, but other manufacturers were thinking along similar lines. In the same year, the Gottlieb company introduced the first machine with button-based flippers, named Humpty Dumpty, which featured three banks of flippers along both sides of the interior of the cabinet, each bank of three operated by a corresponding button on the cabinet's sides. Players reached around the cabinet on both sides as if preparing to nudge the machine, but in fact their actions consisted of the much simpler and less forceful action of pressing a button. In some European nations including Germany, the game that English speakers call *pinball* is known as *flipper* to this day.

Flippers allowed players to intervene more decisively in the action, prolong the game, and demonstrate more skill.[10] A period advertisement for Humpty Dumpty claimed, "Super-sensitive Flipper Button on both sides of the Cabinet controls 6 unique FLIPPER BUMPERS on Playing Field. With SKILL and timing, player can control balls, . . . can send them zooming from the bottom right back to the top."[11] The addition of flippers rapidly changed the complexion of pinball, partly severing it from its association with barroom gambling, even if the most restrictive cities, such as New York, remained unpersuaded.[12] Instead of launching a ball and hoping for a decisive result, as if one were playing a slot machine, the new flipper machines encouraged point scoring competition and prolonged pleasurable play, and some players clearly were more adept than others. Audiences responded enthusiastically. Gottlieb sold sixty-five hundred Humpty Dumpty machines, a level unmatched by any other game until the 1970s.[13]

The buttons on early machines typically took the form of circular discs on the sides of the cabinet, but later adopted the almost universal form of the *arcade button,* about one inch in diameter, often slightly concave and protruding roughly a quarter inch from the surface, spring-loaded, and without any discernable click. When pressed, early flipper buttons closed the contacts of a simple leaf switch hidden beneath. Closing the switch activated a solenoid that rapidly moved the flipper, and when the circuit opened again by releasing the button, a spring returned the flipper to its original position. Even on more recent computerized pinball machines, the player does not typically have any way to influence the flipper's power or speed, because hitting the button more quickly or more forcefully will not make it strike the ball faster or harder.[14] The introduction of flippers also helped the coin-operated amusement industry distinguish pinball from games of chance, such as slot machines or roulette. Because players could intervene in the action continuously, manufacturers and owners could argue that pinball was a game of skill, which they did with increasing success during the 1960s. In the 1970s, Los Angeles and Chicago legalized pinball, as did New York, but only after a proficient player and pinball advocate demonstrated to the satisfaction of the city council that he could call his shots in advance.[15]

Part of the pleasure of pinball involves the skillful management of a complex environment of rolling balls, bumpers, ramps, targets and even hidden magnets with an extraordinarily restrictive binary interface that registers no inputs other than on or off. Even more than the plunger, the flipper buttons organize action in the ways we have discussed in previous chapters: they isolate actions as discrete choices; they shrink the gap between trying and doing; they extend the radius of the will; they amplify physical force; they condense the duration of action to an instant. Accordingly, the kind of mastery demonstrated differs from the analog kind. In billiards, a player must have considerable experience to anticipate all the factors involved, estimate angles, bounces, and friction, make sound judgments about the best plan of action, and finally execute a plan dexterously. None of this is exactly a matter of mathematical calculation, however, but rather of practiced judgment. For that reason, David Hume used the example of billiard balls to insist that causation is purely experiential rather than a matter of a priori reasoning. Even though most people today would probably agree that billiards is entirely describable in terms of Newtonian mechanics, Hume's point was that players do not come to understand it through a priori reasoning. Only by seeing balls collide in the past, both in billiards and in other contexts, do players form judgments about what sort of action is likely to follow any other action.[16] This is why there can be highly proficient billiards players who have no knowledge of Newtonian mechanics at all. But precisely because they are not calculating the physics of bodies in motion according to universal and universally available laws, players will receive a different kind of credit for the results of the causal interactions. Players with sufficient judgment and dexterity born of experience will earn credit for real virtuosity, for they are doing something more than just following a mathematical script to an assured outcome.

In pinball, by contrast, the ways in which a player might try and fail are reduced drastically, and the most conspicuous involves failing to time the pressing of a button precisely. Everything else the machine automates with wonderful reliability, rapidity, and ease. In billiards timing matters very little and dexterity is everything; in pinball dexterity matters very little and timing is everything. Even the player's most deliberative strategies, such as systematically pursuing certain ways of scoring points, must be carried out primarily by managing the precise timing of button presses.

It is easy to discern political meaning behind these differences. Billiards expresses an Enlightenment commitment to rationality that is yet heavily dependent on individual human judgment. It is slow enough to permit, and even require, careful reasoning about the next course of action. It is social, in that individuals compete against each other over a shared table, often facing each other from opposite sides in convivial settings such as public houses. And many different games can be played with the same implements, so players have great freedom to alter the rules of play to create different kinds of games. Billiards is thus a game in which reasoning individuals can test their most deliberate actions against each other in friendly contests, with outcomes minimally attributable to chance, and within a flexible structure that permits the changing of the rules. In these ways, it is a fair model of an enlightened liberal democratic society.

Pinball is quite different. As Warren Susman has pointed out, it "was the ideal toy of the machine age." It has "a complex set of stern rules" scripted entirely into the function of the machine, which limits its flexible reconfiguration by players. Multiplayer games are possible, but players often play alone. In a billiards room, competitors face each other across a table, but in a pinball arcade they turn their backs to each other, hunching over machines that back up to each other or that are lined up against walls. And pinball players must act and react so quickly that there is rarely time for careful, judicious deliberation. For Susman, pinball is a symptom of modern alienation, and even of the contradictions of capitalism. Even though the player is "a convert to, or a true believer in, the vision of greater order and increased rationality," he says, at the same time, "luck, chance, irrationality greeted him everywhere."[17] Pinball might even be seen as a training system that prepares players to endure and even enjoy their isolating exposure to modern mechanized precarity.

This seems true up to a point, but perhaps Susman underestimates the outlandishness of the intensifications of pinball, its wild acceleration of action, and especially its extreme restrictions on manual input. There is something almost parodic about pinball, in that its excesses undermine the mechanical structures it also imposes. Pinball might model modern social and economic life, but it takes its simulation to ecstatic extremes. Outside the machine, the controls approach a minimum of binary functionality; inside the machine, the effects approach a maximum of complexity and sensory stimulation. Inside the glass, an accelerated world of dynamic action and constant change; outside the glass, the crudest possible interactive controls. This contrast between a complex interior and a simplistic exterior virtually defines pinball, intensified over many decades of innovation after innovation. The haptic poverty of touching two crude buttons on the outside nonetheless generated great riches of other forms of sensory stimulation inside. That arrangement affirms the mastery of players over machines, because they can exert their will with the merest touch of a finger. At the same time, it also enthralls them with material processes that are remote from their bodies and never entirely under control. The adjectivity of the player consists of this simultaneous exaggeration and elimination of mastery, of making the machine an extension of the person while simultaneously making the person an extension of the machine.

The broader point is that adolescents and adults learned to play this way, probably for the first time ever, only in the twentieth century. It may be that, as Susman argues, games like pinball are didactic, training players for a life of mechanized capitalist precarity, but this is not the whole story. Plato said that the child who would be a farmer should play at farming, but pinball reverses this logic. For adolescents and adults already acclimated to working with switches and buttons, pinball turns work back into play, and in the process, adult subjectivity back into more fluid and distributed childish forms. Pinball is a machine for turning subjects into adjects, alleviating the stable, bounded forms of subjectivity normalized for most adults most of the time.

Flipper Fingers

Roger Moseley's account of *ludomusicality* asserts that modes of playfulness born from music have infiltrated media forms ranging from computer games to telegraphy.[18] Sometimes this ludomusical legacy appears overtly as musical performance, as in computer games like *Guitar Hero,* but sometimes it simply retains the form of the instrument and some attendant input techniques, as in Baudot's telegraph. One of Moseley's points is that there is more music in technical media than most people like to think. Another point is that music and musical techniques afford critical perspective on digitality in the broadest possible sense, encompassing everything from manual keyboarding techniques to computers' binary code. The prevalence of ludomusical elements in modern media may help explain why a concept album by a British rock band might be the most insightful critical appraisal of pinball ever written.

I am referring to the 1969 rock opera *Tommy* by the British band The Who. The opera's most well-known song, "Pinball Wizard," celebrates the "deaf, dumb, and blind kid" who becomes an expert pinball player as a teenager, and later a messiah figure to his adoring fans.[19] Tommy had lost his sight, hearing, and ability to speak earlier in childhood after enduring several traumas, including seeing his father shoot his mother's lover, being sexually abused by an uncle, and being tied up and tortured by a cousin for sport. Isolated, powerless, and repeatedly victimized, Tommy takes refuge in "poxy pinball," as his father calls it, the great art born of his suffering. In this way, and by the most thinly veiled metaphor, the opera identifies the sources of rock music with teenage suffering and alienation.

The song "Pinball Wizard" makes it clear that Tommy experiences a mystical communion with the machine, reminiscent of that of the rock star with his guitar, but ultimately different from it. For, Tommy is a virtuoso of pinball, a celebrity player whose intuitive command of the machine suggests deeper levels of sympathy and feeling. Eventually, Tommy takes his show on the road for something like an arena tour, where his adoring acolytes seem, by turns, like delirious fans at a rock concert or like fervent members of a religious cult. Overall, the opera is thinking through the ways in which rock music might correlate with other examples of mechanized youth culture, all of which *Tommy* imagines as generating intense mass-cultural affiliations born from the agonies of adolescence.

In *Tommy,* however, the conflation of people and machines is even more profound than the ecstatic transport of a rock musician in the throes of a guitar solo. Tommy "Becomes part of the machine," and he "Plays by intuition" alone. He "Ain't got no distractions / Can't hear no buzzers and bells / Don't see no lights a flashing / Plays by sense of smell."[20] That last line is ironic, spoken naively by his envious rival, because smell has nothing to do with it. Smell stands metaphorically for some more mystical perception, described elsewhere in the opera as feeling. Touch, or some mystical extension of it, is Tommy's second sight, even though touch is the one sense that the glass top of the pinball machine blocks completely. Accordingly, Tommy's hands have "crazy flipper fingers," adept at pushing buttons but also conflated with the flippers themselves. He masters the machine from without but somehow is inside it too, both extended by it and absorbed into it. The opera stages the merging of human and machine into a totality that amounts to the superhuman, an actual *wizard.* But this holds true only as long as Tommy plays the game, because none of his wizardry transfers to other kinds of machines. Tommy cannot drive a car "by intuition," and his adoring disciples have to lead him through the arcade like any other person with a severe visual disability.

Tommy always seems to be either less than human or more than human, either a degraded being who absorbs the most horrifying abuse or a demigod who humbles the mortals around him. As a figurative account of personhood, *Tommy* correlates the process of growth and development from childhood to adulthood with a kind of mechanical augmentation that finally overshoots the mark. Physical vulnerability turns into supernatural ability, no less for Tommy than for Dionysus. Along with that transformation, sadistic abuse also turns into inverted forms of worship. After a cure restores Tommy's senses, he tells his acolytes, "If you want to follow me / You've got to play pinball / And put in your earplugs / Put on your eyeshades / You know where to put the cork."[21] The cork goes in the mouth, presumably, an imitation of his own prior muteness, and an ascetic precondition of his disciples' rebirth. The broader message is clear: apotheosis requires humiliation and denigration, a formula as Christian as it is Bacchic. To rise above the human requires sacrificing the most cherished human capacities. As a result, the crude binarism of the button helps organize an equally crude binarism of two different kinds of personhood, either of which would prove unsustainable on its own, but both of which Tommy manifests. The mastery of the subject gets figured as a wizard's supernatural might, but that master is also another kind of person, a victim disabled by violence. There is nothing in between.

This brings us to the larger point, not just in regard to *Tommy* but in relation to pinball itself as a popular amusement over many decades. It is not just that there are a few extraordinary people like Tommy who achieve (or endure) this denigrating apotheosis, but rather that the binarizing of play makes a roughly equivalent condition available to millions. Tommy's unstable existence as both subhuman and superhuman is a literary dramatization of adjection itself, a fantasy of what it might be like to inhabit that condition permanently rather than as a temporary ludic respite. Put differently, *Tommy* imagines what it would be like if adjection were not just something one *does* but something one *is,* not just a temporary experience but an inescapable identity. In a world that routes so many kinds of action through binary switching for so many millions of people, perhaps it is.

Wii the People

If pinball had been a dead end, a temporary conjunction of binary switching and ludic experience that expired with the arcade, it might have told us something about midcentury life but probably would have little relevance to life today. But pinball was not a dead end. Since the 1970s, computer games have propagated adjection in roughly the same way but more widely than ever, first in arcades where they crowded out pinball machines, later in homes on popular gaming consoles, and later still on smart phones that make gaming available anywhere. With computer games, the political stakes of all this also come more clearly into view. As we shall shortly see, the impossibly high standards of individual autonomy in liberal societies may in fact compel subjects to perform mastery while simultaneously seeking to escape the pressures of such an impossible ideal.

New York City finally legalized pinball in 1976, just one year before the Atari VCS (Video Computer System) home game console inaugurated the second generation of computer games, relegating both pinball and public arcades to an ever-shrinking niche. But the sophistication of modern computer games raises a difficult question. Given that their visual simulations have achieved striking levels of verisimilitude, and given that their audio simulations are more life-like still, why has the haptic interface not developed similarly? The great promise of virtual reality has left the computer-game controller curiously unaltered, with most still heavily reliant on binary switches and only some employing comparatively crude forms of haptic feedback, such as vibration. The convincing simulation of analog forms of sound and vision has only limited equivalents in the realm of touch, where binary inputs stubbornly persist. The reason for this, I want to suggest, is not that technology or human imagination has failed to develop equally impressive haptic simulations, but that computer games have preferred binary input in order to contrive the same adjectifying effects previously afforded by the limiting controls of pinball.

Some will object that recent innovations in gestural control suggest otherwise, but gestural control has not proved as revolutionary or as popular as it once promised to be. It has, however, attracted a great deal of critical interest from scholars of computer games, far more than other kinds of controllers.[22] The gestural Wiimote controller introduced with the 2006 Nintendo Wii drew widespread popular and scholarly praise as a marked improvement over earlier game pads. With the Wiimote, a player swings a tennis racquet within the game by miming the action of swinging a racket with the controller in hand. The Wiimote was hardly the first gestural controller, as Ian Bogost notes, for it was preceded by Sony's EyeToy, Bandai's Power Pad, the Joyboard from Commodore Computers, and Mattel's Power Glove, in addition to earlier light guns and steering wheels, some of which date back to arcade games in the 1930s.[23] For Bogost, gestural control in all these options affords richer kinds of experience, given that gestures are socially meaningful in ways that button pushing is not. I disagree that button pushing is not a socially meaningful gesture—indeed, I have been arguing throughout that it is profoundly so—but it certainly does promote a different kind of sociability. In fact, Nintendo marketed the Wii as a way of fostering lively in-person sociability, and its advertisements frequently depicted active and tightly grouped players viewed from the position of the screen. Some ads did not show the game on the screen at all. The name of the Wii even plays on the English first person plural pronoun, not *I* but *We,* an invitation to collective affiliation.

According to the most enthusiastic advocates of gestural controls, players who use them experience deeper immersion in the game world, more enjoyment, a greater feeling of presence, or a sense of naturalness, not to mention a more life-like simulation of reality and richer social interactions with one another.[24] These encomia often involve a corresponding denigration of mere button mashing. As one critic put it, the button is "sadly inflexible, . . . shows none of the ability of a modern computer to express user intention," and "should be abandoned as soon as possible."[25] Other critics have claimed that buttons hinder learning, or are disembodying, alienating, or complicit with a disciplinary logic of control.[26] In the heady early days of the Wii, traditional game pads appeared as unfortunate historical compromises, endured rather than embraced, perhaps because technology afforded no better option, or because economics made button-based controllers cheaper to manufacture, or because path dependency locked everyone into suboptimal arrangements.[27]

This open association of gestural control with in-person sociability is telling. We already have seen how the binarizing of bagatelle gradually transformed billiards from a social game requiring multiple players to an individual contest against chance. Partly for that reason, pinball machines tended to be legally categorized with individualized gambling devices like slot machines, even after they ceased dispensing prizes and after multiplayer options were available.[28] And unlike billiards, pinball is a game that someone could play for hours entirely alone. So, another part of the story of the binarizing of billiards and bagatelle involves this insistent individualizing of play. When Nintendo attempted to associate the Wii with sociability, it did so not just by devising in-person multiplayer games, but also by replacing binary switches with analog gestures. Socially meaningful gestures fostered sociable play. But this suggests that the opposite also might be true: perhaps individually meaningful gestures like switching foster equally powerful forms of individualization.

There is no space here for a full history of computer-game controllers, but we should note that binary switches have been present from the beginning, starting with the controls for the very first computer game, *SpaceWar!,* which first used test-word switches on the console of a DEC PDP-1 computer installed at MIT. Later, its engineers improvised controllers from Western Electric office buzzer buttons, and finally built custom controllers with toggle switches and buttons.[29] Even so, the line from the switches and buttons of *SpaceWar!* to the buttons on today's console controllers is not a straight one. Dials gained ground in the earliest home consoles, such as Atari's Pong, and the joystick of the Atari VCS dominated the second generation of home consoles after its introduction in 1977. Even though the Atari controller was little more than a stick that activated four binary switches inside its black box, its form promised something like the analog inputs of aircraft controls. At one corner it also presented its alluring orange button, and it is hard to think of a single controller without at least one button since.

The release of the Nintendo Famicom game pad in 1983 made buttons dominant again, in a form that eventually displaced joysticks for roughly a decade. Designed to be held with two hands, the Famicom controller has a directional pad, or D-pad, on the left side consisting of four buttons linked to a single plastic rocker, two small buttons labelled *Select* and *Start* in the center, and two game-play buttons labelled *A* and *B* on the right. Three years later, Nintendo released an American version with a different console design and a different name, the Nintendo Entertainment System (NES), but with controllers that were mostly identical, depicted in Plate 10. The NES proved so successful that other console makers in the late 1980s adapted similar designs, including the Sega Master System in 1985 and its more successful Genesis in 1989.[30] As Nintendo and Sega dominated the industry for the next decade, their entirely button-based D-pad controllers became an industry standard. By the time the fifth generation of home consoles appeared in 1993, all contained some version of the game pad.[31]

By the 1990s, however, small joysticks and sometimes even trackballs began to return to controllers as well, especially after the success of Sony's Play Station in 1994, which added two analog thumbsticks to the basic NES configuration. But even with these additions, button-creep continued. The Sega Genesis added two additional buttons for the right hand, and later two more still for a total of six. The Sony Play Station Dualshock controller introduced shoulder buttons on the front of the controller to be activated like triggers with the index fingers, an innovation widely copied. Later controllers added inputs shaped like triggers beneath those, some of which registered gradation so that they could simulate the variable throttle in a racing game, but even these were probably more often used for binary operations, such as simulating the triggers of virtual guns. Today, the most popular controllers are more heavily festooned with buttons than ever. The Sony Dualshock 4 has fourteen buttons. Nintendo's Joy-Con controller has the same basic format, but users can upgrade to a more familiar game pad with sixteen buttons. Aftermarket controllers are even more baroque. The Razer Wildcat controller for the Xbox One has twenty-two buttons in addition to the standard two thumbsticks, not an unusual number among high-end controllers. Even the Wiimote gestural controller had eleven buttons and could be used like a traditional game pad. Many serious gamers prefer to use a QWERTY keyboard (usually in conjunction with a mouse), instead of any of these controllers, which affords more buttons still.

For all the promise of gestural control, then, players continue to play computer games much as they did in 1968, substantially if not quite exclusively with binary switches. Accordingly, the Wii's popularity proved temporary. The most common home console in 2017, the Play Station 4, sold twenty million units worldwide. The Nintendo Wii U, an updated version of the original Wii, sold fewer than one hundred thousand.[32] After a period of intense enthusiasm, the gaming market eventually gave gestural control a well-known finger gesture of its own.

As a result, for decades gamers have controlled an entire symbolic realm of simulated actions with binary inputs, a far wider range of actions than pinball can facilitate mechanically. Joysticks, some with analog functions, are involved too, especially for controlling simulated movement through space, but a wide range of other activities rely primarily on button pushing, such as jumping, striking, throwing, changing speed, switching views, picking something up, choosing among objects in the environment or an inventory, firing weapons, entering text, and much more. In games without a human avatar, such as strategy games or simulators, buttons activate an equally wide range of options, such as moving an army or building a house, and more easily with a QWERTY keyboard than with a console controller. Switches and buttons are thus not so much baleful and unnecessary compromises, but powerful affirmations that all action can be rendered as a binary equivalent, and that there are experiential advantages for doing so. The embodied action of jumping becomes an automated mechanical routine that one can switch on as needed. Any sort of fluid or continuous action can be translated into the logic of on and off and mapped back onto the body as switching techniques.

It is worth noting that, even though computers today largely conceal their binary function within binary code and a microscopic network of small automatic switches, at the interface with the body the binary frankly and unapologetically manifests as switching. The eyes and ears no longer perceive computational binarism, for only experts can access code at that level. The digitality of audio is now all but impossible for the unaided ear to perceive, and even the digitality of pixels on the screen is harder to discern without magnification. But the fingers enjoy no such illusion. For them, the rigid logic of on and off is inescapable. This is no accident, no failing of technology or imagination. Imposing binary input techniques on the fingers is what a controller is *for*. The controller compels functional binarism just as purposefully, just as insistently, as flipper buttons once did for bagatelle. As a result, the controller does not resort to translating complex gestures into crude binary inputs, but rather accomplishes that translation as one of its core functions. To recognize this function as a success, rather than as a compromise, and still less as a failure, is to begin to appreciate how determined people have been to inject haptic binarism in the experience of certain kinds of play.

And the reason for that, I have been suggesting, is that nothing else so successfully generates that wavering form of subjectivity that I have been calling *adjection*. If young children gradually consolidate stable forms of subjectivity through the pleasures of adjection, adults can mitigate it the same way. By toying with buttons, they can effect a simultaneous dissolution and resolution of the self, a chaotic, unstable affirmation and denial of stable and bounded subjectivity. Like Tommy, they can be both more and less than human, both more and less than the kinds of responsible, autonomous agents that liberal societies expect them to be. Adjection might even be considered a playful exposé of the deepest logic of liberal individualism, one that developed in parallel to the proliferation of switching starting around the second half of the nineteenth century. As switching organized agency as more rational, more individualized than ever, playful alternatives disrupted it from within. To be clear, these disruptions are by no means a radical escape from subjectivity. The adject might be elastic and mobile, wavering between subject and object, but it never escapes the logic of subject–object differentiation.

Because it generates no other settled counterlogic, adjectivity preserves the binary of subject and object intact, regardless of how dynamic their relation turns out to be. For adults, adjection might be a game for passing time in the dualist prison.

But no political system sustains itself by exhibiting its internal contradictions like this. In the next section, we turn more specifically to a "haptic liberalism" through what are generally less playful techniques involving domestic automation, photography, voting techniques, and remote control. When toys become tools, when play becomes something more practical, switching techniques typically produce individual subjects ever more convinced of their mastery.

Part V

»» »» Haptic Liberalism

CHAPTER 13

The Control Panel of Democracy

The switch's most beguiling pretense is its passivity. It appears to wait patiently, inertly, for the activation of human touch, never varying or changing in the meantime. Although previous chapters often have associated switches with action, an equally important function is their *inaction,* their thorough resistance to change absent human contact. After all, a switch must not change states on its own, nor should it fail, function intermittently, or degrade materially. A great deal of engineering went into accomplishing these goals, including plastics that do not erode, springs that endure millions of compressions, and even printed letters that will not rub off from keys.

This passivity proves enormously flattering to the people involved. Every switch is like the sleeping beauty of a fairy tale, dead to the world but ever responsive to the right kind of touch. One of the General Electric Company's most well-known former slogans boasted, "We bring good things to life." In saying so, GE was associating consumer electronics not just with "the good life," its glibbest insinuation, but also with the reanimation of the dead. GE makes everyone into the charming prince who awakens Little Brier Rose with a kiss, or, more fittingly, Viktor Frankenstein, who "might infuse a spark of being into the lifeless thing."[1] To press the *on* button is to quicken the dead. In this way, the apparent passivity of the switch also acts back powerfully upon people. The switch's passivity really amounts to a limited kind of material stasis that is socially, culturally, even ideologically active. When everything works as it should, the changelessness of the switch becomes an active engine that shapes human agency itself.

The previous part addressed toys and button-based games that functioned unpredictably, even randomly, and thereby opened up ludic possibilities. In the process, as we saw, the boundaries between subjects and objects, people and things, sometimes eroded. In the chapters in this part, we turn instead to machines that must not function unless humans compel them to do so, in order to sustain the most fetching illusions of anthropo-exceptionalism. Switches don't just recruit human agency, as if agency were some indwelling human given, but rather generate basic forms of individual agency and autonomy that have long defined liberal subjects. Switching techniques bring agency into existence in many of the forms most recognizable and highly prized in liberal societies today, associated as they are with freedom, autonomy, individuality, and responsibility. The four chapters in this part will trace the production of liberal agency from various switching techniques, and with varying political results, but we begin with what may be the most conspicuously political switches of all, and the ones expected to function most reliably, those employed only comparatively recently on voting machines.

The disambiguation of voting had long been an important project for liberal democracies, and mechanical disambiguation long predates the nineteenth and twentieth centuries. Voting techniques often involve binary alternatives such as *yea* or *nay,* and even in an election with a slate of many competing candidates, voters typically must choose one and reject the rest. There are exceptions, such as ranked-choice voting, but many forms of voting have been organized as binary operations. Whatever ambivalence a voter might feel in choosing one candidate over all others, the choice of the vote itself must be unambiguous. Looked at that way, choice is not at all synonymous with will, and still less with something like desire, for choice by definition involves a structure of alternatives that organizes and constrains the ambivalence of the will. To choose between alternatives is to be limited, yet from this condition of limitation the most compelling demonstrations of agency tend to result. A person who acts inconsistently, ambivalently, intermittently, or with frequent reversals might be said to suffer from weakness of the will, but voting procedures restrict such failures almost entirely. To vote is to have been constrained to select among mutually exclusive alternatives, once and for all, and the autonomous will seems to achieve its highest political expression from this severely limiting condition.

Free and responsible liberal agents expect to self-direct in any number of areas of life, such as entering into contracts, but although it might not be the most materially consequential act for any individual, casting a vote is surely the most politically resonant. Doing so affirms clearly and publicly that the state recognizes the voter's self-sovereignty, and this confirms the voter's full status as citizen. But the act of voting has more private political consequences too, for specific voting techniques can affirm voters' social embeddedness or individual autonomy, their entanglement with other people or their independence from other people. Precisely because voting is so symbolically loaded, its techniques can model the proper constitution of political subjects in a liberal society. The pages that follow show how the mechanization of voting translated a discourse of liberal freedom and autonomy into technical forms, which it then distributed widely but surreptitiously in devices that have political content of their own. Even though the mechanization of voting first arose in Europe and North

America as part of a movement for universal male suffrage, the prevalence of binary input techniques coupled with an imperative for secrecy eventually sustained very different values. In short, the mechanization of voting involved a shift from an earlier phase of socially conscious civic republicanism to a later and much narrower phase of liberal individualism. That general transformation has been described by many historians and political theorists before, but not, to my knowledge, as a function of technical mediation. In the end, this chapter suggests, changes in voting technology are part of the recalibration of liberal democratic society, and not just a secondary reflection of it. How people vote makes as powerful an argument about liberal subjectivity as do new laws, public discourse, or theories about the nature of political subjects. Although inseparable from legal, discursive, and theoretical developments, voting techniques are nonetheless partly parallel to them rather than entirely derived from them. Indeed, because all cultural techniques act so directly on human bodies, new voting techniques may have made especially persuasive arguments to millions of people who had little choice but to acquiesce.

The mechanization of voting had three main consequences for these purposes. First, by mechanizing the act of choice, voting machines obviated the need to register choice through other means, by marking a paper, speaking out loud, signing one's name, or other methods that might be prone to misinterpretation or ambiguity. Once embedded in technology, *yea* and *nay* became equivalent to *on* and *off*. Democracies already have procedures for narrowing down the fullest range of possibilities to a small set of mutually exclusive alternatives, but new voting technologies translated laws and practices into durable techniques. Once embedded in machine-mediated techniques, democratic procedures appeared to be a matter not just of "how we do things," but of "how it is," a mechanical means for naturalizing the way the world works.

Second, this translation of earlier methods of disambiguation into strictly mechanical ones concealed a great deal of social content. For some reformers, this was precisely the point. Voting by voice reveals a great deal about the voter, and when performed publicly, conveys all the social meaning that voters themselves cannot help but exhibit. Written ballots removed some of this social meaning, but handwritten ballots might still signal class or gender, and might be even more revealing of education, especially literacy, than voting by voice. Even the preprinted ballots that nineteenth-century voters might collect from a church or union clearly signaled broader affiliations. In contrast, mechanized voting sheds most evidence of social meaning, thereby ensuring that voting is both maximally accessible to the greatest number of people and also maximally devoid of social content.

Third, and finally, the mechanization of voting is impossible to separate from a parallel project, the popularization of the secret ballot. Stripping away social content could anonymize individual votes, but it also could individualize anonymous voters. As secrecy and privacy became the norm, voting technologies extended the basic principle of "one man, one vote" to a more extreme performance of entirely private self-sovereignty. Liberal theorists understood perfectly well that people are influenced by any number of sources, but voters were trusted to own those influences, and to incorporated them into a self deliberately and rationally. The laudable desire to prevent voters' coercion and corruption thus resulted in dramatic displays of voter independence through physical separation and visual concealment. This culminated in the first voting machine used in an actual public election, the Myers voting machine of the 1890s, which took the form of a metal room closed on all sides, with a narrow double door and a massive bank of pushbuttons within. The obvious techniques of individualist self-isolation effected by Jacob Myers's iron box have, as their complement and correlate, the equally individualizing technology of the switch.

Mechanizing the Vote

The history of voting machines makes one fact abundantly clear: in the material world, binarism is hard to accomplish. In the recount of the 2000 United States presidential election, election officials in the state of Florida famously poured over ballots designed to be maximally unambiguous to try to discern the voter's intent. Like more than a third of all Americans at the time, voters in Florida used punch card ballots, a technology that dates to the 1830s but that became popular in the United States during and after the 1960s.[2] In Florida, the problem was that making a hole in a piece of cardboard was not quite as unambiguous as voters likely believed, which is to say, it was not quite as binaristic as *hole* or *no hole.* The punch tool should have broken a small rectangular piece of paper, called a chad, out of the card entirely, but sometimes it did so incompletely. As a result, election officials had to parse the gray area between the presence and absence of a hole, and the controversy about how to parse the meaning of hanging chads, swinging-door chads, tri-chads, and dimpled chads enthralled and appalled the nation for weeks.

Technical methods of binary disambiguation have infiltrated the language of voting itself. Even the term *ballot* invokes early voting technology. It derives from Middle French *ballotte* and is closely related to cognate terms in Romance languages such as Italian *ballotta,* all of which refer to a "little ball" placed in a box or urn to record a vote. This form of mechanical voting had clear advantages over written ballots, especially during periods when both paper and literacy were in short supply. A further advantage was secrecy, for as William Thomas said in 1549 in his *History of Italie,* in Venice a voter "maie let fall his ballot, that no man can perceiue hym."[3] The *ballotta* had some liabilities as well, not the least of which was its vulnerability to ballot stuffing, for without strict controls, a voter might drop more than one ball into the same urn, and afterward nobody would be able to trace any of them back to the source. Even so, the practice of voting by dropping small balls into boxes survived in Greece until the twentieth century.

Elsewhere, the *ballotta* gave way to a wide range of other practices that developed across Western democracies, including voting by voice and voting by marking or writing on paper, as Malcolm Crook and Tom Crook have shown in several recent essays on the history of the secret ballot.[4] According to Crook and Crook, historians have paid far more attention to who could vote than to how people voted, a tendency that has excluded a wide range of material practices from serious consideration. The English chartists made the first organized push for a compulsory secret ballot in the 1830s, and with it, the first proposal for voting machines. A working-class reform movement, chartism sought universal male suffrage, and many chartists argued that open or public voting was too easily corrupted: those who bought votes could monitor whether the people they paid had complied, including landlords requiring tenants to vote against their own interests.[5] George Grote probably created the first modern voting machine in the 1830s, an important complement to chartist legislation he introduced in Parliament mandating the secret ballot.[6] Using Grote's machine, the voter inserted a punch card into a vertical holder mounted on a table, with guide holes in the wood frame next to candidates' names, and a "steel bodkin" to punch a hole in the cardboard through the guide holes.[7] The apparatus would be arranged so that its back faced the rest of the room, thus concealing the voter's choices.

Grote's punch-card machine was a simple version of the system that American voters later used in Florida, but it too raised problems. For instance, it did not protect against spoiling the ballot by punching holes next to more than one name. More importantly, it was not entirely clear how well it would serve the illiterate or visually disabled, who formerly would have voted perfectly well by voice. That was no small obstacle for reformers seeking universal male suffrage, and they were right to worry that the cost of privacy might be access for illiterate rural peasants and urban laborers. In an 1837 article in *The Spectator,* an author evidently sympathetic to chartist goals anticipated those problems and identified possible solutions. "A blind man might vote, by feeling the holes opposite the names," he notes, though how the blind voter would detect the names themselves is far from clear. He goes on to say that "illiterate persons . . . may discover the names of their candidates by being previously shown their relative position on the card," and boasts vaguely that "some infirm persons" might vote successfully, while without this aid their votes might be lost.[8] None of this seems very convincing, and it convinced too few of Grote's contemporaries to win passage. In the end,

Grote's machine seems never to have been used in actual elections, even though he and his wife prepared "forty or fifty models, in wood, distributed all over the kingdom."[9]

Although advocates of the secret ballot saw it as a defense against corruption and coercion, many others—including eventually Grote's former ally John Stuart Mill—saw public voting as an expression of republican virtue. Arguments against the secret ballot often praised the honorable and manly nature of public voting, and even castigated the secret ballot as a form of "sickly exotic creeping into corners," no fitting conduct for an Englishman.[10] Although the chartists began advocating for the secret ballot in the 1830s, England did not finally approve it until 1872, when the success of the Australian ballot inspired imitators on both sides of the north Atlantic. At that point, the new British law even required a "polling compartment" thirty inches wide and seven feet high, designed to ensure privacy.[11]

For those who have learned to take the secret ballot for granted, it is worth noting what a profound change in voting practices the secret ballot entailed. Many eighteenth- and early-nineteenth-century English parliamentary elections were raucous outdoor affairs, often featuring a fair amount of drinking, singing, and brawling.[12] England's rowdy public elections may have allowed for corruption and coercion, but they also made power arrangements more visible, especially if everyone could see the landlord looming by the polls.[13] If the landlord succeeded in intimidating his tenants, he probably did not succeed in concealing the exercise of his power. Open or public voting also revealed membership in other social groups, signified by one's clothes, songs, banners, or location, and it involved nonvoters conspicuously in the election, as outsiders who could still make themselves seen and heard.[14] Many social constituencies coalesced in the physical space of the commons during an English election, as a public that could both witness and revise its own internal organization. Under such conditions, a tenant might feel pressure to vote for his landlord's interests, but he probably had fewer illusions about the nature of his coercion or the extent of his independence. As Frank O'Gorman has argued, even those excluded from voting were often included in the festivities, making themselves both seen and heard in the larger spectacle, sometimes by disrupting candidates' speeches with their jeers.[15]

How different that is from the current American experience of racing to the polls before or after work, for elections that are not held on weekends when voting might mingle with leisure, in spaces that restrict campaigning, and, most importantly, in the cloistered privacy of a voting booth. I vote in the gymnasium of a nearby high school, where a holy hush hangs over the whole affair. I am usually either late for work or hungry for dinner, and thus little inclined to socialize, but if I happen to meet someone I know, we politely avoid the topic of the election unless we are certain we agree.

Drinking and brawling are not much evident, fortunately, but neither are leisure and pleasure. At the end of this dreary routine, I make my choices at a small plastic table shielded on three sides, usually with the help of a cheat sheet I need to keep track of the down-ballot races. I smuggle this tangible marker of outside influence in guiltily, as if I might be exposed as insufficiently autonomous. Once I slide my ballot into the optical scanner on the ballot box, I receive a little sticker with an American flag that proclaims, "I voted," the only aesthetic remnant of the bands and banners that formerly celebrated on the village green.

In this dismal bureaucratic ordeal, voters have learned to conceal the one thing that motivated them to gather in the first place, their shared fortunes within some form of political organization, and the shared identity that results. In the very place where political choices can be expressed, political affiliations and enthusiasms must remain private at all costs, and some states even forbid wearing clothing that displays a candidate's name. I do not mean to minimize the real problems of coercion and corruption associated with public voting, only to insist that, in addressing those problems, Western democracies have not necessarily reckoned with the social costs. Corrupt though it may have been, open voting allowed the entire public to see itself with its alliances and antagonisms intact, and thus to attain a fuller sense of group affiliations and, one hopes, of shared fortunes. In contrast, the secret ballot conceals that public and its internal organization, and it foregrounds instead the isolated individual voter, whose utterly private act of casting a vote moves to the center of the democratic ritual. Whatever its practical merits, the compulsory secret ballot made radical forms of privacy compulsory too.

Baranowski's *Bouton*

So far, none of these early voting machines involved switches or buttons. In the complicated mix of techniques that included paper ballots, preprinted ballots, punch cards, *ballote,* and more, the introduction of binary switches was by no means a foregone conclusion. Although France had no movement for a compulsory secret ballot similar to chartism in the early- to mid-nineteenth century, it did employ the secret ballot irregularly, and for that reason produced an important early voting machine far more sophisticated than any from England at the time, and apparently the first to register the voter's choice with a switch. Jan Józef Baranowski, a Polish inventor living in France, had devised various meters and calculating machines, which he patented in both France and the United States, before designing several voting machines around 1849.[16] Baranowski promoted his machines as an aid to the secret ballot for general elections,

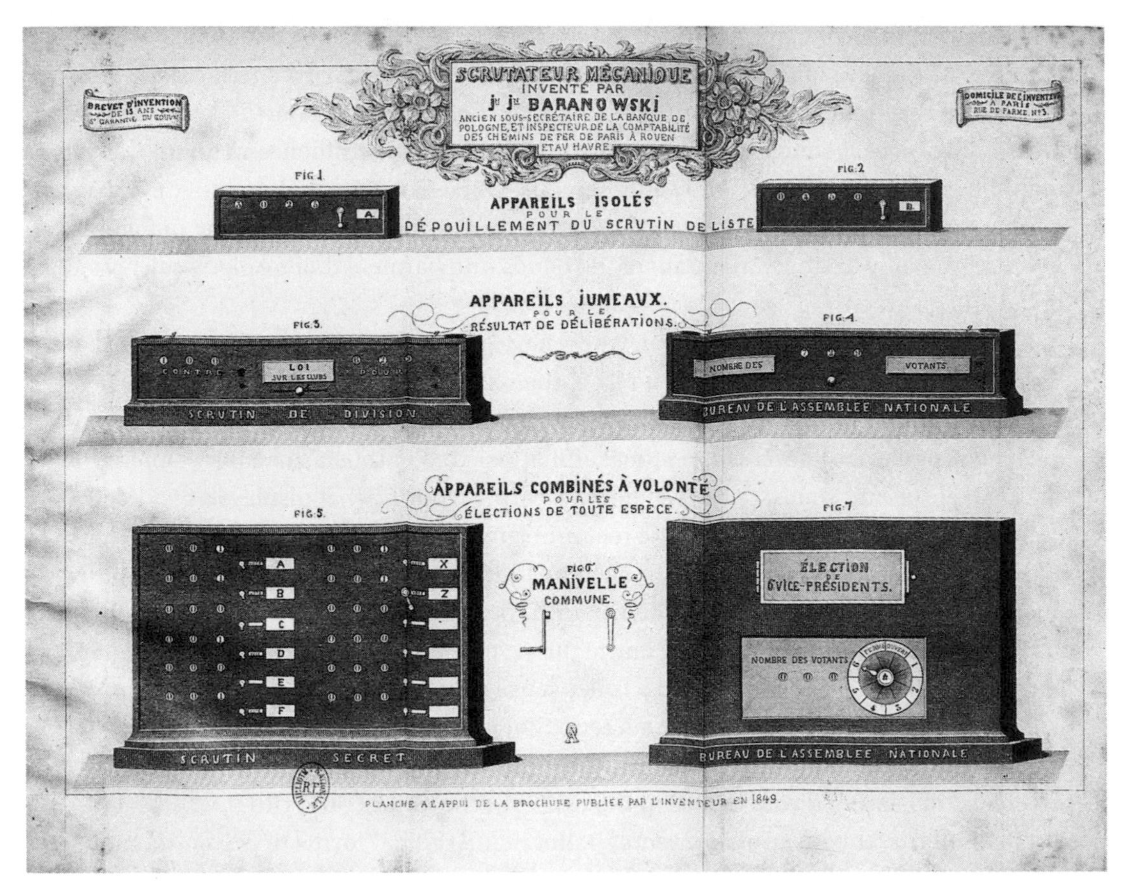

FIGURE 29. Illustrations of three different models of Jan Józef Baranowski's voting machines, from his promotional pamphlet *Nouveau Système de voter au moyen d'un appareil dit: Scrutateur mécanique* (Paris, 1849). The front and back of each device are depicted side by side. The device in the middle row is for recording votes *pour* and *contra* in parliamentary procedures. Courtesy of the Bibliothèque nationale de France.

though he was equally interested in the voting of parliamentary bodies. In elections, he said, his machines would protect against double voting, would speed counting, and would afford each voter "independence of his conscience."[17]

Baranowski's machines seem to have been quite sophisticated, judging by his description, for they protected against both double voting and erroneous nonvoting. A bell rang once the switch had been fully activated, and at that point no further vote was possible until an election official reset the device using a switch on the back.[18] The "twin device" depicted in the second row of machines in Figure 29, used for voting for or against some resolution, also included a mechanism by which voters could place an actual *ballotta* on one of two apertures on top, which would allow only one ball to roll through into the appropriate ballot box once the switch had been properly activated.

Like some of Baranowski's models, many other early voting machines were designed more for parliamentary proceedings than general elections, including Thomas Edison's 1869 electric voting machine, designed to allow legislators to vote by activating electrical switches at their desks. Significantly, that patent not only is Edison's first but also contains an early use of the term *switch* strictly in reference to an automatic device for opening and closing electrical circuits.[19] There does seem to be a rather important connection, then, between the binarism of voting and the machine binarism of switching. Even a paper ballot is a technology of binary disambiguation, but a binary switch is much stricter. For example, Baranowski's device actively prevented voters from spoiling the ballot by voting for both candidates, for the switch could be activated only once. It automates input to defend against other errors possible with paper ballots, such as stray, light, or slightly misplaced marks. It even forbids active expressions of protest, such as marginal notes. With a switch, the voter not only must resolve ambivalence into a strict choice, but is prevented from introducing errors or recording protests in the process.

Most activities are not like that, because they tend to involve all sorts of ambivalences, reversals, degrees of accomplishment, interconnections, or fluid processes, and unlike voting with a switch, they often can miscarry. Still, such activities tend to be described in binary terms; people talk about action in ways that organize it retrospectively as so many discrete choices. This can be useful. Complex, fluid, ongoing, or interactive activities such as swimming are difficult to describe as a sequence of discrete choices. Instead, one can easily say, "I chose to swim," or even "I swam," which collapses all the constituent actions of wearing certain clothes, breathing in certain ways, and moving one's arms around in the water. Something is lost in this reduction, but something is gained as well, which is the brevity of reference, for we assume that swimming involves all the other things that must be done to swim but that would be too tedious to recount separately.

This point is so fundamental that it bears emphasizing. In a society that prizes individual choice, the complexity of whatever happens must constantly be organized as discrete junctures of action, as explicit alternatives that might or might not occur. Actions must be separated from one another, must be bounded and itemized, and the strictest form of this involves organizing them in mutually exclusive binary pairs. We observed a related kind of discretizing function earlier in relation to counting and quantification, but at this stage we turn from the differentiation of objects to the differentiation of actions, a slightly different matter. Discretion always turns magnitudes into multitudes, and by extension, the analog into the digital. This happens prevalently in the language of action, through descriptions that transform fluid process into itemized events. Whenever we talk about joining a meeting, eating a meal, or taking a swim, we bundle a range of processes and interactions under a single name, and thus render the whole ensemble *an* action, a singularity with a beginning and an end, and one that need not be further subdivided.

But there is always something artificial about this bounding of action, and usually it does not withstand too much scrutiny. Although I may have eaten a large meal two hours before swimming expressly to fuel up for exercise, and although I carry that food in my body while in the water and actively derive energy from it along the way, few people would say that I was already swimming when I sat down for lunch. And yet eating before the swim was a necessary part of swimming, deliberately undertaken for that reason. Eating also involves the prolonged process of digestion, which overlaps with the period of being in the water. It might be argued that eating is just as much a part of the act of swimming as moving my arms about in a certain way. But if so, where would swimming end? Is shopping for the food also part of swimming? How about working to be able to afford shopping for food for swimming? It turns out to be easy to imagine cases where such claims make perfect sense, such as the case of an Olympic athlete whose diet is part of a program consciously designed to improve performance in the pool. Such a person is almost always engaged in the larger activity of swimming. The point is that the boundaries of actions are not to be found in the nature of the action itself. Actions are bounded in language, not in life. They are imposed symbolically, and often retrospectively for other ends.

Political stakes are not hard to perceive. Tort law, for instance, has long wrestled with how to bound certain kinds of action, such as fires that spread, in order to fix responsibility.[20] If there were no boundaries to action, no differences between *different* actions, then anyone's responsibility would be potentially infinite. Every action would bleed into the next; every cause would be implicated in all subsequent effects. That would be both practically unworkable under existing legal structures and morally intolerable. Nineteenth-century jurists were eager to set limits on responsibility; otherwise, some feared, people would be so terrified of the damage they might cause that they would cease to act at all.[21] By limiting responsibility based on criteria such as proximity, foreseeability, or negligence, they distinguished between what people might cause and what they could be held responsible for causing. The legal limitations on liability in tort law sometimes have explicitly political motivations as well. For instance, attributing action only to individual human agents can shield other kinds of entities that might also be held responsible, such as factories where the machinery regularly injures employees.

At a more fundamental level, the discretion of action allows for the discretion of human agents too. Without individual actions, it would be hard to discern individual human agents behind those actions. Everybody would be mixed up with everybody else. Methods of itemizing discrete actions are thus also ways of itemizing discrete agents too. And this happens not just in official and institutional contexts, such as law, but also in the customs and habits that people have developed over centuries to describe what happens in everyday life.

I am suggesting that the individuation of action itself produces a specific kind of person, an individual. This happens so pervasively, so subtly, and even so subliminally that it would be wrong to approach it primarily as an effect of institutions or expertise, though they are involved as well, and certainly in the institutionalization of voting. But even in everyday life, an equally important source of action individuation derives from the most mundane switching techniques that appear to be innocent because of the switch's apparent passivity. Passive not active, receiving but never guiding human touch, the binary switch nonetheless draws bright lines between events and defines each as separate and different from other alternatives. In the process, these individuated actions can then be assigned to individual human agents.

For example, switching techniques actively combat the kind of sophistry some will say I perpetrated in my previous conflation of swimming with eating. With a switch, an action appears to start and stop suddenly and entirely, and so action seems less interactive and processual than it otherwise might. In bounding action, the switch also keys it to individual human input so that the action appears to proceed exclusively from people. The switch thus operates in both directions simultaneously, dividing both action and people into discrete, indivisible units. If I were to heat and illuminate a room by making a fire, the process would take long enough that various stages of action and interaction would be unmistakable. Everyone present would be conscious of earlier stages of preparation necessary for success, such as chopping wood and carrying it indoors. Other people may have been delegated to perform some of those actions, and I may have cause to evaluate the quality and quantity of their contributions. If making the fire goes poorly, I could credibly complain that some collaborator failed to select the wood well or dry it properly. This makes it hard to tell where the act of building the fire begins and ends, but it also makes it hard to tell where the agency begins and ends. In contrast, all such interactions are concealed when I illuminate a room by flipping a light switch. With only inputs and outputs visible on either end of the black box, the act of illuminating the room seems to proceed from nothing other than my own act of sovereign will. Of course, fuel was still procured and prepared, but out of view at a distant power plant. Of course, its energy was transmitted to the home, but invisibly through copper wires. And, of course, there was a prolonged process involved, but rapid enough that it appears instantaneous to human senses. Between inputs and outputs, process and interaction disappear into the black box too, and with it any richer, more complex account of the distribution of agency and the social nature of action.

Baranowski's *bouton* is thus just the first of many attempts to mechanize the act of casting a vote so that it appears to consist naturally and necessarily of a similar kind of individualized choice. Between its input and output, process and interaction disappear. The whole active engine of democracy, the whole complex of procedures, laws, arguments, precedents, conflicts, compromises, pressures, and prior election results recedes behind an obdurate mechanical interface structuring a single binary choice. For a pivotal moment, the machinery holds everything steady so that individual human beings can experience their own action as decisive. More compressed and opaque than other technologies of democratic disambiguation, the switch concretizes theories of human agency, translates them into technique, and situates them at the ritual center of democratic experience. In the process, the intrusion of switches into the procedures of democratic decision making also has had further effects on liberal individualist subjectivity, especially once combined with complementary techniques of visual and physical isolation.

Inside the Nonopticon

Despite their technical accomplishments, voting machines like Grote's or Baranowski's did not play a significant role in general elections in England or France during the nineteenth century. Most voters in Europe and North America continued to vote as they always had: *vive voce,* with paper ballots, or with *ballotte.* Not until more sophisticated voting machines appeared in the United States in the 1880s and 1890s did residents of Western democracies vote mechanically in significant numbers, and when they did, they used devices made by inventors who returned to earlier chartist arguments but who also preferred Baranowski's interface of binary switches. By the time England adopted the secret ballot in 1872 and renewed the law permanently in 1880, some Americans had begun calling for similar reforms. Also by 1872, all but one American state had laws requiring the use of either written or printed ballots.[22] Massachusetts passed the first compulsory secret ballot law in 1889, followed by seven other states that same year, and twenty-four more over the next two years.[23] Precisely at that moment, other American inventors began producing the first successful and widely used voting machines. Thus, for the second time—but now a half-century later—advocates for the compulsory secret ballot attempted to mechanize the vote.

Given the length of ballots in the United States, Americans had extra incentive to automate voting. Nineteenth-century English voters often chose the occupant of just a single office, but American elections typically included separate contests at various levels of government (local, state, federal), and for various branches within each level (executive, legislative, judicial). In 1875, an English inventor named Henry Spratt patented a Voting Apparatus in the United States suited to long American ballots. It promised improvements in secrecy, prevention of fraud and tampering, and efficiency of tabulation.[24] In Spratt's design, the voter opened a small door over the name of the candidate he wanted to select, then pulled *handles* or *knobs* to register his choice. Alexander Roney patented a similar device in 1878 meant to "avoid trouble, . . . prevent intimidation, . . . guard against fraud, . . . save the cost of elections, . . . and preserve the purity of the franchise."[25] The voter operated his machine with knobs or handles that, when rotated, brought into view the name of the preferred candidate for each office in a small window. Once the voter properly set the dials, he pressed another knob or handle to record all selections at once. These were significant advances over earlier and more rudimentary ballot boxes that either counted paper ballots or stamped them as they were inserted (to guard against stuffing multiple ballots at once), but neither Spratt's nor Roney's machine seems to have been used in an election. On the many machines that followed, the kinds of handles and knobs on Spratt's and Roney's devices gradually became smaller, simpler, and one presumes easier to operate. Anthony Beranek's patent of 1881 seems to have been the first to employ what would now likely be called push buttons, but which Beranek himself still referred to as *knobs*.[26] Beranek's elegant design made choosing a candidate simple, and the inner workings of his machine effectively prevented double voting for the same office, an important automation. Even though his machine also never saw service in an election, it influenced subsequent user interfaces and methods of guarding against spoiled ballots.[27]

Following Beranek's patent of 1881, most new voting machines in the 1880s used some version of a push-button interface to register choices, including the first machines used in actual elections.[28] The first was the Myers Automatic Voting Machine, patented in its earliest form by Myers in 1889 and first used in New York in 1892.[29] Although many earlier American machines had to observe prevalent laws requiring some form of paper ballot, the Myers machine simply tallied votes on mechanical counters concealed on the back of each machine.[30] In form, the Myers booth was a formidable room, roughly five feet square and seven feet high, made of metal and entirely enclosed on all sides, including the top.[31] It had double doors for privacy and an array of buttons covering the back wall within, all illuminated by oil, gas, or electrical lights. In his promotions for his machines, Myers echoed many of the chartist positions from a half century earlier, which were in turn informed by his own active political allegiance to the Democratic Party in upstate New York. Born in Pennsylvania in 1841, Myers moved to Rochester, New York, as an adult to work for the Rochester Agricultural Works, where he improved self-raking mechanisms. When that business collapsed, he built bank vaults and safes, which seems to have influenced the form of the voting chambers he began designing around 1888.[32] Myers was the rare Democrat in the solidly Republican state of New York, and predictably failed in his own attempt to win election to the state senate. Like his chartist predecessors, he sought to safeguard the ballot against corruption and coercion. He marketed his new device with a pamphlet proposing a new law to make the secret ballot compulsory, and recommending his machine as the best means for doing so. His pamphlet entitled "Arise, Americans, Arise!" proclaimed that his "Marvelous Ballot machine" would eliminate the "bribery, intimidation, and corruption, which has now become the rule instead of the exception."[33] He called his device the "The Poor Man's Voting Machine," and insisted that it "prevents intimidation and prevents the bribery resulting from capitalists contributing to the corruption fund."[34]

Even though his arguments closely resemble those of the chartists in England a half century earlier, the politics of the secret ballot had become more complex. As John Crowley has argued, its success was hardly the fulfillment of radical democratic values.[35] Indeed, each political party hoped the secret ballot might expand turnout for its side and depress that of the other. Myers hoped the secret ballot might embolden working-class voters, but his opponents hoped it might instead deter illiterate voters and those who could not read English. Crowley also shows how the secret ballot did disenfranchise Black voters by functioning as a *de facto* literacy test. In some southern states, officials used the argument for privacy to create discriminatory restrictions on requesting help from election officials.[36] Under open or public voting, those constituencies could simply pick up a preprinted ballot from their union, club, party, or church, but the Australian system asked them to complete the long, complex American ballot by themselves. And although the chartists had been concerned that landlords might coerce the public votes of tenants, New York Republicans were equally concerned that unions might do the same. The secret ballot thus found defenders on all sides, largely because skepticism about corruption and coercion had become more widespread, and because the secret ballot itself could function as an active engine of discrimination.

The Myers voting machine promised to alter the nature of elections, but it also altered the experience of voting. Imagine entering such a formidable hull. The voter would admit himself to the small antechamber through a narrow door and close it behind. He would then open another door into the voting chamber, lit by a gas flame or dim electric light, and confront a grid of a hundred or more buttons covering the back wall entirely alone. In this imposing chamber, a new kind of individual subjectivity might be formed.[37] As Michel Foucault argued, Jeremy Bentham's panopticon was a device for convincing prisoners to regulate themselves, once they internalized the surveillance of an observer who might or might not be watching. But Myers's box is a *nonopticon,* a zone of state-mandated privacy that produces radically individualistic modes of political subjectivity. In that crucible of privacy, the voter has no clear view of the public, and the public can have no clear view of him. Foucault famously claimed that "visibility is a trap," but invisibility can be a trap too, a way of confining people to a model of agency that seems not to involve anybody else.[38] In Myers's nonopticon, the social disappears from view, even as the isolated subject faces extraordinary expectations of autonomy. Any real gains in autonomy come at a cost. Although the voter may not be coerced by others, he cannot be encouraged or assisted by others either, nor can he perceive a group whose shared fortunes his vote might represent. The machine itself structures his conduct as a matter of purely private conscience. Designed to protect the prerogatives of individualism, Myers's voting machine made them mandatory too.

Inside the machine, the method of registering choice was similarly individualizing, through a vast elaboration of Baranowski's *bouton.* Myers experimented with several interfaces and in 1889 received two different patents for similar devices, one of which used metal tokens as *ballotte,* the other "push-keys" that resemble buttons.[39] He preferred push buttons in all designs that followed and improved them significantly over the next several years.[40] Myers's first version of the push key featured a small round touch surface mounted orthogonally to the machine's back wall by a thin rod surrounded by an exposed spring.[41] Durability may have been a problem, for just one year later he redesigned the interface completely with a new push key that slid forward and backward in an open slot beneath a rigid protruding cylinder. In this new design, the rigid cylinder concealed the spring within and the traveling touch surface hung beneath it. By 1893, Myers had settled on a third design that was as simple as his first version but as durable as the second. It concealed the spring inside the shaft of the button, not wrapped around it, and the user depressed the entire protruding device (see Figure 31).

The switches inside the Myers machine complement the individualizing effects of its confining metal shell. The switch itself is highly individualizing, a way of casting votes one at a time, in contrast to a parliamentary voice vote in which the cry of *yea* or *nay* from all voters could sound simultaneously. Any residue of the social has been concealed behind the action of the switch, lest individual action devolve into interaction. Reporting on early uses of the Myers machine, the New York *Sun* announced in a subheadline with evident wonder: "The freeman presses a knob and his will is registered."[42] That telling sentence bears closer scrutiny, for it reveals the real stakes of this method of voting. The voter is identified archaically as a *freeman,* so by definition a creature of liberty. This freeman is the grammatical subject of the sentence, the one who actively "presses" the knob. Action emanates from him, and only in one direction.

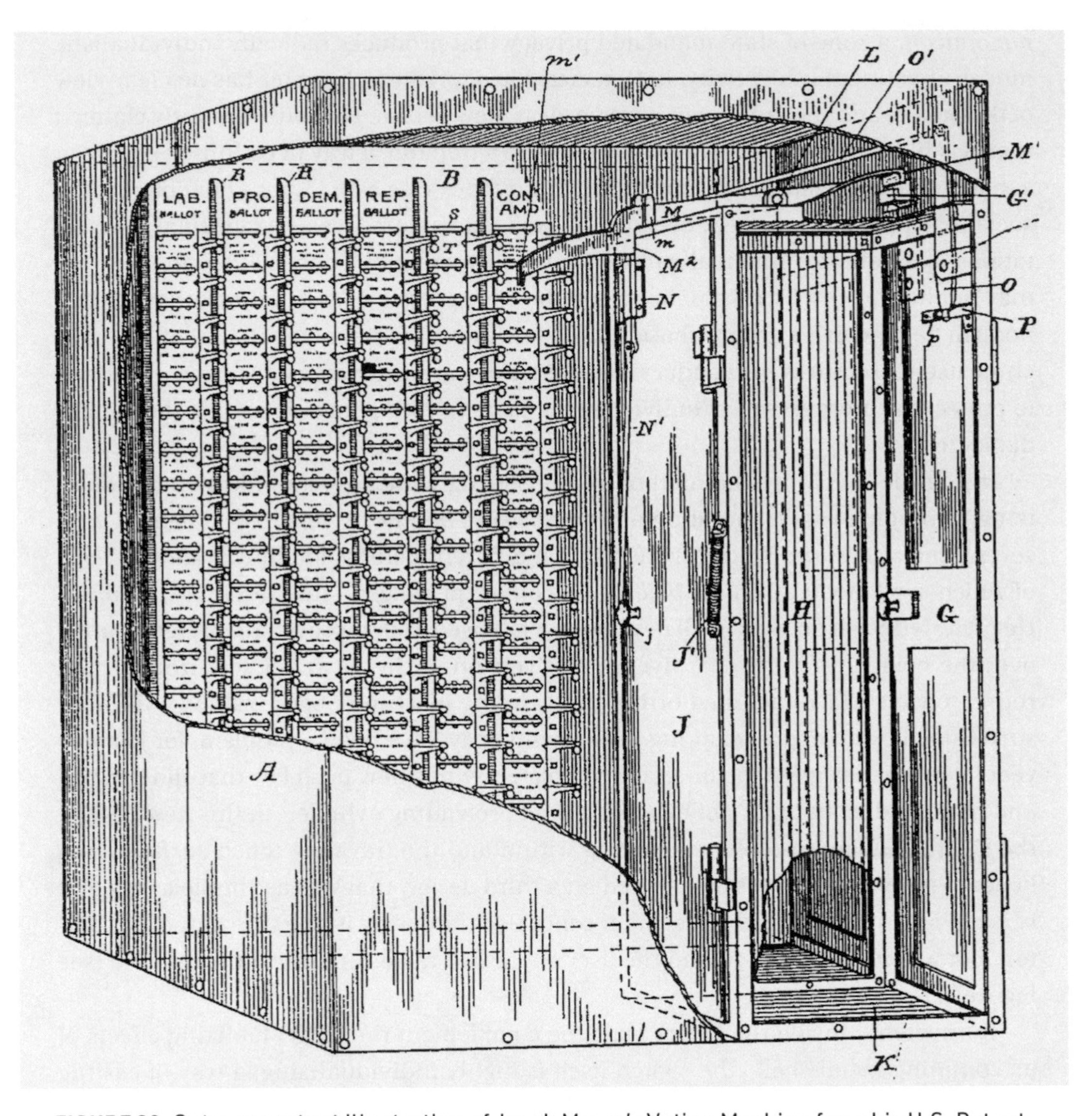

FIGURE 30. Cutaway patent illustration of Jacob Myers's Voting Machine from his U.S. Patent 424,332, issued March 25, 1890. The diagram shows roughly ninety individual "push-keys" arrayed from floor to ceiling, with space for more available behind the door. Courtesy of Google Patents.

In contrast, the machine itself operates in the passive voice, a target of action rather than a source: the result of the press "is registered." Finally, the headline insists that more than just the freeman's "choice" or "vote" has been registered: nothing less than "his will" has registered too. More than just a vote recorder, then, the machine func-

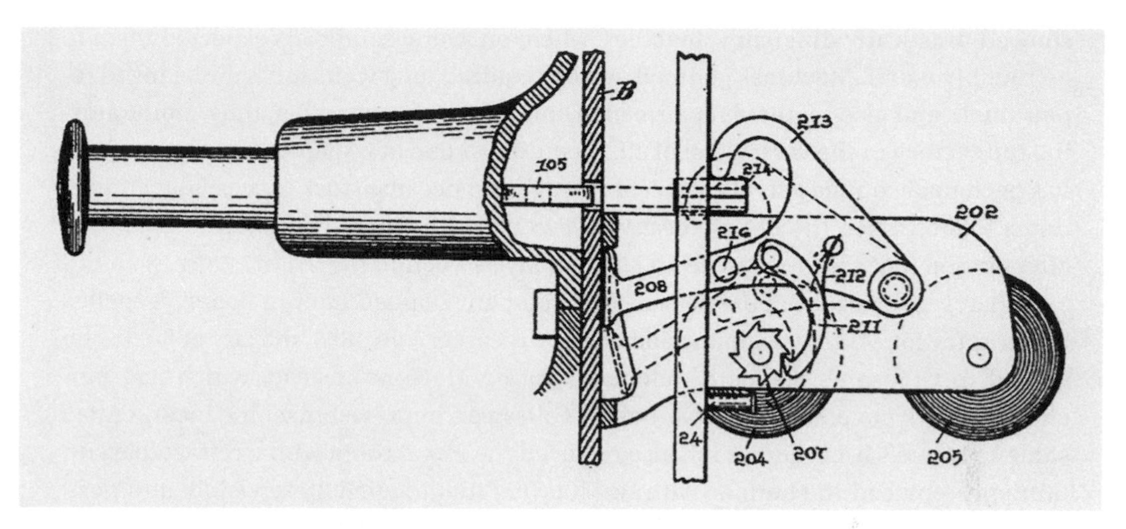

FIGURE 31. Patent illustration of the "push-key" of Jacob Myers's 1893 voting machine, from U.S. Patent 494,588, issued April 4, 1893. Courtesy of Google Patents.

tions as a will detector too, or better yet, as a will validator, a machine that configures action so that it can be maximally attributable to individual people. Myers's voting machine may be imagined as one of those fantastic chambers in science fiction films that transform people into different kinds of beings. People enter and emerge moments later fully human. Purified of dependency on others, even if only for a few moments, the voter stands confirmed as the highest and purest sort of liberal subject of all.

The New York Senate approved the use of the Myers machine in town elections in 1892, and the city of Lockport employed them that same year, with relative success. Other towns in upstate New York installed the machines in 1893, and by 1894 New York voters approved a constitutional amendment that permitted Myers's voting machines to be used in all state elections. Myers had competitors by this point, especially Sylvanus Davis, one of Myers's former employees, whose machine employed "moveable slides or pushes" that also resemble buttons, and from Alfred Gillespie, whose much smaller voting machine employed rotating levers. Both would have been easier for election officials to manage than Myers's iron room.[43] Despite such advantages, the city of Rochester adopted Myers's more proven machines citywide for the high-turnout election of 1896, a presidential election year. The city agreed to buy the machines after the election if they proved satisfactory, and Myers must have sensed that he was poised to triumph over his competition.[44] Nobody had to wait long for the verdict. Three days later, the *Rochester Democrat and Chronicle* reported, "Myers Machines Far from Perfect," and lamented a "Sad Lack of Secrecy." On some machines, the number of votes recorded differed widely from the total number of voters. Adjacent machines showed drastically different vote tallies, when, on average, officials expected them to be roughly equal. Machines jammed, which required long waits for technicians to repair them, and also required repairmen to enter the voting chamber, thus compromising the secrecy of the vote. Worst of all, according to one newspaper, "much complaint has been made on the part of corpulent voters," who became stuck in the eighteen-inch doors.[45] Soon after, the *Buffalo Evening News* blared a two-word headline in edition after edition that sounded the death knell for Myers's company: "Didn't Work."[46]

Whatever the cause of the debacle (the company blamed uneven floors), it spelled bankruptcy for Myers's company and ended his career.[47] In 1898, the city of Rochester turned to Gillespie's Standard Automatic Voting Machine instead, which had purchased the Myers patents the year before. Gillespie's improved machine incorporated some of Myers's innovations, but also replaced the metal room with a retractable curtain and replaced the buttons with small levers that toggled more visibly into position.[48] Gillespie's company eventually merged with Davis's, and the resulting enterprise dominated the industry through the 1960s, at which point more affordable punch-card systems and, eventually, optical scanners began to compete. Only recently have buttons returned to voting again, now in the form of virtual buttons on touch screens.

I offer this history at some length to show how intimately the values of liberal individualism were entangled with the technical history of mechanical voting. We have already seen how the early chartist values for the secret ballot were not the only ones, or even the primary ones, that eventually led to its nearly universal adoption. What New York Democrats and Republicans had in common in 1892 was a roughly equal paranoia about threats to voters' individual autonomy from the other side. One feared corporations and the other unions, but both agreed that the sovereign will of the individual voter must be protected. From that consensus, the secret ballot became part of the guiding mythology of liberal individualism, which paved the way for even more practical ways of enforcing individualist values through voting technology. A fuller history of these developments might show how the Myers machine was part of a broader shift in Western conceptions of liberty from communitarian modes of civic republicanism toward more highly individualistic modes of liberalism. In that transformation, republican theories of liberty as nondomination by arbitrary power gave way to a different theory of liberty as strict noninterference with individual will, or what Isaiah Berlin famously termed "negative liberty."[49] Civic republicans were comfortable shaping individuals socially and institutionally in order to make them fit for democratic decision making, as through mandatory public education. Liberals, in contrast, were prone to see such efforts as interference with liberty, and over time they preferred an economically inflected model of more isolated selfhood prevalent in many forms of libertarianism and neoliberalism.

Arguments for the liberalism of noninterference, of negative liberty, were not just rhetorical but also technical. The voting chamber is an incubator of negative liberty, first of all through its enforced isolation. Myers's chamber is an iron fortress designed to shield the voter from all interference and to permit him—but also to force him—to act entirely on his own. The switches inside compound these effects. Their seeming passivity affirms that the people who use them are all the more agentive and free. There is no action in this room other than what the voter inaugurates, and no interaction with anybody else. Registering votes happens with an absolute minimum of ambiguity, and with no allowance at all for any kind of human ambivalence that might connote weakness of will. The machine is even designed to prevent the most common kinds of human errors. But the machine itself must be entirely reliable if it is to help make the voter experience himself as entirely reliable too. If its keys stick, gears jam, or wires cross, as happened with Myer's machines in the disastrous election of 1896, everyone wonders what other agencies lurk within. Action devolves into interaction after all, a terrible betrayal in what was supposed to be a citadel of human self-sovereignty. Accordingly, the possibility that computerized voting machines might conceal the surreptitious agencies of others threatens far more than just free and fair elections. It threatens a primary engine of liberal individuation itself, without which liberal citizens might stand exposed as rather less autonomous than they are supposed to be.

If voting chambers were the only place where this sort of political effect occurred, it would be richly symbolic but still limited to one kind of activity. However, many other devices have similar properties, less conspicuous but sometimes more insidious. As we turn to still other switching techniques of haptic liberalism such as those used in photography and video gaming, we will see how the seeming passivity of the switch configures human agency in even more extreme forms. But, before turning to those other technical contexts, we need to broaden our conception of the term *agency,* a slippery term that has been treated too generally so far. The next chapter will address just what we mean by *agency* by attending to an influential strain of mid-twentieth-century action philosophy that exhibited its own peculiar dependency on the binary switch.

CHAPTER 14

Switching Philosophies

The more scholars try to pin down the term *agency,* the more slippery it starts to seem. Few disciplines can agree on even a basic definition. For some, agency is the vaunted liberal ideal of individual autonomy, and that which too often has been denied to women, colonial subjects, non-whites, the disabled, and many more. For others, agency is rather a collective or social phenomenon, never precisely isolated in any individual. For still others, agency is an illusion of self-sovereignty, and thus, strictly speaking, an effect of ideology. Some draw the definition narrowly and see agency largely as a matter of legal or ethical responsibility, a way of acting within a system of norms and laws. But some draw the definition much more broadly and view agency as not exclusively human at all, but equally present (though perhaps differently constituted) in animals, machines, and other nonhumans. Presumably the kind of agency attributed to liberal subjects is different from the agency of human collectives, which is different from the agency that Bruno Latour sees delegated to technologies, which is in turn different from the agentive "thing-power" that Jane Bennett discerns in a bottle cap or a dead rat.[1] Still, nobody has been very clear about how they differ. Bennet herself rightly observes, "No one really knows what human agency is, or what humans are doing when they are said to perform as agents."[2]

These matters will not be resolved here. I take it for granted that all these competing definitions serve different purposes, so none should be considered impartial attainments of truth, even though most present themselves that way. Asserting, contesting, and reformulating theories of agency is essential political work in liberal societies, for there is no getting around the shibboleth of agency. Although my own critical sensibilities incline toward theories of distributed, delegated, and nonhuman agency, what follows does not develop those arguments further. Rather, it attempts to understand how more traditional, individualistic models of human agency came to be so widely accepted within popular culture. How, in other words, have millions of people come to take it for granted that a single human being should be endowed with this capacity commonly termed *agency,* and further, why does it seem so strange to so many people to imagine that animals, plants, and machines might be agents too? The dominant model of individualized human agency has been naturalized by centuries of liberal philosophy, and in popular culture by a steady barrage of mass media, from sports broadcasts to Sunday sermons, legal proceedings to television drama, most of which presume the freedom, autonomy, and efficacy of individual human agents. In this chapter, I suggest that an equally important form of popular media for naturalizing individual agency is the binary switch.

Today, few historians of science would dispute Langdon Winner's influential claim that machines have politics, but not many have considered political effects as fundamental as the production of individual agency.[3] Switches familiarize a certain structure of action for millions, and thereby make it seem inevitable. In the process, switches also help produce certain kinds of agents, people whose decisive intervention starts and stops so much of the action in the modern world. Indeed, they establish that action results from an individual person with a finger on the switch, for as we saw in the previous chapter, a central function of almost every switch is to prevent action from happening any other way. Moreover, machines are not just invested with political ideas, scripted into them at the design stage or improvised by users later, but may in fact be sources of political ideas too. Even though theories often get formalized as practices, sometimes technical practices generate new theories. Some theories are second-order ideations that elaborate in language what first took shape in the realm of technique. I suggest in what follows that switching techniques modeled forms of individual agency that philosophy later expressed more abstractly and critically.

Accordingly, I am not interested in starting with any of the theories of agency mentioned above and using them to explain the practices of liberal subjects. Instead, I will be looking at how switching itself may have shaped an influential body of theory known as "action philosophy." Action philosophy dates to Elizabeth Anscombe's *Intention* in 1957, at which point Anglo-American philosophers, primarily, began to treat issues related to action and agency with greater rigor. Action philosophy partly diverged from two prior subfields that had previously dominated the conversation, ethics and moral philosophy, on one hand, and the liberal political theory of individual autonomy and responsibility, on the other. The break was never complete, however, and many of the most important voices in action philosophy, such as Joel Feinberg, were active legal and political philosophers too. Nonetheless, action philosophy usually wore its commitments to liberalism more lightly, as it engaged in fundamental debates about just what constitutes *an* action in the first place. If, as these philosophers have since assumed, an action can be distinguished from an event by its connection to an agent, a whole host of questions arose about the nature of that connection, questions that the field has pursued energetically ever since. The one philosopher whose work is most important for these purposes was also a leader in the field of action philosophy for decades, Donald Davidson.

In Davidson's work we encounter the switching technique in question too, the simple act of turning on a light switch. In a justly famous example of how agency works, one that has been referred to by thousands of other books and essays, Davidson uses the example of flipping a light switch to establish the broad outlines of large parts of his philosophy. In a deceptively simple thought experiment, Davidson says, "I flip the switch, turn on the light, and illuminate the room. Unbeknownst to me I also alert a prowler to the fact that I am home. Here I need not have done four things, but only one, of which four descriptions have been given."[4] We will discuss this short story in due course, for Davidson condensed much of his philosophy of action into it and discovered problems that would occupy him and others for decades. In addition to his reinvocations of this same scenario in later work, Davidson also relied on many other examples of other binary inputs, such as buttons, gun triggers, and the primitive action of moving a finger. His repetition of these examples amounts to a preoccupation, and it is hard to escape the conclusion that, for Davidson, switching served as the normative model of action by which all others could be judged. Davidson's philosophy may be seen as a more detailed and rigorous account of those assumptions about individual agency commonly held by millions of people, people who might not be turned on by philosophy but who undergo a more practical kind of liberalizing enlightenment every time they flip a switch.

Philosophy in Action

Davidson wrestled with some of the most vexing questions related to agency in a remarkable series of papers published primarily in the late 1950s and 1960s. Even more than Anscombe's influential volume, this sustained body of work profoundly reoriented the philosophy of action and agency, even if it also proved inconsistent and even contradictory in its conclusions. Only the narrowest account of Davidson's philosophy of action is possible here, one that gives too little credit to the many other theorists of action working around the same time and ever since. Moreover, even though Davidson was surprisingly preoccupied with switches and buttons, he never acknowledged that fact, and to my knowledge, none of his readers have noted it before. To philosophers, this must seem like an idiosyncratic way to approach this influential body of work, but my goal is not to advance or contest Davidson's arguments about action and agency in themselves. It is rather to show how an entire culture of switching made its own mark on action philosophy, and by extension on philosophical accounts of liberal individualist agency. That said, Davidson's body of work is large and complex enough that we must proceed patiently if we are to understand how such a mundane interface relates to such an ambitious philosophical project.

Let us observe just a few preliminary appearances of switching in Davidson's philosophy. Davidson writes in one place, "If the officer presses a button thinking it will ring a bell that summons a steward to bring him a cup of tea, but in fact it fires a torpedo that sinks the *Bismarck,* then the officer sank the *Bismarck.*"[5] In a slightly later essay: "A man might have the power to destroy the world, in the sense that if he were simply to push a button before him, the world would be destroyed."[6] And also: "We must suppose the agent intended to kill the victim by pulling the trigger because he reasoned that pulling the trigger would cause the gun to fire, which would cause the bullet to fly, which would cause the bullet to penetrate the body of the victim, thus causing his death."[7] This last example is one of many that use the gun as a mediating device by which agency is both organized and affirmed: "Before I can hit the bullseye, I must load and raise my gun, then aim and pull the trigger";[8] "Suppose Jones intentionally causes Smith intentionally to shoot Clifford to death."[9]

Davidson does use many examples of more analog kinds of actions too, such as swimming, but often these are problematic examples that he clarifies with scenarios involving switches and buttons. Binary switches turn out to be useful to Davidson because they organize action as a discrete and bounded event, as *an* action. And often, that action has only two alternatives, such as the light that must be either on or off or the torpedo that can be only launched or not launched. We can still perceive the shadow of ethical philosophy in this decisional binarism, which involves fundamental questions about what human agents could or should do, narrowed down to a stark, mutually exclusive pair of alternatives. But many of Davidson's questions are not ethical, and although there are possible ethical dimensions to many of them—such as Smith shooting Clifford—he takes little interest in ethical considerations. Still, this decisional binarism has more than a little in common with one kind of especially pressing ethical choice, the ethical dilemma, which involves a conflicting set of mutually exclusive values. In an ethical dilemma, there are only two alternatives to choose between, and one generally cannot choose both or commit only partially. For instance, some would find aborting a fetus to save the life of the mother an intolerable ethical dilemma, for in either event a life would be taken. Faced with such a choice, the consideration of other values would have to come into play, such as whether it would be worse to actively take a life or to passively allow a life to be lost, but even then the result of the choice might involve a sense of compromise or regret.

During the 1960s, when Davidson was working most avidly on the philosophy of action, other ethical philosophers devised a thought experiment that can help us understand why the switch may have infiltrated action philosophy from the direction of ethics. First articulated by Philippa Foot in 1967, the thought experiment known as the "trolley problem" sets up an ethical dilemma: "It may be . . . supposed that [a person] is the driver of a runaway tram which he can only steer from one narrow track on to another; five men are working on one track and one man on the other; anyone on the track he enters is bound to be killed."[10] The arrangement is variable. One track might have the prime minister, a child, or your mother, while the other might have a great many elderly people, an escaped criminal, or an agile athlete. Many people make the utilitarian choice to kill one person rather than five, but other forms of ethical reasoning are common too, such as the choice to refrain from any overt action that will cause harm, even if doing nothing causes more people to die. Foot's thought experiment was part of her own analysis of the ethics of abortion, and it has proved extraordinarily influential, appearing regularly in scholarly work and the popular press. In 2011 psychologist Carlos David Navarrete created a virtual reality trolley test to assess a more realistic version of the scenario, complete with videos of screaming victims.[11] Wearing a virtual reality headset, test subjects could turn and look one way at the approaching rail car, then the other way at the people on both tracks, and they could flip a switch to change the position of the switch track. In both Foot's imagined scenario and Navarrete's virtual reality one, the rigidly binaristic mechanical situation creates a pressing ethical dilemma in a very short time period of a kind otherwise hard to imagine.

We already have noted that the term *switch* developed metaphorically from railway switch tracks, like those beneath Foot's trolley. To switch is already to participate in a binary operation first conceived by analogy with the junctions of a railway network. For Foot, the utter inflexibility of the switch track forbids any other forms of creative problem solving beyond just two alternative results, so that the choice is as locked into mutually exclusive alternatives as the trolley itself. The trolley problem mechanizes choice in this way, eradicating third, fourth, or fifth best options, and permitting no partial compromises or gradations. The trolley test also removes complexities related to such things as trying and failing, or making mistakes, since the act of switching appears to require no special strength or skill. In other kinds of situations, one might choose to kill but fail to accomplish it, or perpetrate violence intending only to injure but kill instead. In the trolley test, in contrast, such complexities give way to the iron binarism of two and only two choices, each of which has an entirely certain outcome, and which admit no compromise, no middle ground, no negotiation, and no going back.

It is not so much that such strictly bounded choices were inconceivable before the industrial revolution, for even Plato talks about ethical dilemmas, but surely they were harder to imagine in such mercilessly inexorable forms. The binarism of the switch track gave people a new metaphor for choice, different from other common examples such as murdering or marrying. One cannot murder or marry halfway, but it takes considerable effort to ensure that the act of marrying is utterly unambiguous. Certain rituals must make intentions clear and public, and certain ritual actions specify the moment when there is no going back. Murdering and marrying are also rare kinds of events, loaded with social meaning and carefully policed by the societies in which they occur. However, as electronic and mechanical switches proliferated, strict binary choices proliferated too. The methods of disambiguation were not ritual or even rhetorical, but technical. We could even say that the technology of binary switching created new choices, new structures of mutually exclusive alternatives mediating more and more aspects of everyday life. Choices are thus not something that humans make with the inert materials around them, but something that humans and their environments make together, and that technologies can structure in especially assertive ways.

Intentions

Action philosophy remained largely indistinguishable from ethics until Anscombe's *Intention* in 1957. Anscombe and those she influenced were less concerned with how people act or should act in ethical situations, and more concerned with what an action is in the first place. If an action is a kind of event that belongs to an agent, as she concluded, what defines an agent? Anscombe's answer was that an action must be intentional, but that basic claim turns out to be more complex than might first appear.[12] With astonishing compression, relentless intellectual energy, and admirable style, Anscombe dismantled the term *intention* and clarified its importance for considerations of action and agency. By disambiguating intentions from predictions, commands, motivations, and causes, among other easily conflated concepts, Anscombe did more to clarify the stakes than anyone before, and she established the parameters of a conversation that Davidson would inherit and expand.

Davidson's first essay on action and agency was also his most famous, "Actions, Reasons, and Causes" from 1963, which stirred a wave of investigation into the philosophy of action and framed key issues in the debate. Assessing its influence thirty-five years after its first publication, Antonio Rainone noted that the essay contains an incipient "extensional theory of causal sentences, a theory of events, a philosophy of mind and a theory of practical reasoning," no small accomplishment for fewer than sixteen pages.[13]

It is also a complex argument, so much so that some preliminaries are in order. Anscombe's *Intention* had argued that actions must be intentional, but she also insisted that intentions are more than just internal statements one makes to oneself about a future course of action, such as private resolutions: "I'm going to lose weight," or "I'm going to quit my job." Instead, Anscombe claimed that intentions consist of reasons for action, the beliefs and desires that motivate a person to select one course of action over another. If I am sawing a plank of wood, the action is intentional if I have a plausible answer to the question "why?" Maybe my roof has a hole and I need to cover it, or I enjoy sawing, or desire revenge on my neighbor who owns the plank. Any of these reasons would render the sawing intentional, and so configure me as the proper agent of a proper action. But, by this logic, some events that people cause do not qualify as actions, and accordingly the person will not qualify as an agent. If a man knocks his glass off a table because he was startled by a face in the window, Anscombe concludes that he is not the agent of the action, because being startled does not emanate from reasons such as desires and beliefs. The falling glass is just an event, and the man who truly did cause it to fall remains something less than a proper agent after all.

Davidson and Anscombe would agree about the falling glass, but Davidson's later example of entering a room, turning on a light, and unwittingly alerting a prowler is a more complicated matter. The man who illuminated the room had reasons for doing so. Perhaps he wanted to see where he was going, or he had a settled belief that it was right and proper to illuminate rooms, or he was acting under some injunction from another person who required him to turn on the light.[14] But, even though the man had no expectation that a prowler might be present and the prospect of alerting a prowler played no part in his reasoning, Davidson still considers the man who turned on the light the agent of all the actions that result. For Davidson, the man is the agent of the action of alerting the prowler too. This proposition is sometimes called the Anscombe/Davidson thesis, and amounts to the basic claim, expressed logically, that: "If a person *F*s by *G*ing, then her act of *F*ing = her act of *G*ing." Or in more familiar terms, if he alerts a prowler by intentionally turning on the light, then he was the agent who alerted the prowler.[15] This was a surprising claim precisely because it dispensed with mental actions like willing or trying, and substituted for them only the causal properties of reasons themselves. In Davidson's view, all that matters is that an agent acted for reasons and not necessarily for a specific reason keyed to a desired or expected result.[16]

Davidson sometimes referred to this as *intentionality* at this early stage, but he seemed to mean something different from what Anscombe did. Influenced as she was by Ludwig Wittgenstein, Anscombe was resistant to cluttering the stage with mental activities that somehow determine whether some event amounts to an action. In Wittgenstein's well-known example, he asks, "When 'I raise my arm,' my arm goes up. And now a problem emerges: what is left over if I subtract the fact that my arm rises from the fact that I raise my arm?"[17] According to Michael Scott, the answer is decidedly not something like the *will,* a surplus that some invoke to confirm that a happening truly was agentive. For Wittgenstein, actions are determined by contextual factors, not by special mental prerequisites like willing.[18] Similarly for Wittgenstein's pupil Anscombe, reasons for action are not ways of thinking that accompany or give rise to an action, but retrospective contextual tests, assessments of surrounding conditions that involve asking *why* some action was performed. Moreover, for Anscombe, reasons are not exactly causes either, not in the way they would be for Davidson. If the answer to "why?" takes the form of reasons, then the event is an action after all, but that is not quite the same as saying that reasons themselves have causal properties.[19]

But, for Davidson, at least at this early stage, reasons were causes too. Turning on the light switch causes the prowler to be alerted, even though that result did not figure into the thinking of the person who flipped the switch. All that matters is that there were reasons for turning on the light. This is Davidson's major departure from Anscombe, and it leads to his further conclusion that reasons for action need not be overidentified with the person acting. The reasons themselves are causal, even though they are still somehow associated with a person. As a result, Davidson says, "some causes have no agents," by which he means that some causes have no human agents, or even that causes should not be seen as equivalent to human agents. He goes on to say, "Among those agentless causes are the states and changes of state in persons which, because they are reasons as well as causes, constitute certain events free and intentional actions."[20]

This extraordinary proposition detonates in the last sentence of his essay, but seems to this reader to raise more problems than it solves. Leaving aside the fact that the return of the term *intentional* sows nothing but confusion, what does it mean that these reasons are "in persons"? Davidson seems clear that they are not equivalent to persons; but, if that is the case, how do they relate to the people they apparently inhabit? One possibility is that the reasons determine the conduct of the people who contain them, which may be why Davidson obscurely insists that these are *free* actions as well; however, that is not necessarily the same as saying that the people themselves are free. All of this easily becomes circular: a person's reasons cause actions, but what caused the person to have those reasons? If we answer that the person is in control of those reasons, then it is not clear why we need to refer to reasons in the first place. We would be back to sourcing action from something like executive will. But if the person is not in control of the reasons, then reasons might be acting through people who play no real role in the process. By this line of thinking, people are in danger of becoming passive vehicles for reasons, which do all the work. But Davidson rejects these and other objections with brusque efficiency, simply insisting that we must not assume that a "cause demands a causer, agency an agent."[21] He had written himself into a difficult position, which he would spend the next decade trying both to defend and to escape.

In asserting a causal theory of agency, Davidson edged close to a nonhuman theory of agency too, though he resolutely avoided considering that possibility explicitly. His reasons for action somehow reside inside a person, he insists, yet he seems aware that they might derive from a wide range of sources, some personal and some impersonal, including "desires, wantings, urges, promptings, and a great variety of moral views, aesthetic principles, economic prejudices, social conventions, and public and private goals and values."[22] None of this necessarily means that the reasons are not entirely one's own, for liberal societies do not require agents to be without influence, only to own those influences by having selected and retained them voluntarily. But if causation is all that matters, why should we halt our inquiry into the chain of causation when we reach a person in whom these causal reasons apparently reside? Of all the possible repositories for reasons, what is it about the figure of a person that forbids us to inquire any further, that throws up an impassable blockade to the search for prior causal reasons still?

The question is even more difficult if we consider several other kinds of reasons for action that Davidson does not mention: habits, laws, material constraints, scripted procedures, cultural techniques, and much more. Is an intention to turn off a light different from an ingrained habit of turning off lights? If the habit is just one of many different kinds of intentions, as Davidson would likely agree, then is a habit of turning off lights different from an official sign posted at a door that instructs employees to turn off the lights? That would supply a reason too, but now one that is located outside the person touching the switch. Such a sign would effect, through people, what an ingrained habit otherwise might have accomplished. But further, how is such a sign different from the aggravating motion sensor in my office that automatically turns off the lights when it detects too little motion for too long? With Latour, one might argue that these all represent similar reasons for action, but installed in different locations: the first is in the habits of a person, the second in an official sign backed by institutional authority, and the third in an automated technical procedure.[23] Reasons for action can lodge in all sorts of places, in people but also in other institutional and technical resources. If Davidson is serious that "some causes have no agents," there would seem to be no reason why agency could not reside just as sensibly in machines.

Accordion Effects

Perhaps the threat that agency might escape the human entirely eventually led Davidson to moderate his attempt at a causal theory of action by incorporating more Anscombian views of intentionality. Five years after "Actions, Reasons, and Causes," he returned to the example of the light switch in "Agency," an essay that elaborates on the problem of infinite causal regress and much more. Midway through "Agency," Davidson announces, "At this point I abandon the search for an analysis of the concept of agency that does not appeal to intention," and he later confirms that causality "yields no analysis of agency."[24] Causality in action is simply event causality, he says, so there is no specific agent causality as he had supposed five years before. In truth, his conversion was partial, for he retained parts of his causal theory of action, though qualified by new recourse to intentionality. Intentionality consists of what he would call "pro-attitudes" toward some course of action, attitudes that should not be confused with the mysterious faculty of the will, which he continued to reject. Intentions equivalent to these pro-attitudes are similar to certain kinds of wants or desires, but not exactly equivalent to them. A person might add sage to the stew to improve its flavor, Davidson explains, but though improving the flavor is the *reason* for adding the sage, the reason itself is not the source of the action, as he formerly could have claimed. That reason was guided by an intention, an "attitude of approval" of improving the stew, of wanting to improve the stew.[25]

The precise merits of Davidson's claims about intention are less important for these purposes than the fact that he had retreated from what was initially a much stronger causalist position. By reintroducing intention, he made individual human agents superintendents of the reasons that formerly seemed to operate more independently. This later view is much more compatible with traditional liberal humanism, partly because his pro-attitudes edge dangerously close to the kind of volitionism he otherwise rejected. But if Davidson's intentionality risked smuggling *will* back in under another name, it also defended against an even greater threat, the danger of infinite causal regress that his earlier model invited. With events, as opposed to actions, causal regress is the norm. One event happens (snow builds up on a branch) that causes another event (the branch breaks). But what caused the snow to build up? And what did the breaking branch cause next? It is hard to know where to stop. Then again there is much less pressure to stop, for most people are more comfortable with a complex account of causation than they are with a complex account of agency. Moreover, one can break down any of these events into component subevents, such as counting each falling snowflake as a separate event, or their interactions that made the snow stick, or the effects of cold temperatures on the strength of the wood, and much more. Few people find this troubling until questions of responsible agency intrude. If the branch fell on a child, suddenly people might want to isolate starting points in the causal chain precisely to transform the event into an action. Where were the child's parents, and why did the property owner not trim the branches? But under a strictly causal theory of action, it becomes difficult to isolate any action from a potentially infinite causal regress of prior actions. Why, for instance, stop at a reason for action without inquiring into the *reasons* for those reasons? Not only is it hard to argue that one particular reason must be the cause of some subsequent action, but it also proves difficult to limit the effect of any one reason to a single culminating action.

Another part of Davidson's solution to the problem of causal regress was to argue that the stages of an action are not separate actions after all, but just different descriptions of the same action, some more detailed than others. And it is the intention, the guiding or organizing pro-attitude, that determines where the action begins and ends, not the reasons themselves. By this logic, descriptions of any action can "accordion" or "puff out" the action to a large size, or shrink it down to the "primitive action" of the movement of the body.[26] But these are just ways of talking about action. The action itself is bounded by the intention that gives rise to it. Even in Davidson's early essay on strictly causal agency, he was cautious about the possible devolution of one action into many separate actions, each of which might require a different agent. "During any continuing activity, like driving, or elaborate performances, like swimming the Hellespont, there are more or less fixed purposes, standards, desires, and habits that give direction and form to the enterprise," he says.[27] Swimming the Hellespont is not a set of actions (planning, dressing, and performing identifiable movements repeatedly, such as the strokes of the arms) but just one irreducible action. "Fixed purposes" seems to mean something like "pro-attitudes" here, though the concept is undeveloped. Similarly, when a person stops a car, according to Davidson, one need not itemize all the stages between pressing the brake pedal, compressing fluid in the master cylinder, transferring that pressure through lines, and creating friction on the rotor that transforms kinetic energy into heat. The act of pressing the pedal just *is* the act of stopping the car, no less and no more. The example of swimming the Hellespont is arguably more complex, given that it is rather harder to reduce the prolonged activity of a human body submerged in a fluid medium to a single act. As I suggested in an earlier chapter, swimming surely tempts us to think about the distribution of agency between a wide range of human and nonhuman interactants, an assemblage of people and things that all contribute to the result. But this Davidson could not allow, for by that route agency threatens to detach itself from the human altogether, which a liberal humanist theory of agency cannot permit.

Sometimes during this period, Davidson seems to be combining two actions of his own: having his cake and eating it too. On one hand, he insisted on a causal theory of action, but on the other, he argued that one must not inquire too minutely or extensively into causes, otherwise a single action is liable to fracture into multiple ones, and who knows where that might end. He settled this conflict largely by returning to Anscombe's claim that action must be "intentional under some description."[28] So, even though it would be easy to describe opening a door as a series of different actions, Davidson would say that the reasons for grasping, twisting, and pushing parts of the door are the same as the reasons for merely opening the door; one intends to go through, which later means one has a pro-attitude toward going through. Only when there are separate and different intentions behind the reasons for action are there separate and different actions.

If the seemingly multiple stages of opening a door can be condensed into a single action, the seemingly multiple stages of moving the body can be too. One no more need itemize stages of brain chemistry, blood flow, nerve impulses, and muscle contractions than one need itemize turning a key, twisting a knob, applying pressure, and swinging a hinge. All of it is part of the same action, just described differently depending on whether one expands or contracts the rhetorical accordion. "We never do more than move our bodies," Davidson concludes. "The rest is up to nature."[29] On that basis, Davidson argues that the "primitive action" of some body movement is equivalent to a much more extensive action as a whole. "Doing something that causes my finger to move does not cause me to move my finger; it *is* moving my finger."[30] In other words, moving a finger is not a separate action at the start of a chain of other actions, nor is it an effect of some earlier action, such as a prior act of willing, but is tantamount to the entire action itself. Now we are getting closer to the importance of switches and buttons, for if moving a finger is not a stage in a chain of actions but the sum total of some much larger action, then flipping a switch is both the minimum and the maximum of action all at once. Unlike swimming the Hellespont, the primitive action of moving a finger is substantially contracted already in space and time, so there is not a great deal left for the rhetorical accordion to compress. One might need to argue at some length that swimming the Hellespont is a single action, against skeptics who would itemize many separate stages. But how many stages of action can one perceive in the split second when the light turns on? Or within the black box of the switch itself? Very few. Tellingly, Davidson's use of "moving a finger" as a "primitive action" is by far his most common one. In just this one essay, he refers to: "pointing my finger," "cause my finger to move," "moving my finger," "moving his finger," "moves his finger," "move a finger," "moving one's finger in a certain way," "moving my trigger finger," and more.[31]

In most of these references, Davidson is not talking about wagging one's finger about in the air, but specifically about moving it to operate a switch. He does refer to the indexical gesture of pointing early in "Agency" to isolate the movement of the body on its own, but more often fingers press buttons or pull triggers, especially as the essay proceeds. His real interest lies in "what relation an agent has to those of his actions that are not primitive, those actions in describing which we go beyond mere movements of the body and dwell on consequences, on what the agent has wrought in the world beyond his skin."[32] Why should the primitive movement of a finger attach itself to switching at precisely the point where Davidson yokes body movement to the world beyond the skin? Because switches tightly couple extensive and seemingly instantaneous actions at a distance to the body in ways that are difficult to pry open, break down, itemize, sequence, or analyze. They contract the accordion not in the rhetoric of description, but in the rhetoric of procedure. They condense what otherwise might be mistaken as so many separate actions into a unit so small, rapid, and opaque that it would be hard to perceive it as multiple actions after all.

It might be assumed that Davidson uses examples of switches because he has discovered the deep truth about action, but it makes more sense to say that his theory of action idealizes and generalizes a kind of compression already accomplished by switching, which then becomes the model for his own theory. Davidson even concedes that a single action is more liable to appear to be multiple actions when there is a time lapse, as with "an action that begins when the movement of the hand takes place but ends later."[33] For instance, he says, "Suppose I pour poison in the water tank of a spaceship while it stands on earth. My purpose is to kill the space traveler, and I succeed: when he reaches Mars he takes a drink and dies."[34] The question is whether he killed the traveler. The intuitive answer is *yes,* in accord with the Anscombe/Davidson thesis. However, as Davidson points out, if that is the case, it seems he killed the traveler before the traveler died. These kinds of paradoxes crop up when temporal duration separates the primitive action from the end result and, according to Davidson, sorely tempt everyone to muddy the waters by dividing one action into two. But "we feel less paradox in saying, 'You have killed him,'" he says, once the possibility of death becomes more certain, and especially once it is in the past.[35]

Many probably feel even less paradox when there is no perceptible passage of time at all. After all, shooting someone with a gun is just a more high-speed version of poisoning the water that the space traveler takes to Mars, because the shooter also kills the victim before the victim dies. The timescales are just shorter, almost but never quite instantaneous, which makes it unlikely that other intentional actions might be perceived, or that anyone would perceive temporal paradox. The briefer and smaller the action, the more resistant it is to rhetorical elaboration, and the less likely it will appear to consist of a chain of separate actions after all. For Davidson, switching is a better example of action than swimming precisely because it resists extended descriptions.

To summarize, switching conceals process in three different ways. First, in some cases, the action happens so rapidly that other agents have fewer options to intervene in the ongoing process. Other functions that might be regarded as agentive are thus harder to discern. Second, the concealment of action inside the black box of the machine renders mediating processes mostly invisible to the observing eye. The linkage between primitive action and result thus appears to be unmediated and direct. Third, automation can make an action functionally irrevocable even when the action is of longer duration, such as the trolley crossing the switch track. Even if the results are delayed, there is no going back, so the action is as good as accomplished as soon as it is commenced. Many of Davidson's examples have one or two of these three qualities (instantaneity, invisibility, and irrevocability), such as poisoning the astronaut on Mars, which is irrevocable and invisible but not instantaneous. But his examples involving switches usually have all three, as if to foreclose any possible route by which other sub-actions might intrude. That makes it appear as if the human agent is immediately coupled to the desired result, without any coparticipating agencies in between.

Enlightenment at Home

Davidson is fond of illustrating his arguments with short narratives. Some of these are sensational and dramatic, like pressing a button to launch a torpedo to sink the *Bismarck,* but many others involve familiar domestic situations. Like his example of returning home and turning on the light switch, these examples align philosophy with everyday experience. In many of Davidson's illustrations, the most mundane episodes of daily life take place at the switch. In "Agency," for instance, Davidson begins with a short narrative of his actions during his morning routine. "This morning I was awakened by the sound of someone practising the violin. I dozed a bit, then got up, washed, shaved, dressed, and went downstairs, turning off a light in the hall as I passed."[36] There are many artifacts in this story, and some are substantially analog, like the violin, but he takes comparatively little interest in those. However, the automatic, habitual act of turning off a light switch registers clearly, even though I suspect not many people would mention such an act in their own account of their morning routines. The act itself has none of the analog extensibility and complex material interactivity of actions like washing, shaving, and dressing. And yet, of all these actions, turning off the light attracts a little extra attention from Davidson later too, as he goes on to clarify that it would have been different if he had "turned off the light by inadvertently brushing against the switch."[37]

In another article from the late 1960s in which Davidson attempts to translate action sentences into first-order logic, he betrays an even more telling consciousness of switching. In a similar inventory of everyday actions, he reports, "I sleep, I snore, I push buttons, I recite verses, I catch cold."[38] These expressions denote familiar activities conveyed in active sentences with verbs that will interest Davidson later, but for our purposes it matters only that button pushing finds a place even in such a pedestrian list. Two of the actions involve the body alone, one involves an interaction with an infectious agent, and one combines speech with remembering, but the only action that represents the physical manipulation of the material world involves pushing buttons. Most philosophers of action have tended to focus on simple actions in which bodies have effects in the material world, such as Anscombe's spilled glass or sawed plank, and they have tended to find solitary or mental events like snoring or remembering inconveniently complex. For that reason, it is all the more surprising that Davidson ends up using nothing but buttons to represent the class of action most commonly considered by action philosophers. Far from standing in for an especially mediated or problematic form of action, pushing buttons registers here as action's most normative form.

Davidson is profoundly at home in a world of binary switching, as are most people in the developed world today. The advent of binary switching in the mid-nineteenth century had already begun reconfiguring agency in ways that Davidson would make philosophically explicit only in the mid-twentieth century. Theory caught up to practice. Ideas reexpressed established techniques. But the underlying assumptions were already comfortably situated in millions of mid-century European and American homes, and in many other places too. Like it or not, at an intuitive level, most people today are probably Davidsonian theorists of action, simply because the sheer ubiquity of binary switching has been shaping their assumptions about action for more than a century.

Ironically, the more actions involve binary switches and buttons, the more agency comes to seem exclusively human after all. In a predigital world, everything is more like swimming the Hellespont, fluid and uncertain, drifting and prolonged, and rendered as a single action primarily through rhetoric. It could be experienced as totality, as Davidson argues, but also as a tangle of other interacting factors. And there is no reason at all why entire societies might not conceive of action in these other ways. Indeed, many already do. Mario Blaser has pointed out that "indigenous ontologies" take for granted a distribution of agency across a wide range of human and nonhuman locations, assumptions that non-Indigenous scholars trained in different expectations have typically derived from actor-network theory, process philosophy, or other academic sources.[39] But Blaser helpfully reminds us that, in some societies, these assumptions are foundational norms. Even Jean Piaget recognized that what he called the "adherences" of subjects to objects involve dozens of different causal logics, and with them, different models of agency. My point is simply that widespread assumptions about individual agency in liberal societies have to be demonstrated, and have to be continually demonstrated, if they are to overcome competing conceptions like these. Precisely because there are viable alternatives at all levels, liberal individualist agency had to be hard-wired into the interfaces of the modern world.

Switches are potent little machines for shaping a kind of human subject long thought to be central to liberal democracies, one that is individual, rational, autonomous, and free. In a world festooned with switches and buttons, a great deal of action already has been organized as so many latent binary alternatives, ready and waiting for the active intervention of human touch. What a flattering world to live in. How rewarding to experience the world as *off* until my touch turns it *on,* as passive until contact with me brings it to life. Precisely because these switches and buttons conceal internal complexity, people tend not to perceive action as prestructured, still less a collaboration between themselves and technical systems, but solely as their own capacity to inaugurate action well beyond themselves. Those self-sovereign liberal individuals, imagined as free to select a course of action and responsible for the consequences, needed machines to be equal to their Enlightenment description.

CHAPTER 15

Pistolgraphs

In her 1973 essay on photography, "In Plato's Cave," Susan Sontag writes that "a camera is sold as a predatory weapon," and that, "just as the camera is a sublimation of a gun, to photograph someone is a sublimated murder—a soft murder, appropriate to a sad, frightened time." Similarly, she says, "the old-fashioned camera was clumsier and harder to reload than a brown Bess musket. The modern camera is trying to be a ray gun."[1] In linking cameras to guns, Sontag was part of a long tradition. Christian Metz says, "The snapshot, like death, is an instantaneous abduction of the object out of the world into another world, into another kind of time. . . . Not by chance, the photographic act . . . has been frequently compared with shooting, and the camera with a gun."[2] And Friedrich Kittler has argued at some length that "the history of the movie camera coincides with the history of automatic weapons," and that "the transport of pictures only repeats the transport of bullets."[3]

The metaphor of the camera as a gun derives at least partly from a prevalent sense that photography is closely related to death, as many theorists have claimed. André Bazin compared the photograph to a death mask, and concluded that it "embalms time."[4] Siegfried Kracauer observed early in his career that the photograph "destroys the human being" and "gobbles the world" and that, in it, "the present seems to have been snatched from death but in reality is delivered up to it."[5] The metaphor of the camera as a gun expresses similar sentiments in a more popular and accessible idiom. It has other less lethal associations as well, not least of all in relation to photography's function as a scopic regime of power, an instrument of policing and other more subtle forms of discipline.

Significantly, all these theoretical contexts are firmly grounded in strictly visual practices, from Bazin's visual arts to Michel Foucault's visibility traps. As a result, the most prevalent theories of photography as death or discipline tend to be partial, for they leave out equally compelling haptic dimensions by which the camera mediates not just looking, but also touching. In fact, the full meaning of camera guns seems to derive from a combination of haptic and optic properties, from the configuration of looking *as* a kind of touching, and even from the extension of the haptic to optic scales. This occurs especially in the operation of that particular kind of binary switch commonly termed a *shutter button,* but what was also called, in its earliest Victorian example, the *trigger.*

In chapter 1, I noted that the binary switch has two especially eminent ancestors, the musical keyboard and the trigger of the gun. If the keyboard is primarily associated with mathematical rationality, as we saw in earlier chapters, the gun trigger is associated with more direct expressions of mastery and violence. Just as the language of musical *keys* made the analogical leap to the buttons of writing machines, so too the *triggers* of projectile weapons transferred readily to the inputs of cameras. In both cases, the language of *keys* and *triggers* reveals the lingering significance of the original technical analogy, long after changing machine forms made it harder to perceive. To speak of *shooting* a photograph is thus to remember that some early cameras seemed rather more like guns than they do today, and even imitated actual guns with surprising fidelity. As we turn to this next iteration of haptic liberalism, we find that metaphors of shooting photographs reveal latent fantasies of mastery and violence that are also activated by the switch.

Metaphors of cameras as guns have a much longer history than twentieth-century theoretical sources suggest, for photographers had been comparing their practice to shooting since at least the 1850s. In fact, the metaphor of the gun was present at the very birth of instantaneous photography. The following pages attend specifically to still photography and to cameras designed to look and function like guns starting as early as the mid-nineteenth century. A media archeology of early camera guns can help us better appreciate the values invested in this durable and influential metaphor, especially those values that shooters were able to claim for themselves. In the previous two chapters, we witnessed isolated individuals performing sanctioned forms of private choice that had been configured for them in advance by the binary switch. In the context of shooting photographs, however, we encounter a different situation in which liberal subjectivity emerges not from privacy and isolation, but from the kind of asymmetrical power relations created when one person is holding a gun. This is an individualism born of domination, sustained through inequality, and validated in the most extreme cases by actual violence. Obviously, this is not how most liberals prefer to think about themselves, but I hope to show how a certain kind of liberal individualist subjectivity derives not just from the avoidance of being dominated (always regarded as illiberal), but also from sanctioned forms of domination, among them the practice of photography.

And yet, precisely because the camera is not usually an instrument of actual bloodshed, we must distinguish between two different theories of the violence of photography, one of which will be more important here. The first and less important theory proceeds from a general sense that the photographer's gaze amounts to a kind of violence against the photographed subject, as a violating intrusion or as a vector of coercive control. The gun thus stands in for more subtle but still dominating systems of visual discipline, frequently tied to race, nationality, or gender. This version appears in Sontag's essay but shows up more dramatically in works she cites, such as Michael Powell's 1960 film *Peeping Tom,* in which a male serial killer photographs his female victims at the moment he murders them. But this will not be my primary concern here, partly because it has been so well documented already. It is closely related to more than a half century of theoretical accounts of the optical production of subjectivity, from Jean-Paul Sartre's "look," to Foucault's "panopticism," to Laura Mulvey's cinematic male gaze. All of these and more share a pronounced interest in the targets of visual attention who are variously produced, disciplined, or dominated in the process.

The second and more important theory is more evident in the quote from Metz above and proceeds from a somewhat different tradition of thinking about the photographic image as deathly or corpse-like, a claim Sontag also entertains, but one made more influentially by Roland Barthes's *Camera Lucida.* In this mortuary theory of photography, the image of an utterly still human body functions elegiacally as a kind of death mask, preserving not just a person who died, but any photographed moment that is always irretrievably lost.[6] Looked at this way, the camera-as-gun abducts the living subject into the silent and timeless mausoleum of the photographic medium, still as a corpse. Even so, the living subject has not been killed, for the photographer has only created a simulation of death, a surrogate version of another human being now rendered as a lifeless thing. The effects, in this instance, are not principally on those who are photographed, but rather on the photographers who establish themselves as the living masters of the dead. In the act of stilling life, the photographer creates a new relation of subject to object, of living agent to lifeless body, of active shooter to passive target, of person to thing. By abducting apparently lifeless corpses into the timeless crypts of the print, photographers establish all the more surely their own status as agentive human beings. To excavate the practices of firearms from the history of photography is to perceive how the camera projects not only a field of possible domination in front of it, but also an envelope of liberal mastery behind it.

Active Shooters

Before addressing camera guns we had better start with guns themselves. Modern firearms date to the fourteenth century, but the nineteenth century witnessed a revolution in firearm technology. Although rifles had existed since roughly the seventeenth century, technological developments made them much more powerful and accurate by the century's end. In eighteenth- and early nineteenth-century battles, most soldiers would have used smooth-bore muskets, far less accurate guns that compelled soldiers to fire from ranks at close range. Starting around the time of the U.S. Civil War, more powerful, accurate, and affordable rifles drastically altered military tactics. Once armed with rifles, soldiers would have found it suicidal to fight in tight ranks and fire at close range, so they spread out, operated at a greater distance from one another, concealed themselves from view, and exercised more initiative during battle.[7] One historian has traced the rise of more individualistic tactics to the technical capacities of the rifle, a development I have elsewhere termed "rifle individualism."[8] Rifles made it not just possible but also necessary for soldiers to abandon the tightly coordinated, closely ranked firing practices of the eighteenth and early nineteenth centuries, and thus permitted and even required much more individual autonomy. Soldiers exercised more discretion about where to move on an expanded battlefield, and more judgment about where, when, and whether to fire. From new technologies came these new techniques, which had broader social and cultural repercussions still. As early as the 1890s, for instance, long-range, accurate, and plentiful rifles had made officers' earlier practice of exposing themselves to enemy fire utterly impractical, so officers began to conceal themselves in ways that would have registered as cowardly just a generation before. Under changed technical conditions, the performance of masculine courage changed too.

Rifles had other social effects, as modern versions became more affordable and plentiful, especially in the United States. Hunting grew more popular in the United States after the Civil War, even as commercial food production became the norm. During this period, Americans placed new emphasis on their constitutional right to "keep and bear arms" as a foundation of liberty, encouraged by the National Rifle Association, established in 1871. That organization, tellingly, was not called the National *Arms* Association, but specifically invoked the precision and power of rifles. A musket would have been a terrible metaphor for individual liberty: cumbersome, slow, and imprecise, it suggested that the outcomes of any intentional action were unpredictable and liable to miscarry. Indeed, the inaccuracy of the musket required overtly collectivizing tactics such as firing volleys under the command of an officer who called the shots. In contrast, a late nineteenth-century rifle was incredibly accurate at much longer ranges, and once jacketed ammunition, breech-loading, and magazine storage became the norm, soldiers could fire rifles quickly, easily, reliably, and with great precision at long range. As a result, rifle firing could serve as a compelling demonstration of individual autonomy, self-sufficiency, and responsibility. In fact, rifles proved so reliable, accurate and predictable that it became harder for anyone to claim that their effects derived from any source other than the person who pulled the trigger.

For these reasons, the political effects of modern rifles derive less from the actual violence they cause than from the structure of action they make possible. The threat of violence is always present, but even the most fervent adherents of American gun culture rarely employ their weapons against people. Today, Americans buy ten to twelve *billion* bullets every year—more than one bullet for every person on the planet—but despite the fearsome levels of carnage caused by guns nationwide, the vast majority of those bullets are either not fired or not fired at living bodies.[9] They do not need to be, because the real goal is to possess that potential power and to be able to exert precise control over it even at great distances. For the vast majority of gun owners, then, the violence is mostly imaginary, possessed as potential, rehearsed in fantasy or enacted in target practice, but not projected against people or animals. Ideologically, then, the gun transmits its most powerful effects not through the barrel but through the trigger, which acts on the shooter by enlarging and refining an agency that otherwise might seem too puny or diffuse.

In his reading of the American gun-rights slogan "People kill people; not guns," Bruno Latour points out that people should really attribute the killing to the "composite agent" of the "citizen-gun" or "gun-citizen," rather than to either taken alone.[10] But, as noted already, the power of the gun often has nothing to do with actual killing, but rather with the fantasy of having the power to kill. From the front of the gun, a field of domination envelops the target; from the back of the gun, a field of mastery envelops the shooter. In this way, the mere presence of a gun establishes a relation of domination in which one person enjoys a pronounced surplus of authority, the other a decided deficit, and all of this usually happens without anybody being shot. I am reminded of Richard Wright's great short story "The Man Who Was Almost a Man," about a seventeen-year-old son of a Black sharecropper named Dave who is desperate to buy a gun. When he does, he "held it loosely, feeling a sense of power. Could kill a man with a gun like this. Kill anybody, black or white. And if he were holding his gun in his hand, nobody could run over him; they would have to respect him."[11] Wright captures the experience of bearing arms on subjectivity, a fraught source of political power, as the story makes clear, but a seductive one nonetheless. Just holding the device makes Dave feel more powerful than anyone, a turning of the tables in a social system that, Wright makes clear, is entirely stacked against him. This fantasy eventually turns into disaster when Dave accidentally shoots his own mule, after which he fantasizes about shooting his employer too. The story brings us close to the meaning of bearing arms in Wright's broader culture. To bear arms is not just to make a claim for equality, not to demonstrate a preexistent sovereignty of self, but to affirm outright superiority, to effect a righteous domination by the weak of the strong, and thus to become the strong in turn.

Bearing arms recapitulates the dynamics of subject formation from Hegel's account of "lordship and bondage," in which the lord and vassal, or master and slave, mutually and reciprocally constitute each other. As Hegel argues, both of their identities are relational, interactive, comparative, and thus minimally individualistic, a fact that the master can find especially painful to concede. Donna Haraway associates Hegel's dialectic with the power disparities of all dualisms and binary oppositions. Socially, Haraway argues, the domination of "women, people of colour, nature, workers, animals" involves the "domination of all constituted as others, whose task is to mirror the self." By this logic, "the self is the One who is not dominated," Haraway observes, for "to be One is to be autonomous, to be powerful, to be God." However, she goes on, this is also an illusion, for the one who dominates depends on the other through this "dialectic of apocalypse."[12] Judith Butler has similarly suggested of the master's apparent autonomy that "Self-consciousness learns this lesson first in the context of aggression toward the other, in a vain effort to destroy the structural similarity between the two and restore itself to a sovereign position."[13] Violence is the master's attempt to beat back its double, the slave who makes mastery manifest but who also exposes the master as anything but independent. Butler goes on to say, "Violence is the act by which a subject seeks to reinstall its mastery and unity," a task that can never be accomplished once and for all, which is why it has to be violently transacted in the first place.[14]

But as we have already seen, actual violence need not be involved. The mere presence of the gun structures forms of social domination simply by making the person on one end of it commanding and the person on the other end compliant. Heralded as an epitome of liberty in American gun culture, the gun also generates instantaneous inequality, a zone of asymmetrical power that distorts everything within it. The gun is a mobile boundary that sorts out masters from slaves, hunters from quarry, subjects from objects, and hence what all of these distinctions finally amount to: people from things. Even when wielded defensively, the gun affirms the bearer as nobody's fool. But it thereby also turns other people into things—obedient servants but also corpses if necessary—in a rehearsal of mastery that validates the shooter as apparently self-sovereign and autonomous after all.

Although most of this tends to be accomplished in the visual domain, the gun's latent power finally derives from its potential for violent touching. To touch the trigger is really to touch the target with a bullet almost instantly, with great force, and from a distance. The bullet is a deadly fingertip, about the same size and shape but vastly exceeding the reach of the hand. Accordingly, the gun is a machine that expands the power of the liberal agent by making touch possible at visual scales. Conversely, the figure targeted by the gun cannot touch, indeed must not touch anything at all, hence the standard imperative issued to one facing the barrel of a gun: "Hands up!" The raised, separated, isolated hands confirm that the power to touch lies entirely on one side. The bearer of arms might even issue another equally telling command: "Freeze!" Only the body behind the gun may move, for agency lives there exclusively. Before the barrel, we thus find a living body already pantomiming the attitude of death, a vertical corpse. Deprived of motion, forbidden to touch, the person before the gun resembles a rather different kind subject, also frozen, also deathly: a figure in a photograph.

Snapshots

In photography, there is no kind of violent touching exactly equivalent to being shot, but photography can still be coercive, even violent. It is not that the photograph directly impacts the body, like a bullet, but rather that it promises longer-range coercions generally derived from violations of privacy. The photograph can function as a source of shame, an instrument of blackmail, or a mode of policing. Because such effects can be anticipated easily enough, even aiming a camera can be as coercive as aiming a gun, but in both cases this matters only because people rightly anticipate being touched materially in other ways later.

The aggrandizement of the photographing agent occurs more squarely in the realm of touch. Cameras needed finger-activated triggers only once instantaneous photography became possible, with exposure times much shorter than a single second. From the birth of photography in the late 1830s until the late 1850s, photography required long exposures, typically around a half minute for mid-century portraiture, depending on conditions. There were only three ways to improve exposure times: create more sensitive chemical emulsions, illuminate the subject more brightly, or admit more available light through the lens. Most of these options were limited in the 1850s, but some early attempts seem to have produced the earliest association of cameras with guns.

Well before the dry-plate technology of the 1880s, there were at least two wet-plate camera guns. The less significant but more beautiful was Thompson's revolver camera from 1862, a compact pistol camera that made four exposures on a single revolving wet plate behind a barrel-mounted lens.[15] The other early wet-plate camera gun was slightly earlier and far more influential. Thomas Skaife's Pistolgraph of 1858 is often treated as a charming oddity in camera encyclopedias and technical histories, yet the Pistolgraph almost certainly marks the first moment when the practice of photography could be experienced as something similar to firing a gun, an association the inventor encouraged with the camera's unusual name. Designed with a lens protruding from a compact brass box, and with the shutter release positioned either above or below the lens (surviving models differ), the form of the camera resembles a pistol in a rudimentary way. Still, to those accustomed to early box cameras, the pistolgraph's size and shape must have been far more suggestive.

Technically, the pistolgraph was a remarkable machine. By the time Skaife perfected it, the camera used a very small plate about 1.5 inches in diameter in order to permit a very large aperture of about f1.1, fast even by today's standards.[16] A surviving example is reproduced as Plate 11. The wide aperture allowed for a rapid shutter speed that Skaife estimated at under a tenth of a second, a time span too short for people to perceive as duration, and which Skaife termed the "twinkling of an eye."[17] At a moment when most photographic portraits required fixed expressions and metal headrests to prevent inadvertent movement, Skaife's London studio promised stop-action portraiture of the kind that people have taken for granted for well over a century.

Skaife encouraged the association of the Pistolgraph with guns in his promotions. In an 1859 lecture at the Royal Pavilion in Brighton, Skaife said that its "size and form . . . were determined by the same rule which decided the size and form of the Colt's revolver." Moreover, the Pistolgraph "is to photography what the revolver is to gunnery, and bears the same relation to a twelve-inch plate camera as a pocket pistol does to a twelve-inch mortar."[18] Many journalists discussed Skaife's device through the metaphor of guns as well. One wrote that, when preparing to take a picture, Skaife "simply dips his pistolgraph into a species of elastic bag and charges it with something—for anything the spectator can see to the contrary it might be powder or shot."[19] And in an anecdote that Skaife himself seems to have relayed, police once detained him for pointing his Pistolgraph at Queen Victoria, and only released him when he opened his camera to show that it was loaded with a glass plate rather than a bullet.[20]

However, the pistolgraph was not Skaife's first conflation of shooting with photographing. He already had gained public prominence for taking a remarkable photograph in 1858 that he alleged was the first to show an artillery shell in flight. Even though his photograph received widespread attention in the popular and professional press on both sides of the Atlantic, it is not clear what kind of camera he used, and the original plate now seems to be lost. Only a printed *carte de visite* survives, reproduced as Figure 32, which Skaife sold through advertisements in the newspaper.[21] It looks rather painterly and may have been retouched significantly. The image emphasizes the equivalent speed of the shutter and the shell. By equaling the speed of the gun, the camera could make moving objects perceptible again. In a letter to the *London Times* celebrating his own success, Skaife said that photography "appears to eternize time," and he claimed that it "promises to do to epochs of time what the microscope already does to small objects, and the telescope to distant ones." As an instrument of demystification, however, the camera exceeded even material constraints, according to Skaife, for he also wrote to the *Times* to note the presence of "the likeness of the human head" in the smoke from the mortar, a "phantom" similarly invisible to the naked eye.[22]

FIGURE 32. Thomas Skaife's *carte de visite* showing a mortar firing a shell on June 28, 1858, on or near Woolwich Common, London. Courtesy of the Royal Astronomical Society, RAS MS Add 91, vol. 1, no. 109.

Although I find no casual references to *shooting* photographs in British or American photography journals in the late 1850s or early 1860s, some were beginning to venture similar metaphors, possibly influenced by Skaife. In 1860 the astronomer John Herschel applied the term *snapshot* to a quickly composed instantaneous photograph, thereby likening it to firing a gun quickly from the hip.[23] Literary references also began to appear. In an 1858 English stage farce, a woman having her portrait taken exclaims, "Oh, lor!—it's a gun!" and "he's going to fire!"[24] The date coincides with the start of Skaife's active promotion of the pistolgraph in London. Similarly, in Irish playwright Dion Boucicault's *The Octoroon,* which premiered in New York in 1859, characters frequently conflate a camera with a rifle.[25] The camera belongs to Salem Scudder, a plantation overseer in the American South, who makes a photograph of a young belle early in the play. As he begins the exposure he calls out "Fire!" He also notes, "I've got four plates ready, in case we miss the first shot."[26] Later scenes make it clear that the camera requires a long exposure, so the metaphor seems not to lie in equivalencies of instantaneous operation, as it did for Skaife. Perhaps Boucicault was aware of Skaife's invention from just a year before, given its attention in the press, but was unacquainted with the details of the device. In any event, the fact that Boucicault could refer to exposing a plate as a *shot* early in the play and without explanation suggests not just that audiences could be expected to grasp the metaphor, but also, perhaps, that they already did.

The metaphor of firearms is even more prevalent in Boucicault's play when non-white characters interact with the camera. Later in the play, a "quadroon boy" named Paul (likely the son of the recently deceased plantation owner) and an Indian named Wahnotee happen upon Scudder's camera and attempt to take a picture.[27] Wahnotee fears the camera is a "carabine," and Paul despairs that "he tinks it's a gun."[28] Eventually Paul determines to take a photograph of himself, and asks Wahnotee to expose the plate. To time the exposure, Paul asks Wahnotee to uncover the lens, run to a distant tree and return before covering it again. While sitting still and alone before the open lens, Paul is murdered by a third man, a former plantation overseer conspiring to steal the plantation through various stratagems. When Wahnotee returns and finds Paul dead, he imagines that Paul really was shot by the camera, and smashes it with his tomahawk. However, unbeknownst to anyone until later in the play, the long exposure had captured the identity of the murderer on the plate's new "self-developing" emulsion.[29]

All of this is interesting enough as an early instance of photography as surveillance, even as criminal forensics, but it also establishes the relationship of photography to power disparity more generally. When Scudder photographs the young woman early in the play, the photographic scene takes familiar form, with a man operating the camera and calling the shots, and a woman mostly passive before it. Later, when Paul attempts to appropriate the master's instruments for himself, Wahnotee is right to perceive the camera as a dangerous vector of power, controlled by the same person who oversees the plantation. Plantation overseers were often seasonal contract laborers charged with maximizing slaves' productivity during the planting and harvest, and many American slave narratives identify overseers as sources of great brutality. That the man who *oversees* the slaves also operates an instrument for mediating sight further associates the camera with violent domination. When Paul is killed by a white man during the exposure—and a former overseer at that—the play superimposes the impunity of white domination on the uneven power relations of photography. Paul's stillness before the camera even amounts to a premonition of his imminent death. In the end, the murderer is caught, but even in that instance the camera is an instrument of even higher power, as it keeps a "vile ruffian" and low-class usurper in his place.

Skaife's instantaneous Pistolgraph brought similar power differentials into play.[30] Few images from the Pistolgraph survive, but one striking plate from 1865, reproduced as Figure 33, depicts two girls from the Brighton Deaf and Dumb Institution and confirms the camera's considerable abilities. One can see the candid attitude of the subjects, still rare in photographs from the 1860s, especially in portraits taken from such

FIGURE 33. Pistolgram by Thomas Skaife of two girls from the Brighton Deaf and Dumb Institution, England, 1865. Courtesy of the Science Photo Library.

close range. Neither child is posing for the camera, and both appear to be occupied with objects or people outside the frame. Moreover, these are not just young children, but children with serious communication disabilities that could render them unphotographable by other contemporary means. After all, the subject of a photograph with a half-minute exposure has considerable responsibility for making the photograph too, mainly by sitting still. But with instantaneous photography Skaife could capture images of even the most unwilling children, conceivably without their knowledge. The instantaneity of the process concentrates agency in the hands of the photographer and further dramatizes the distance between him and subjects, who no longer need to cooperate or even acquiesce.

Five years before photographing the girls at Brighton, Skaife was already advertising "Pistolgrams of Babies" in London newspapers, and suffering pointed mockery from *Punch* magazine as a result.[31] Around the same time, he began licensing others to set up Pistolgraph studios and directly marketing the camera itself to amateurs, specifically to women. One advertisement read, "'Skaife's Ten Guinea Pistolgraph,' by which a Lady can take a beautiful locket likeness of her baby, at her own residence, in the twinkling of an eye, without staining the fingers."[32] In an 1864 address to the London Photographic Society, Skaife recommended his camera as part of the standard equipment of the Victorian household, because it was small in size, easy to operate, and suited to low light levels in the home. He specifically recommended it for women, who might employ the Pistolgraph "to take a likeness of a baby or other domestic pet with as much ease and comfort to herself in her own drawing room or boudoir as she could sit down to do a piece of embroidery or other needle-work."[33] It is more than a little absurd to shape a camera like a revolver, name it after a pistol, and then market it to the angel of the household to aim at her children. And yet there is some cruel candor in that arrangement, given that women could be trusted to use the camera only when aiming it at less powerful subjects still.

Instantaneous photographs also brought other hidden facts to light, including discomposed facial expressions, furtive glances, disarrayed clothing, and even blinking eyes. In such moments of sudden capture, the balance of agency shifted decisively toward the person behind the camera, exaggerating power disparities that already existed and confirming the photographer as the primary locus of agency.[34] This is evident enough in the late 1850s, but as the century progressed, camera guns like Skaife's approximated the form and function of firearms even more closely, and the language of "shooting" photographs became the norm. In the process, the functional equivalence between cameras and guns began to take on an even more pronounced inflection of violent mastery.

The Quick and the Dead

The invention of dry plates and, later, film, made many other camera guns possible during and after the 1880s. In London in 1885, the Sands and Hunter camera company created a compact brass camera mounted on a rifle stock, containing a magazine of twenty-eight dry gelatin plates. In Switzerland in 1888, Albert Darier's Escopette (musket or carbine) consisted of a small box camera with a protruding lens and a pistol grip with a trigger. And in France, Theophile-Ernest Enjalbert's camera firm produced a highly realistic Revolver Photographique around 1882.[35] A trigger operated the shutter, and ten small plates slid into position for each subsequent shot, as illustrated in Figure 34. One journalist describing Enjalbert's camera wondered "why on earth apparatus-makers manufacture instantaneous cameras in the form of a gun or

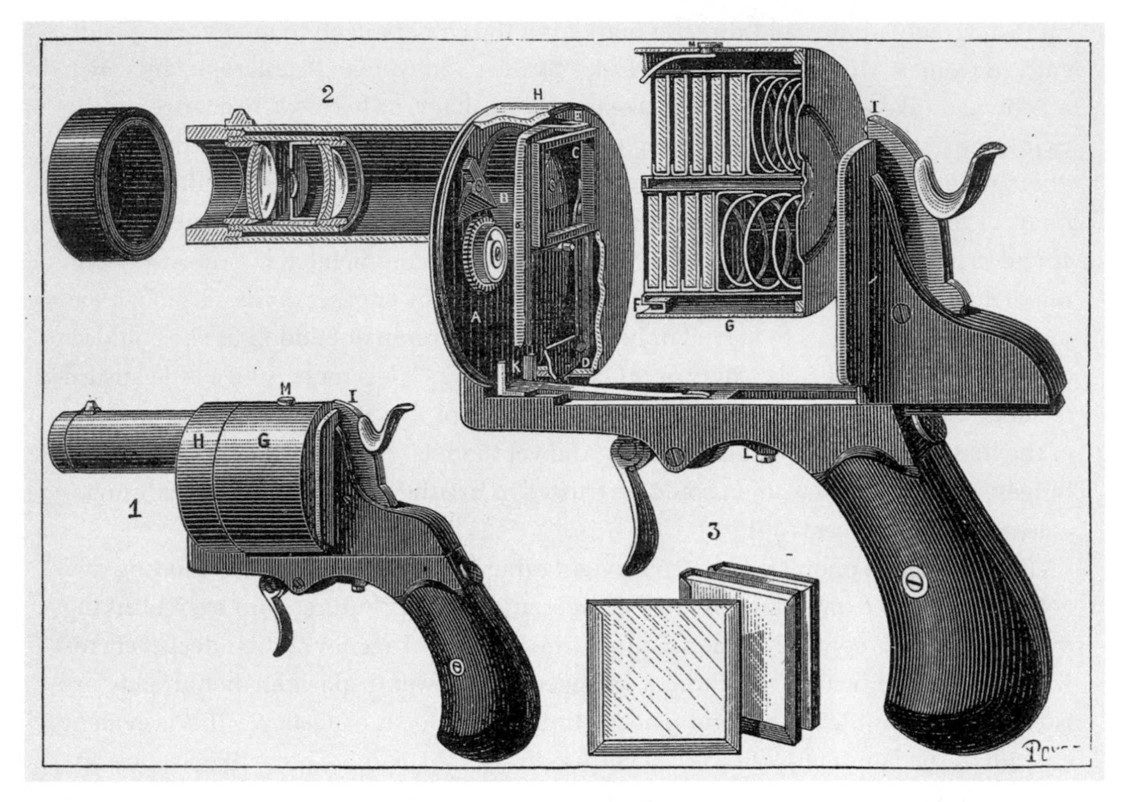

FIGURE 34. Theophile-Ernest Enjalbert's Revolver Photographique of 1882, which housed ten small dry plates in a camera body shaped like a pistol. Illustration from Josef-Maria Eder, *La Photographie Instantanée,* trans. O. Campo (Paris: Gauthier-Villars, 1888), 31.

pistol? . . . It is all very well to go skylarking about with a mock sort of pistol of this kind, but one fine day a fellow-creature may misinterpret your motive, and it is always difficult to explain matters after you have been knocked down."[36]

The most famous camera gun of the 1880s is surely Étienne-Jules Marey's chronophotographic gun, a rifle-shaped camera that could record up to twelve images a second on a single rotating plate, and an important technical precursor to cinema. Marey designed and used the chronophotographic gun for his animal-motion studies starting in 1882. Unlike his American precursor Eadweard Muybridge, whose motion studies used multiple cameras, Marey's chronophotographic gun exposed multiple images on a single plate that could yield strikingly beautiful images, especially of birds in flight. Like many other camera guns of the 1880s, Marey's and Enjalbert's cameras more fully associated photography with actual firearms practices, such as hunting or policing, and more extensively imitated those techniques.

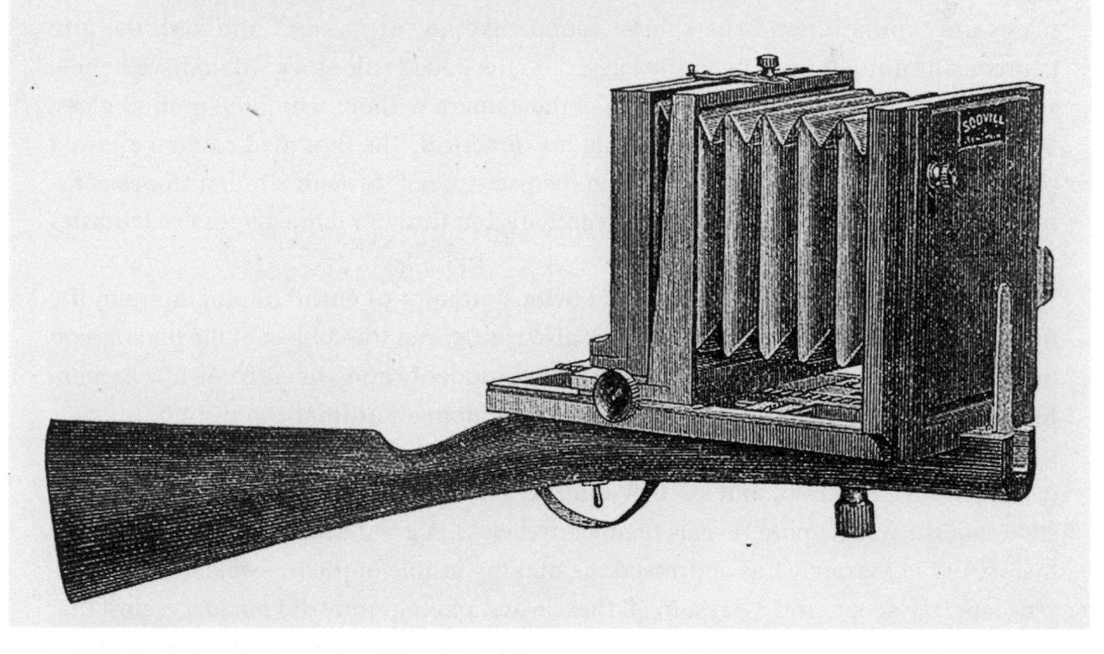

FIGURE 35. Advertisement for Kilburn Gun Camera, a small view camera mounted on a rifle stock. From J. Traill Taylor, *The Photographic Amateur*, 2nd ed. (New York: Scovill Manufacturing, 1883), appendix page 22.

There were more practical reasons for building camera guns in the United States, where the culture of hunting differed so drastically from that of Europe, and where a burgeoning gun culture developed rapidly after the Civil War. In that climate, the Scovill Manufacturing Company introduced the Kilburn Gun Camera in 1882, pictured in Figure 35, the same year Marey created his chronophotographic gun. Consisting of a 4×5 camera mounted on a rifle stock, and with a shutter speed up to 1/700 of a second, the Kilburn Gun Camera took advantage of more stable and sensitive dry-plate emulsions to allow photographers to freeze their quarry in flight. It is not entirely clear how many Kilburn Gun Cameras were produced, but the device was certainly cheaper and simpler and enjoyed much broader appeal than any of the other nineteenth-century gun cameras mentioned so far. It also was one of the earliest indications of what Matthew Brower has identified as a vogue in "camera hunting" in the United States around the turn of the century, in which hunters pursued wild game with cameras rather than rifles.[37] Accordingly, Scovill marketed the device to hunters who wanted to participate in the manly rites of outdoorsmanship while avoiding "the maiming of fish, flesh, or fowl." One period advertisement for the Kilburn Gun Camera could not be clearer about the fantasy of gun violence at the heart of photographic practice. The plates are "ammunition," the shutter sounds like an "explosion," and with the gun camera, any animal "may be easily bagged."[38] The added rifle stock would have helped stability and allowed for rapid aiming of the camera without using the ground glass, but presumably the goal was more than just practical. The modified camera allowed photographers to experience nature, and themselves, not through familiar tropes of romantic contemplation and aesthetic receptivity, but through more aggressive fantasies of violent capture.[39]

Beyond the Kilburn Gun Camera's obvious attempt to enroll photography in the masculine rituals of big-game hunting, it also configured the subject of the photograph as a fugitive that the photographer must seize suddenly and violently. As one camera hunter wrote, "The hunter never realizes how seldom an animal comes into full view until he has followed him around with a camera."[40] Skaife's artillery shell, quick and uncooperative children, and skittish animals are all of a kind: reclusive from human sight, but the gun camera reveals them nonetheless. A great deal of recent work on the history of photography has addressed the making visible of the invisible, through microscopy, telescopy, and X-rays. In all these ways, photography did not just record that which could be seen, but also revealed that which never had been seen before.[41] Skaife himself claimed that his instantaneous camera "promises to do to epochs of time that which the microscope already does to small objects and the telescope to distant ones."[42] As Jimena Canales has shown, even Marey's camera gun was part of a more general nineteenth-century preoccupation with the mysterious content of the tenth of a second. To photograph a bird in flight is not just to hunt for the bird, but to hunt for the elusive motion of the bird, that part of the animal that resists the unmediated eye. The instantaneous camera roots these images out of their hiding places between imperceptible instants. It captures them deep inside the thickets of rapid change. To shape a camera like a gun is to acknowledge the fleeting and evasive nature of action itself and then to reassert human mastery over it.

Crucially, these activities were not limited to a small number of wildlife photographers or camera hunters, but by the end of the 1880s had gained mass-market appeal. When Kodak introduced its first camera in 1888, it marketed it with a slogan that paid no attention to the visual whatsoever: "You press the button, we do the rest." Preloaded with rolled film for one hundred exposures, the camera automated everything except aiming the camera and triggering the exposure. This is the logical culmination of decades of shooting metaphors coupling power to primitive movement, even though shooting had retreated from technical to purely linguistic analogies. Once that happened, the association of cameras with guns began lodging not just in the vernacular of "shooting" photographs but also in the names of popular Kodak cameras. By 1895, Kodak named two roll-film cameras the Bullet and the Bulls-Eye and played up the firearms connotations in their advertisements.[43] The company used those names for other models in both the 1930s and 1950s. Despite the fact that these cameras were shaped like simple rectangular boxes, Skaife's original synthesis of firearms and photographic practices persists in their names, along with the resonant fantasy that making pictures is still a form of domination akin to shooting.

As a result, the camera is not just an optical instrument for the practical control of others, but also a device for generating individual self-sovereignty through this rehearsal of machine-mediated mastery. This mastery can arise in different ways. The camera can produce mastery, by exerting its own disruptive force. It can affirm mastery, by mediating existing inequalities. And finally, we are now in a position to see, it can simulate mastery, by capturing that part of the subject that would otherwise escape control, sometimes at no real loss to the subject photographed. The still body in the photograph is everything the photographer is not, frozen in time, changeless, immobile, without agency. Accordingly, the act of pressing the shutter button, which produces this simulation of death, is already an anticipation of mastery. It is mastery's founding moment, the instant when the master can bring another to a standstill, or so it seems to the photographer with a finger on the shutter button. Triggering the shutter is all the more compelling as a demonstration of mastery because the action involved is so slight. To touch is to still and kill, without any moderation of force or duration, without special skill, without even the perceptible passage of time. Like a gun, the camera turns whatever is before it into things; it turns whatever is behind it into human beings. Or rather, it turns whatever is behind it into that particular kind of being that liberalism regards as fully human: individual, autonomous, and free.

This is a liberal subjectivity born from a particular kind of domination, the interring of the other in images, or so it seems from the perspective of the photographer. To some photographed subjects—the infant, the animal—there is likely no harm done by this capture, but then again their experience is typically deemed irrelevant. They are subordinate beings from the start. In this form of domination, photographers achieve relation and nonrelation at the same time, domination and nondomination. The animal may go about its business, and the camera hunters can pride themselves on their respect for its independence. Meanwhile, they can also take home its semblance as trophy, adequate evidence that they mastered it after all. This deniability helps make domination tolerable for those otherwise convinced that domination itself is illiberal. If practiced in ways that are too violent and bloody, domination can affirm mastery only within the most comprehensively dehumanizing regime, such as slavery. Absent a consensus that those dominated are not fully human, and so deserve to be dominated, modern liberal domination must be rehearsed in more subtle ways. Photography raises domination to its highest art, by miming a form of violent capture that can confirm and deny mastery at the same time.

This helps explain why bearing arms and taking photographs function so powerfully as techniques of liberal individuation. The master must both preserve and destroy the other, must maintain the life of the slave and also demonstrate the power to end that life. Both options are necessary, and both intolerable. To preserve the other is to continue the relation by which mastery is constituted, but also to confess that the master is dependent on it. To destroy the other is to demonstrate mastery in its most extreme form, but also to eliminate the relationship on which mastery depends. Photography admits both of these possibilities at the same time. To shoot pictures is to effect a soft, simulated killing in which nobody dies, while also deploying photography as a real vector of power. The same subject can even be interred in the photographic crypt over and over again, sometimes without even knowing it. For liberal subjects who need to dominate in order to demonstrate autonomy, but who also have come to think of all domination as illiberal, a camera is the perfect weapon.

If the voting booth forged individuals in a crucible of solitude, still photography forged them from aestheticized rehearsals of domination accessible to millions. But firearms turn out to have influenced another technique of individualization that migrated to binary switching: remote control. As we shall see in the next and final chapter, the remote control of other kinds of visual media, from television to desktop computing, also derived in part from analogies with shooting. Freed from the narrow medium of photography, these later elaborations of shooting have given itchy trigger fingers to millions more people, and in media technologies that are now extremely difficult for anyone to escape.

CHAPTER 16

First-Person Shooters

Firearms techniques seeped into other contexts too. Although the first camera guns issued from Britain and France, it should surprise nobody that the next generation of optical devices influenced by firearms appeared in the United States, where a burgeoning late-nineteenth-century gun culture made firearms into a central symbol of individual liberty. Starting in the mid-twentieth century, activities as mundane as changing a television channel, operating a computer, or playing a video game all incorporated shooting techniques into new methods of remote control. The linkage between optics and haptics first established through "shooting" photographs migrated readily to these more interactive, screen-based forms of visual media.

As Caetlin Benson-Allott has shown, television remote controls have long mediated not just between users and broadcast media, but also between users and their consumption practices, leisure conduct, gendered identities, and family dynamics.[1] The television remote control made new kinds of image consumption possible, such as viewing in bed, but also fostered new social interactions, such as the battle to control the remote, as well as activities that some researchers have called "aggressive play," such as the desire to "annoy or tease family or friends." It also afforded individuals the simple haptic pleasures of "playing with buttons" or of having "something to do with my hands."[2] Remote control is thus socially and culturally complex even in its most mundane forms, and it turns out to be closely related to that one social entity that the chapters in this section have been tracing, the autonomous individual so central to liberal societies.

Ideal liberal subjects first must be individuated, as we saw in the previous chapter on voting technology, by insulating them from certain kinds of sociality so that their actions can seem to emanate from an isolated self. They also must experience a degree of mastery over other people and things, as we saw in another chapter on camera guns, so that they can enjoy their apparent sovereignty. But beyond even that, they also must be endowed with that mysterious property called agency, a liberal condition of selfhood without which one would not quite qualify as human. This final chapter attends to the history of remote controls modeled on firearms, and especially the ways in which they function as technologies of self-production. Remote control itself is a technique for generating selves that are not just properly integrated and individuated, but also properly agentive.

As a technique of self-production, remote control entails a peculiar relation between distance and touch. The unaided eye can focus on objects at infinite distance, but no closer than about six centimeters away, while the unaided hand must be in direct contact with what it senses. But remote control conflates visual and haptic ranges, so that mediated touching can happen at optical distances. To touch *here* and contact *there* with minimal exertion, duration, and skill is not just to extend the range of the haptic to the scale of the optic, but also to make touching almost as effortless as looking. And unlike camera guns that might propose that taking pictures is like shooting bullets, the remote controls considered here actually do make a material impact somewhere else.

In fact, they make an impact in two different places at the same time. A master who rings a servant buzzer is effectively acting in two places at once, in the comfortable parlor where the servant button is pressed but also downstairs where a bell summons the servant. Whatever else might be involved, the master experiences a displacement, as agency in one place accomplishes action in another. One of the privileges of mastery is to be in two places at the same time, spanning distance while compressing time. This is even more evident when the function of the servant is delegated to automated systems. When I press a button to open my garage door, I move the surface of a button a millimeter or two with my finger but my agency is confirmed by a larger action that occurs almost immediately but dozens of feet way. Where formerly the servant would have stood, I now seem to serve myself, both here and there, near and far, bounded by my body but also extending well beyond it. And in liberal societies, this chapter will suggest, agency consists substantially of such transitive properties and their aggrandizing effects. To be an agent, I will argue, is always to be displaced in order to act in two places at once.

One immediately thinks of actions that seem not to be distanced in the least because the target of agency is the agent's own body. To act back on the body would seem to foreclose all distance between self and other, contracting agency to a hard core of purely individualized personhood. For instance, is it not the case that physical exercise undertaken for the sole purpose of strengthening the body is an intransitive agency, an agency that has no target beyond the self? Perhaps. But, then again, the body itself becomes other, a piece of warm nature to shape just as one would manage any other part of the environment. The weight lifters at my gym admire themselves in mirrors not just out of narcissism, but also to effectively *other* their own bodies in a visual register, to make their bulging arms external targets of what they can then better experience as indwelling agency. The mirror makes more than just their muscles show up. It makes the *agency* that makes the muscles show up too. This may explain why people often involve mirrors when ministering to the body most privately, as many do before their bathroom sink each morning. The vanity mirror has its practical utility, no doubt, but while agents are combing, brushing, washing, or shaving, they can behold a second self on whom those actions land, one who appears to be looking back from just a few feet away.

Still, perhaps there are purely mental actions that are never externalized, and thus never distanced from the self, such as willing oneself to resist an intense desire? For instance, in moral contexts being a kind of person who abstains from drugs or sex might be an especially important method of self-production. Indeed, for those who identify selfhood primarily with moral self-regulation, the agency of abstinence might be the most important agency of all. Then again, to think of the self this way is already to divide it into an oppositional other and a superintending self, an animal desire controlled by a reasoning will, an id governed by a superego, an inner demon resisted by better angels. Some philosophers have even suggested that humans are not distinguished by the mere fact of wanting, which animals also do, but of *wanting* to want *differently*. Only these "second-order desires," as Harry Frankfurt called them, are characteristic of humans. To be a person, to have a self, is to establish a critical distance between first-order desires and the more important, more human second-order administration of them: not just wanting to drink, but also wanting not to want to drink as much.[3] And yet, by this route, Frankfurt defines human agency specifically in terms of transitive action after all, now mapped onto an inner psychodrama. Agency must operate on something other. To avoid being controlled by one's desires, one must control those desires instead.

If agency is transitive in this way—if it must always target an other, however constituted—firearms and the remote controls that imitate firearms are technologies of selfhood, much like mirrors, but ones that operate haptically rather than optically. Because touching the trigger also means touching something else at a much greater range, the gun, like the mirror, organizes the self through an interaction between proximity and distance. In both cases, a person seems to be in two places at once, but far from dissolving the self into multiplicity or distributing it across multiple locations, this extension helps bound and individuate the self. To venture a familiar analogy that we will return to in due course, Jacques Lacan described the production of the self as an effect of mirroring. Only by apprehending the self as other can the self cohere as an integrated whole, Lacan claimed.[4] Guns and the remote controls that imitate them produce selfhood similarly. I reclaim my agentive selfhood only from afar, after it has been doubled and distributed. Or it might be better to say that, through action at a distance and in two places at once, a solidly agentive *I* comes to be organized and naturalized. At that stage, this *I* appears as an appropriately autonomous, bounded, and empowered liberal self, and one that seems to have resided within a consolidated *me* all along.

We can even wonder whether there might be such a thing as haptic first-personhood, a mediated way of defining the self not through grammar (the pronoun *I*), narrative (a *first-person* narrator), or visual representations (mirroring or cinematic point-of-view shots), but through manual techniques, including remote control. In fact, remote control establishes first-personhood just as surely as those grammatical conventions, narrative forms, and visual practices that most of us take for granted. In arguing that there is such a thing as haptic first-personhood, and that it is more prevalent than commonly supposed, the following pages will proceed by degrees through examples of remote controls configured as firearms: first in early television remote controls, then in the "light gun" of a Cold War radar system, and finally through that popular genre of video games known as "first-person shooters." No less than the ubiquity of the mirror, the ubiquity of the remote control has made it a ready instrument of self-production well suited to the values of liberal societies. Many people, perhaps all people living in modern, technologically advanced societies today, are first-person shooters, because they have all internalized techniques of selfhood influenced by the remote control and the legacy of firearms behind it.

Screen Shots

The first wireless television remote control looked nothing like those most common today, typically small rectangular boxes with an array of buttons to turn the device on and off, change channels, select components, and adjust volume. Introduced in 1956, the Zenith Flash-Matic was shaped like a small pistol, emitting a beam of visible light when the user pressed a small electrical button configured as a trigger. Light-detecting sensors on the four corners of a specially designed television registered the flash of light and operated one of four controls: power on or off, channel clockwise, channel counter-clockwise, and sound on or off. If users preferred to walk to the television they could also change channels by "push-button operation" using interfaces located just above the light sensors on the top two corners. In advertisements for the Flash-Matic, the depicted viewers—always women—typically sit comfortably while operating their sets from across the room by firing a Flash-Gun at the television.[5] One advertisement promises that "A flash of magic light from across the room . . . turns set on, off, or changes channels, . . . and you remain in your easy chair!" But the same advertisement also anticipates another range of possible meanings when it reassures consumers that the Flash-Matic is "Absolutely harmless to humans!"[6] It was harmless, but the disclaimer reveals that its creators understood that shaping it like a pistol suggested other possibilities.

The difference between shooting a camera and shooting a remote control shows up in the posture of the those wielding the weapon. In photography, an empowered, usually adult, often male photographer turns his gaze on subjects who often inhabit the postures of traditional Western portraiture: seated, reclining, passive, still, nude. With the media remote control, however, the still and passive figures who had been the targets of others' agency take control of the action. With a remote, a woman formerly represented reclining on a couch becomes a "couch potato" instead, still immobile but now actively changing channels with her trigger finger. Active photographers shoot immobilized subjects, but with the Flash-Matic, immobilized subjects could actively control moving images on the screen. This fantasy of immobile agency is not as unusual as readers might think. As we have seen, switching had long immobilized the body, channeling action through smaller and smaller primitive movements, as if to narrow the gap between willing and doing to an absolute minimum. One magazine article about the Flash-Matic calls its beam a "finger of light," a veritable extension of the hand operating from afar.[7] The more negligible the exertion required, the more completely intention and action seemed to coincide. Shooting was already a model of this kind of agency, a reduction of killing as much as possible to the choice of whether or not to kill.

Many early advertisements for television remote controls depict immobile women, including the Flash-Matic advertisement in Figure 36, which shows a woman curled kittenishly in her easy chair. No doubt gender dynamics are in play, perhaps related to the prosthetic thinking so prevalent in women's typewriting, but labor dynamics seem to be important too. According to television historian Elana Levine, many at midcentury complained that women neglected their housework in order to watch soap operas. Unlike earlier radio serials, which women could listen to in the background, television commanded their visual attention and kept them in one room.[8] Perhaps remote controls promised that women could stray further from the television and accomplish more, but in the advertisement for the Flash-Matic, the woman appears to have no housework to do. Or perhaps other automated devices are doing it for her, acting elsewhere on her behalf. Rather than suggesting indolence, her leisurely remote control of the TV suggests that she is already remotely in control of the rest of the household, regardless of whether she delegated that work to a maid or to a Maytag. Needless to say, the fact that women's immobility could be glossed as evidence of their agency no doubt contributed to their predicament.

THE SATURDAY EVENING POST September 10, 1955

YOU HAVE TO SEE IT TO BELIEVE IT!

FLASH-MAT

A flash of magic light from acros

off, or changes channel

YOU CAN ALSO SHUT O

WHILE PICTU

Here is the most amazing television development since the invention of the picture tube—and only Zenith has it!

Without budging from your easy chair you can instantly turn your new Zenith TV set with Flash-Matic Tuning *on* or *off* or *change channels*—or even shut off the sound, eliminating annoying commercials while picture remains on the screen. If the program is dull, your flash-gun will give you another channel. All you do is aim your Zenith flash "gun" at a corner of the set and touch the trigger. A beam of magic light does the rest. There are no wires of

NEW! **Flash-Matic Tuning.** Enjoy the ultimate in TV viewing comfort, convenience. Flash-Matic remote control tuning available on **X2674EQ**, above and four other models.

NEW! **24" Table Model TV.** **(X2640R)** Genuine Zenith quality in 24" table model. Cinébeam, Ciné Lens for clearer pictures. Zenith Top Tuning. Also in blonde **(X2640E)**.

YOU HAVE TO SEE IT TO BELIEVE IT!

THE SATURDAY EVENING POST

...C TUNING BY ZENITH

ONLY ZENITH HAS IT!

... room (no wires, no cords) turns set on,

...d you remain in your easy chair

...NG, ANNOYING COMMERCIALS

...MAINS ON SCREEN

...ny kind to trip you or tangle you. No mechanisms or contrap-...ons to clutter up your living room. This is not an accessory. This ... a built-in, integral part of several new Zenith Royalty TV receivers.

Let your Zenith dealer show you the famous Zenith features ...ch as Cinébeam,® Ciné-Lens, Top Tuning, 20,000 Volts of Picture ...ower, and this latest achievement in creative electronics—Flash-...latic Tuning. It makes your television set a home theater! But ...ou have to see it to believe it. Stop at your Zenith dealer's soon.

The Bismarck. (Model X2264EQ). 21", Flash-Matic Tuning. New Royal "X" Chassis. 10" speaker. Cinébeam. Ciné-Lens. Spotlite Dial. Blonde grained finish cabinet on casters. Also in mahogany color (X2264RQ). As low as $399.95.*

...lso see the other outstanding new Zenith models just announced

IF IT'S NEW, IT'S FROM ZENITH!

...EW! **Over-and-Under 21" TV-...honograph. (X2280E).** In one hand-...me cabinet, superb picture quality; ...clusive Cobra-Matic® record player. ...lso in mahogany color (X2280R).

NEW! **Jet Tuning Table Models.** The set that brings television to every room in the house! Compact, convenient, colorful. Two-tone color combinations. (X2222Y, shown.)

The Royalty of TELEVISION and radio

Backed by 36 years of leadership in radionics exclusively

ALSO MAKERS OF FINE HEARING AIDS

Zenith Radio Corporation, Chicago 39, Illinois

*Manufacturer's suggested retail price. Slightly higher in Far West and South.

FIGURE 36. Advertisement for Zenith Flash-Matic remote control in *The Saturday Evening Post*, September 10, 1955, 14–15.

Class and labor issues are surely in the background too. Rachel Plotnick has recently shown how electrical servant buttons actively distanced masters from servants by making the act of summoning remote. According to Plotnick, "the button pusher could 'toggle' the servant's presence 'on' or 'off' at will so as to minimize any 'unnecessary' interactions."[9] Eventually, servant bells and buzzers did even more, as they automated not just the summoning of the servant but also the service itself. The history of the thermostat is an instructive example. Warren Johnson, a professor of natural sciences in Wisconsin in the 1880s, became frustrated that the temperature in his classroom periodically dipped too low, a situation rectified only when a custodian made the rounds each hour to adjust dampers and stoke the furnace. One solution would have been to install electric servant bells to summon the custodian as needed, ensuring that the master could delegate the adjusting of the furnace with little more than the touch of a finger, and that the two would not have to interact. But an even better way was to automate the master's already minimal act of summoning. For that, Johnson devised a thermostat that rang a bell when the temperature dropped too low, which summoned the janitor to his task. The company he founded, the Johnson Electric Service Company, later became Johnson Controls, today a large multinational conglomerate that manufactures everything from thermostats to fire detection systems.[10] Eventually, the same principle of automation controlled the furnace directly. When I press a button to raise or lower the temperature on the digital thermostat in my home, I forget that a servant ever was involved, not least of all because the work of stoking furnaces now happens even further away at an electrical power plant, and behind many other layers of mediation.

Though it is tempting to regard remote controls as labor-saving devices, they are really labor-relocation devices, and by extension, labor-concealing devices. They hide labor by shifting it out of view, and thus help those in control of the interface imagine that they compel the results by themselves. These absent laborers are what Bruno Latour called the "missing masses," social beings whose functions have been delegated to machines but whose social meaning persists.[11] When I use my washing machine, I can say that I "washed the clothes," even though my labor involves little more than dumping in the laundry and operating a single switch. But how would it seem if I said, "I washed the clothes," after handing them to a maid who then washed them for me? An aristocrat might not smirk, but the maid probably would. And yet, even in the case of my washing machine, the labor of the maid has not been eliminated, but just redistributed, shifted to the coal mine to extract fuel, the rail yard to transport it, the power plant to convert it to electricity, and the factory to build the washing machine. And it is not just that such commodified machines abstract and alienate labor, as Marx claimed, nor just that social meaning survives in machines, as Latour argued. The concealment of labor within technical systems also consolidates an apparently individual agency in the person at the switch. The alienation of labor, followed by the concealment of labor, finally results in individuals' ability to experience the sum total of action as their own. Marx was not wrong that alienated labor appears as a certain magic property of the commodity, but in the context of many kinds of remote controls, it also appears as a certain magic property of the people who use them.

Firearms provided one early version of these dynamics. A person could conceivably make a longbow from scavenged materials, but for most of their history, guns and even crossbows have been produced by specialists and necessarily involved an extensive division of labor. Just as every shot releases the stored energy of the powder, so too it releases the stored energy of labor gathered from far and wide, all of which the apparatus concentrates in the shooter's trigger finger. As an energy-storage system, the gun made violence more potent and portable. As a labor-storage system, it concentrated a different kind of power—the power of others—in shooters who could experience the entire arrangement as an extension of themselves.

By mid-century, the Flash-Matic's fantasy of such surplus power was being marshaled against a formidable intruder, the flood of televised sounds and images invading domestic space. A television screen could suddenly beam all kinds of unwelcome visitors into the home, to say nothing of other assaults against manners or morals, especially in commercials. Earlier wired remotes were already marketed as defensive measures, including the Blab-Off, which consisted of a simple switch attached to a cable, and which allowed viewers to "Stop annoying TV commercials from your easy chair."[12] Zenith played up the Flash-Matic's symbolic assault on unwanted intruders. One Flash-Matic advertisement invited viewers to "shoot off annoying commercials from across the room."[13] An article in *Popular Science* declared that users could use their "pistol" to control the set, and that a "bullseye on the slot at lower right tunes out the sound on commercials."[14] The Flash-Matic was part of a much broader attack on television commercials in these early years, and Benson-Allott notes that the first television function available exclusively on the remote control was *mute*.[15]

The Flash-Matic had its drawbacks. Because it used the visible spectrum, it suffered from interference by ambient room light. It also required a large battery, and users had to aim it with some precision. By the late 1950s, the next generation of remote controls remedied these shortcomings by using sound frequencies inaudible to humans, but in doing so they changed the form and function of the remote as well.[16] Because these sonic devices did not need to be aimed, Zenith abandoned the Flash-Matic's pistol-shaped form in favor of the rectangular shape familiar to this day. However, Zenith did not abandon the martial connotations, nor did television remotes cease to promise protection against televisual intruders. A more muscular name, Space Commander, offset the innocuous new form, later shortened to Space Command, part of a fantasy of militarized high-tech power. Other manufacturers embraced space-age and military rhetoric too. Motorola introduced the Golden Satellite remote controls in 1958, four years before the first communications satellite was launched, so likely more suggestive of the militarized space race. Roughly one year later, Admiral created the Son-R, later changed to Sonar, a reference to submarine warfare.[17]

With the abandonment of the Flash-Matic's form, one might think that the association of television remote controls and guns was subsiding, subsumed within a more general rhetoric of militarism. That is partly true, but at precisely that moment, firearms analogies reemerged in an unexpected place, the other side of the screen. The ray guns of 1950s science fiction clearly inspired the Flash-Matic; but their energy beams were not yet called lasers, which were not named until 1959. Inspired by that new technology, the *Star Trek* television series imagined a directed energy weapon called a "phaser" in the mid 1960s, which looked nothing like a gun but a great deal like the second generation of television remote controls. Grasped by the hand and operated by the thumb with a switch on its top, the iconic *Star Trek* phaser fired a beam of light more than a little reminiscent of depictions of the Flash-Matic, though also clearly indebted to earlier science fiction precedents, as depicted in Plate 12.

The phaser's resemblance to mid-century television remote controls is more than just coincidental. At a 2009 roundtable on *Star Trek* set and prop design, John Jefferies, one of the phaser's designers, confirmed that they loosely based it on a Magnavox remote control marketed around the same time.[18] He was probably remembering the Magnavox Phantom, pictured in Figure 37, which started appearing in Magnavox advertisements around 1964, two years before the series premier. The Phantom shares certain formal features with the phaser, including narrow width, buttons on top for the thumb, and a slightly tapered shape. It also had the same unusual combination of letters at the start of its name: the *phaser* is half *phantom* and half *laser,* a hybrid device with a hybrid name, and the deadly double of the device some viewers were using to operate their own sets.

None of this is entirely surprising, given that *Star Trek* has always been one of the most self-referential television programs, even before *Star Trek: The Next Generation* made the "holodeck" a metacommentary on digital media.[19] Just as the holodeck explored the forms and functions of interactive media, so the technologies of the original *Star Trek* series sometimes reflected on the still novel medium of television. For instance, the "transporter" broadcasts human bodies through space, delivering them in a flicker of televisual static. At this time, television itself was beaming all sorts of intruders into people's living rooms, and also transporting viewers to strange new worlds. Similarly, on the bridge of the USS *Enterprise,* the crew members spend much of their time gazing

FIGURE 37. The Magnavox Phantom remote control from 1964. Pressing the buttons compressed a small bellows inside, which forced air through two ultrasonic whistles, one of which triggered the television set to change channels, the other to change the volume. The Phantom was likely the model for the *Star Trek* phaser ray gun. Photograph by Patrick Manning.

at an enormous video screen, as if they were a large multiethnic family watching television in their flying white house. The family patriarch, Captain Kirk, even sits in a futuristic easy chair, appropriately appointed with another remote control on its armrest. The meaning of that might not have been lost on viewers. Whoever controls the remote really is the space commander.

The phaser is part of this televisual metacommentary too. Both weapon and remote control, it maintains a perimeter around the sovereign self against all sorts of space invaders. When fired at full power against a living body, the phaser causes a victim to glow brightly then simply disappear, dissolving gradually like a fading picture on a cathode ray tube. In that sense, the phaser is a gun that makes violence as easy and bloodless as changing the channel. It turns anything into television, even bodies, which become as ephemeral as pictures on a screen. As such, it also represents a kind of transport different from traveling in a spacecraft or beaming one's body from place to place. With the phaser, shooters could reach out with a finger of light that gave them power over the presence and absence of others. To touch here is to kill there, instantly and all at once, and so completely that not even a corpse is left behind. Plotnick notes that the servant buzzers could toggle the servant off or on, but the phaser has just one of those functions: instead of summoning, it only dispatches. It is a remote control for switching human life off.

Soldiers and Secretaries

Around the time Zenith released the Flash-Matic, the U.S. military developed a different way of shooting at a different kind of screen, but with much higher stakes. The Semi-Automatic Ground Environment early-warning radar system of the late 1950s, better known as SAGE, devised a novel graphical user interface for a powerful new computer, and an innovative way of interacting with it.[20] SAGE was a colossal and colossally expensive Cold War project that monitored airspace across North America, centralized radar readings to track targets, and coordinated military responses. The twenty-two completed direction centers distributed across the United States each housed two IBM AN/FSQ-7 computers, at that time the largest ever built. They took up forty thousand square feet, drew three million watts of power, and occupied an entire floor of a windowless concrete blockhouse.[21] Using twenty-five thousand vacuum tubes each and communicating with other installations over telephone lines, the SAGE computers could track two hundred enemy aircraft in addition to two hundred defensive interceptors.[22]

Thomas Hughes has argued that the real significance of SAGE lies in the complexity of its systems approach to design, production, and management, but I am concerned with just one small part of that system as it interacts with human users, the OA-1008/FSQ situation display, which was just the second graphical display ever made for a computer.[23] More than ninety of these display consoles in each direction center were connected to the operating mainframe, each relaying maps of North American terrain, indications of aircraft above it, and information about each flight. At a point when computers still operated primarily with punch cards or magnetic tape, and a decade before Douglas Engelbart would popularize "interactive computing," the SAGE situation display already employed a configuration familiar to most personal computer users today, with a vertical screen and a graphical user interface for operator interactions.[24]

Although many historians of SAGE have termed its graphical input device a "light pen," the 1959 patent refers to it exclusively as a "light gun," and refers to the push button on its front as a "trigger."[25] Pictured in use in Figure 38, the SAGE light gun developed directly from its simpler predecessor in MIT's Project Whirlwind, which one member described as "shaped like a backward pistol."[26] There had been earlier light pens

FIGURE 38. An operator using the light gun to operate the IBM OA-1008/FSQ radar display console of the Semi-Automatic Ground Environment (SAGE) radar defense system. Reprint courtesy of and copyright IBM Corporation.

for interactions with analog oscilloscopes, but for the computerized defense-industry projects Whirlwind and SAGE, these were refashioned in the forms of firearms.[27] On the SAGE light gun, a spring-loaded spacer protruded to help the operator align the device properly in relation to the screen. Just under this spacer, a series of lenses in the barrel focused light from the screen to sensors in the gun, and light from the gun onto the screen.[28] Once the operator selected the representation of an aircraft with the light gun, he could label it appropriately, and the computer could calculate an intercept position. Interceptor aircraft or missiles could then be sent to those coordinates. In such situations, the light gun appeared to do far more than just manage data. Through many subsequent stages invisible to the operator, it was also the first step of firing back.

Within a military context, SAGE translated the computer mediation of air defense into the familiar martial terms of aiming and shooting. For the men who operated these displays, the metaphor of shooting may have helped affirm their masculinity, given that other interfaces such as QWERTY keyboards were still strongly associated with women's work. Even swapping a light pen for a light gun might have helped men feel more like soldiers than secretaries. Beyond avoiding secretarial techniques, the light gun also helped characterize the operators' work as material causation rather than as symbolic communication, as fighting rather than writing. In chapter 14, we noted that Donald Davidson endorsed Joel Feinberg's "accordion effect," which holds that the results of an intentional action, even at distant stages, are all part of the same action. To identify stages of a single action is not to identify additional actions, Davidson argued, but only to expand a purely rhetorical accordion by which the action is described. By this logic, according to Davidson, if the SAGE operator aims his light gun at a representation of an aircraft on his screen, and designates it in such a way that it is sure to be shot out of the sky, then he has shot it out of the sky. All the other mediating technical and labor systems that bring this result about are just stages of a single intentional action.

The accordion theory of action is harder to apply to acts of communication, even extremely rigid ones such as the issuing of a military order, but it is at least possible to conceive of communicative acts in Davidson's terms. If an officer instructs the terminal operator to designate an aircraft to be shot out of the sky, presumably the intention and the reasons behind the intention would lie with the officer. But, at this stage, we can perceive the decisive role of the metaphor of the gun in the SAGE interface. Soldiers may regard themselves as having no agency if ordered to do something trivial or foolish, something they would never choose on their own, like marching in circles, but it would not be quite so easy to disavow agency if they were ordered to shoot someone. Even when acting under a direct order, a soldier would likely experience some amount of pride or shame for having done so, depending on the situation, and thus would likely claim some responsibility for it, especially if aiming were involved. Similarly, the pilot of the interceptor who fires the missile designated by the SAGE operator will not likely experience himself as just one more fold in the accordion. The act of triggering is difficult to disown, partly because switching is so widely established as the starting point of action, and partly because guns are so deeply associated with individual autonomy. Even if skillful aim were not involved, as in the case of a soldier ordered to execute a prisoner, the soldier would still probably feel some measure of responsibility. So, even though, in Davidson's terms, the officer giving the order to execute the prisoner might count as the agent of the prisoner's death, the soldier who pulls the trigger is not likely to feel innocent. In contrast, the secretary who typed the order probably does not feel quite so agentive. Guns generate a different kind of agency than pens.

Needless to say, metaphors of writing and shooting are both prevalent in computing, and remain so to this day, along with many others.[29] When people say they are "writing" to the hard drive or "reading" a file, or even when they use a QWERTY keyboard for computer interactions, they frankly inhabit the role of secretary rather than soldier. In some programs such as email or word processing, computers openly invoke the organization of a Victorian office with all its obsolete analog functions, including clipboards, paint buckets, brushes, scissors, rubber stamps, paper clips, and more. The actions implied are equally archaic, including opening, cutting, copying, pasting, filing, highlighting, and labeling. References to paper are everywhere, and with them, suggestions of secretarial labor.

But alongside these imitations of obsolete clerical functions, residues of soldiering remain just barely visible. As Thierry Bardini has shown, contemporary graphical user interfaces owe more than a little to the SAGE light gun.[30] Moreover, the iconography of shooting persists in visual representations that most users rarely recognize as deriving from projectile weapons. In most operating systems and in most applications, the mouse controls a cursor shaped like a stubby arrow, reminiscent of the bolt from a crossbow. Computers could have used other graphic indicators—a pen, an *X,* a typesetter's registration mark, a simple geometric shape such as a circle—but the arrow was present from the beginning at the top level of the interface. It served as the cursor in all the major graphical user interfaces since the 1960s, including Engelbart's cursor for his Augmenting Human Intellect in 1968, the Xerox PARC interface, Apple's interface for its earliest Macintosh computers, and Microsoft's Windows. Substituting a projectile for a projectile weapon captures another important shift. As Bardini notes, the mouse unlinks the hand and eye in a way different from a light pen or light gun, which must be held in contact with the screen in the line of sight.[31] The arrow, in contrast, is the marker of separation from the agent's hand, a shot already underway.

An arrow is not fired from a gun, but crossbow bolts were triggered in similar ways. They used kinetic rather than chemical systems to store energy that could be released at the touch of a finger. When the arrow emerged as a popular graphical marker in the eighteenth century, archery was already obsolete, but rifle bullets were still spherical, so could not connote direction.[32] By the eighteenth century, arrows began to appear in printed books to indicate direction of motion, such as the direction of the flow of a river on a map.[33] These earliest printed arrows tended to be realistic, with barbed heads, long shafts, and fletching, but even by the mid-nineteenth century, they were being truncated and stylized into the forms common today. At some point in the eighteenth century, arrows also began indicating some part of an illustration for special consideration, without connoting motion or direction. The arrow cursor of contemporary computer interfaces generally functions this way, because it indicates elements on the screen for possible interaction but never direction of movement. More recently, the "bullet points" list has become a mainstay of corporate communication, in which a small round dot reminiscent of a rifle ball accomplishes something similar. The mouse pointer is more explicit still. As a machine-mediated symbol, its analogy with shooting remains clearly discernable. Not quite a "finger of light," it is nonetheless a finger that acts at a distance from the body, a fugitive action, an intention in flight.

In fact, when arrows began appearing as indicators in eighteenth-century printed materials, they replaced an earlier graphical marker, the *manicule,* a small illustrated hand with a pointing index finger, like the one that appeared in chapter 6 as Figure 9. Some of the earliest manicules appear in the eleventh-century *Domesday Book,* but one historian notes that they were among the most common symbols in the margins of manuscripts and printed books from the twelfth to the eighteenth centuries.[34] When arrows began indicating parts of an illustration for special attention in the eighteenth century, they replaced earlier representations of human hands making indexical gestures, which had previously been used for the same purpose. So arrows and index fingers have been substitutes for a long time. And they remain substitutes in personal computers today. When browsing the internet on both Mac and PC computers, the default arrow cursor turns into a hand with an extended finger when it hovers over a hyperlink, in a form still strongly reminiscent of medieval and early modern manicules. That gesture is virtually identical to the primitive movement used for pressing a button. As the finger replaces the arrow, the threat of violence disappears; shooting turns back into touching again. But that substitution also affirms that touching was shooting all along, a displacement of action beyond the body. The interchangeability of the arrow and the manicule in modern computers reflects in iconographic form the highest truths and deepest contradictions of agency: the simplest primitive movement of the body counts as agentive only when it overshoots its limits, happening somewhere else at the same time.

Thus does action at a distance become agency at a distance too, not just in science fiction, as we saw in the previous part, but in computer interfaces used daily by hundreds of millions of people. The arrow and the manicule are icons of autonomous selfhood, a code that confirms that some kinds of agency should be in touch but also never too near at hand. All of this signals a deeper logic of liberal selfhood at work, embedded especially in switching techniques, but most visible in those switching techniques that claim affinity with shooting. Who can say what kind of ideological influence such iconography has had on what are now several generations of personal computer users?

The arrow and the manicule are the sacred symbols of an information age, a profession of faith in individual agentive control over the material world for those who traffic mainly in information. It remains to be seen how something like a self might derive from that arrangement.

Haptic First-Personhood

The Zenith Flash-Matic and SAGE light gun both have precedents in early coin-operated shooting gallery games that also used light guns, such as the Seeburg Ray-O-Light from the mid-1930s, which used the first light-sensing vacuum tubes. Later video games employed light guns too, including some of the earliest home consoles, such as the Magnavox Odyssey Shooting Gallery from 1972, which featured a large, realistic light rifle that had to be cocked between shots. Similarly, the Coleco Telstar Marksman from 1978 used two dials for multiplayer games like pong, and a militaristic light pistol for single player shooting games.[35] Given the association of shooting with individualism observed in the previous chapter, it matters that more social, interactive, multiplayer games like pong employed the analog interface of dials, while the single-player games employed the light gun. In later years, Sega, Nintendo, Atari, and others continued to produce light guns for their own systems.

Today, however, light guns for home consoles are comparatively rare, but video games involving shooting are more popular than ever. The most popular video game genre for the last few decades has been the "first-person shooter" (FPS), in which players operate an avatar through whose eyes they perceive an interactive environment.[36] Genre boundaries are not easy to fix in video games, but it would be hard to think of another genre that has sold better or been played more since about 2000.[37] In many games, part of the gun protrudes into the bottom of the screen, as if glimpsed in the lower peripheral vision. Visually and even audibly, these can be strikingly realistic games. And yet the haptics of input remain crude, depending as they usually do on controllers that still make heavy use of binary electrical switches. We noted earlier how curious it is that, unlike the audio and video components of games, the haptic components have settled for a comparatively crude verisimilitude, and that gestural controls, though intermittently popular, have not yet seriously rivaled button-based play. But there is one major exception to that claim: push-button game controllers have always been highly realistic gestural controllers when the gesture in question is shooting a gun.

In that sense, the button actually might count as the first gestural control. The first video game was a shooter, though not yet a first-person shooter. *SpaceWar!* from 1962 was made at MIT on the Digital Equipment Corporation's PDP-1, a small but sophisticated interactive computer with a cathode ray screen and optional light pen. The history of *SpaceWar!* is sufficiently well known that I need not recount it here, except to say that, in the game, two players attempt to shoot down each other's space ships.[38] In early versions of the game, the controls consisted of some of the forty-one toggle switches on the PDP-1, which both players would use to control thrust, rotation, and firing of weapons, but programmers later improvised remote controllers from a four-button office buzzer, and later constructed custom controllers out of wood, wire, bakelite, double-throw toggle switches, and a momentary push button.[39] The point is that, from the beginning of video gaming, programmers simulated an activity, shooting, that was triggered by a similar kind of finger-based activity, switching. They thereby established an equivalence between the binary interface of computers and the binary interface of guns, and so too of the interactions those interfaces facilitated. Even though the human operators of interactive computers like the PDP-1 worked less directly with binary code than their predecessors, their computer interactions required a growing number of binary manual actuations.

At the very moment when binary code began receding into the machine, then, binarism itself was reconstituted as interface. A functional binarism suited to the fingers migrated to the outside of the box, culminating eventually in the widespread adoption of QWERTY keyboards. Eventually, binary inputs dominated the interface of desktop computers almost completely, alleviated today only by the simulated analog movements of the mouse. And none of this seems to have been forced on unwilling users who endured a sadly restrictive interface. Users seem to have preferred binary switches for their computer interactions, even in games that imitate analog fluidity at every other level of sensation and simulation. One reason for this, I suggested in an earlier chapter, is that switching was a pleasurable and playful way of both affirming and alleviating individual subjectivity. But we now are in a position to see that the kinds of triggers considered so far might generate something like first-personhood *itself,* the sense of having a self, of being a localized *I* with a distinct perspective. But the term *perspective* is inapt, implying as it does only visuality. I am after some haptic equivalent of visual perspective, a method of constituting first-personhood through touching rather than seeing, through manual rather than visual techniques. Most people are so deeply invested in visual descriptions of selfhood that such an idea may seem strange, so to explain what I mean, let us approach the matter more slowly by returning to the familiar visual simulation of first-personhood in the first-person shooter.

The visual perspective of the FPS has clear cinematic precedents. Alexander Galloway has associated the FPS with the film convention of the "subjective shot," in which the camera inhabits the limited visual perspective of a single person.[40] From a perspective roughly five or six feet off the ground, with an angle of vision that corresponds to the human eyes, and often with simulated bouncing to indicate walking or running, the subjective shot implies the situatedness of a single person's body, sensory capacities, and consciousness. By this visual grammar, viewers understand that subjectivity itself is being signified. But Galloway distinguishes the meaning of the subjective shot in film from its correlate in the FPS, because in film, he argues, the subjective shot is an othering view, associated with drunkenness, madness or monstrosity, or literalized as voyeurism or predation, such as those that simulate a view through binoculars or a rifle scope. In the FPS, in contrast, the perspective is far less othering, and in fact tends to be so thoroughly normalized that it no longer represents a dangerously deviant perspective at all. It may in fact be the only perspective the player can occupy.

The reference to the "first person" in FPS games was borrowed from grammar, where the term refers to the first-person pronoun, and by extension from narrative theory, where it indicates narrators who speak from the position of a single limited consciousness, indicated with the pronoun *I*. To read a first-person novel is to inhabit the consciousness of another person rendered grammatically and narratively, but typically without the pressing sense of othering that Galloway observes in film. In fact, however unreliable first-person narrators are in fiction—and they are all unreliable to some degree—the reader tends to be coaxed into a too-comfortable intimacy. Literary narrative usually naturalizes the other's perspective so completely that it can subsume the reader's own independent perspective. To be alert to irony (the distance between what the narrator says and what the novel is saying), readers must maintain critical distance from a perspective the novel forces them to inhabit. The results of this narrative arrangement can be just as disturbing as a film's subjective shot. In Vladimir Nabokov's *Lolita,* for instance, the pedophile narrator seduces readers no less than the teenage girl named in the title, and their captivity within his perspective, and the fact that they take aesthetic pleasure from it, can also feel shameful and violating. In fiction as in the FPS, the problem with the first-person perspective is sometimes that it is not quite othering enough.

But let us remember something crucial about all first-personhood, which is that it always involves the displacement of subjectivity into some other location, the symbolic substitute *I* that stands in for an otherwise fugitive self. First personhood should not be confused with some natural, self-sufficient subjectivity, a Cartesian *cogito* that is given, not made. First-personhood is always a production in which a seemingly natural self is inhabited from afar. Even to use the pronoun *I* is to occupy a place in a symbolic order in which selfhood registers as a separate symbolic substitute. To be a *self,* one must first be an *other.* As noted at the outset, Lacan influentially imagined subjectivity specifically as a visual displacement, a mirroring by which subjects perceive themselves as objects, as integrated, bounded beings *over there* in their reflections, and for him this was a version of what happens through linguistic displacements too.[41] Whether in mirrors or in words, dislocation is a prerequisite for integration as a self.

However, organizing the self in a mirror is somewhat different from inhabiting another first-person perspective, as if seeing through other eyes. Inhabiting a simulated first-person visual perspective layers one first-person view within another. When playing an FPS, I do not see a body from a distance that registers as mine, as I would in a mirror, but I see a screen depicting a simulated perspective that I can imaginatively inhabit. Yet this too is a dislocation, a constitution of selfhood at a distance, even though no mirroring is involved. As a result, the player's first-personhood does not come first, for selfhood in any form is compounded from these kinds of secondary representations. After all, nobody referred to a "first-person perspective" until it was simulated at one remove. And nobody would say, "I am a first person." First-personhood is by definition the name of a grammatical, narrative, or visual offsetting from which a sense of the seeming naturalness of the self can later arise.

Whenever there is a representation of first-personhood, we can always ask what shows up on both sides of the mediation, at the intradiagetic and extradiegetic levels. This is obvious enough in the visual dynamics of FPS games, which reproduce a field of view that roughly corresponds with the field of view of human eyes. But such repetitions need not be purely visual; literary narratives often rely on nonvisual repetitions both inside and outside the representation. Romantic and specifically gothic literary narratives, for instance, often reproduce scenes of silent and solitary reading, precisely the activity required to consume them, as a way of aligning the reader's consciousness with the consciousness represented in the text. Edgar Allan Poe's "The Raven" starts on "a midnight dreary, while I pondered, weak and weary, / Over many a quaint and curious volume of forgotten lore." Any reader pondering over Poe's own quaint and curious volume finds in the very act of reading the grounds for first-person identification with the narrator, above and beyond that narrator's grammatical first-personhood. Similarly, the first-person narrator of Charles Brockden Brown's *Wieland* imagines she hears footsteps in her home while in bed, thus priming readers to share her fear, given that they might be imagining the same footsteps while reading in bed too.[42] These duplications of the scene of reading involve much more than just *perspective,* though we sometimes refer to it that way. They involve whole ensembles of bodily positioning, tactile experience, technical interactivity, and social forms, even in such a familiar activity as reading in bed. Where the intradiegetic and extradiegetic coincide, we will find the most important parameters of any production of first-personhood. The combining of the intradiegetic and the extradiegetic puts a person in two places at once. Being scared while reading in bed about being scared in bed links self and other in ways that privilege imagination under conditions of silence and solitariness, key components of the romantic model of selfhood. When we can identify the common element reproduced inside and outside the diegesis, we are on our way to identifying the core claim about first-personhood.

In the FPS, the common elements between the intradiegetic and extradiegetic layers of first-personhood are also haptic, specifically involving that one part of shooting strictly associated with triggering, the pressing of a binary switch. A certain configuration of touching, like a certain configuration of seeing, can generate the kinds of distancing required for first-personhood. Through the reproduction of haptic experience on both sides of the mediation, the player can accomplish the displacement necessary to become a first-person. The game is thus not just *simulating* a preexistent first-personhood, but rather confecting it out of multiple sensory displacements. In FPS games, this includes both haptic and visual displacements, but other video games may depend exclusively on the haptic. For example, *SpaceWar!* has no first-person visual simulation at all. Visually, it occupies a third-person perspective, a view from above. And yet, even with its earliest and crudest controllers, touching the switch outside the game registers as firing a weapon inside the game. The players are visually outside but haptically inside; they are visually alienated but haptically identified. In that way, even shooting games that do not simulate a first-person visual perspective might be first-person shooters too, but *haptic* first-person shooters exclusively.

Accordingly, not many have noticed that the square game pads of the 1980s and 1990s have been gradually morphing into imitations handguns. Many recent FPS games map the switch used to fire guns onto the controller's trigger buttons by default, which is shaped and positioned like a gun trigger. The rest of the controller is starting to feel like a gun too. The most common controllers for the Microsoft Xbox and Sony PlayStation now have "grip handles" that fit snugly in the palm, wrapped by the three outer fingers of each hand while the extended index finger curls around the trigger and the thumb operates controls on top. This is the manual posture of holding a revolver, the grip in the palm, the index finger on the trigger, the thumb on the hammer. I invite readers to heft one of these game controllers and to *feel* whether there is any closer analogy than holding a pistol. The controller is becoming a flattened pistol, or rather two pistols joined together, so that, even in games that do not simulate firearms techniques, the player can feel like a shooter nonetheless.

Because first-personhood is not imprisoned in the visual, perhaps we need a term like *pertactive* as an alternative to the ocularcentric term *perspective*: a "touching through" rather than a "looking through." We might entertain a thought experiment about some strange beings in a haptocentric society that esteems touch as highly as sight. For them, first-personhood might be experienced differently. In their peculiar society, inhabiting the subjective position of another would require equivalencies at the level of touching rather than just looking, and so too would require self-production through haptic displacements. For them, a first-person *pertactive* would not involve just seeing like another person, but feeling like another person too, and they would design their technologies accordingly. As a result, they would also experience selfhood not solely through the visual, as theorists from Sartre to Lacan have supposed, but through haptic dislocations and displacements that make selfhood palpable.

We are those strange beings, but we are mostly in denial about it. Alongside the screens and mirrors that shape the visual domain of many societies today, a haptic domain is no less formative. Ocularcentric societies have successfully repressed or concealed the mechanics of haptic first-personhood, even though the machines designed to generate it are all around. They are most conspicuously those first-person shooters played by hundreds of millions of people, and which are now among the most popular forms of mass media on the planet. They are also those other arrangements entirely devoid of visual first-personhood, such as games with third-person visual perspectives, television remote controls, or desktop computers' graphical user interfaces. Decoupled from visual simulations of first-personhood, these machines permit their users to touch at different locations, right *here* and over *there,* inside and outside the diegesis, at the boundary of the body and far beyond it, and thus to gain enough distance from the self to make it tangible.

These dynamics of self-production may feel humbling to liberal subjects who prefer to think of themselves as naturally free-standing, autonomous, and individuated. Skeptics might even be beside themselves, in that ancient idiom for intense feeling: jarred out of the comfortable occupancy of self. But if they are beside themselves at the thought, they are only recovering what it really means to be a self, for everybody is beside themselves, given that selves must be constituted at some remove. If this is most evident in remote controls, and especially in those remote controls that imitate the material power of guns, it is not so alien to other binary switches, including all those others we have seen in previous chapters, from telegraphs to typewriters to light switches to musical instruments to computer controls to interfaces of even the most pedestrian home appliances. In truth, every switch is a remote control, for every switch distances, however slightly, the input from the output, the touch of the finger from the result. Because all of them effect action at a distance, all of them generate agency at a distance too. Mediated though my actions are by machinery, alienated as they are by distance, they are yet part of a digital technique more fundamental than any other in this book, for they trigger effects that a good liberal subject can subsequently detect, collect, and designate with that privileged title of *me.*

Epilogue

Self-Destruct

Where will it end? From the mid-nineteenth century to the present, binary switches have proliferated massively and continuously, in almost every area of life. Actions that had been analog for all of human history now rarely happen any other way: illuminating rooms, writing letters, making images. Before the mid-nineteenth century, these acts required different skills and dexterities, but today they all can be accomplished through remarkably similar motions of the hand, most of which require minimal training. How odd it is that millions of people use the same gesture to write, cook food, open doors, summon people, watch television, heat and cool the home, calculate sums, exchange money, send messages, wash dishes and clothing, and much more. It is no exaggeration to say that people have never before altered the relation of their bodies to the material world so rapidly and so completely as they have over the last century and a half. This has been more than just a matter of technical change, as noted frequently in the chapters of this book. It amounts to a structural transformation of action itself, accomplished through technologies but not entirely determined by technologies. For it was no foregone conclusion that the aftermath of the industrial revolution would necessarily privilege binary inputs suited to the fingers across so many different technical and social domains. One switch at a time, people made a new world to live in, and in the process also remade themselves.

The many technologies encountered in previous chapters share a set of related techniques, often similar but rarely identical in form or social function. A typewriter key is not quite the same as a light switch, which is not quite the same as the shutter button. Some are more expressly mathematical, others more strictly causal; some more rationalizing, others more ludic; some are keyed to aesthetics, others to violence. And yet the extensive analogies that have marked the history of switching also testify to the softness of these categories, and to their constant reformations and deformations as input techniques have slid from one application to another. This is a history in motion, one that has settled only into temporary patterns so far, and one that continues to change. If past is prologue, we should expect more extraordinary analogies still.

Not long ago, it would have seemed absurd for toothbrushes and toilets to have push buttons, but here we are.

Although this has been a story of rapid and drastic change, it also has been a story of historical persistence that continues to the present. In chapter 10 we considered the "fossil utility" and "zombie utility" that characterize many of the most familiar technologies today, and which take the form of centuries-old techniques that survive within the haunted crypts of our machines. For that reason, computers today are as *steampunk* as any retrofuturist science fiction fantasy, for their keyboards preserve the form and function of nineteenth-century writing machines mostly intact. It takes effort to see these historical relics in technologies so widely glorified as new. It takes still more effort to recognize that the digital techniques of the people using them are even older, older than the earliest computer or than anyone alive today. At the extremities of the body, and through the most advanced digital techniques, we cannot fail to remember the past.

Part of what the digits of the hand remember is the installation of new political values after the revolutions of the late eighteenth and early nineteenth centuries. New philosophies of individualism gave rise to new political systems, new societies within those systems, and new kinds of subjects within those societies. Ever since, those renovated values have persisted not just in political theory, but in technical media too. A liberal philosophy that shaped new political systems also shaped new kinds of machines, especially machines adept at validating philosophies of individual human autonomy. The eighteenth century described people in a radically new way. The nineteenth century built a world in which that description could seem credible and true.

After all, it takes a great deal of work to make it seem as if action normally emanates from individual people. That idea is so implausible that it must derive from a truly comprehensive system of reinforcement. For individualism, as we have seen, is not the absence of social organization, but just another mode of social organization, an orderly method of fostering experiences of autonomy and of limiting or concealing experiences of other kinds. This social shaping of individuals typically happens subtly, even surreptitiously, in order to conceal the painful fact that it is happening at all. For that reason, switches are among the most effective engines of individualism, operating quietly but relentlessly to help everyone experience themselves as maximally autonomous and free. This is true even within that pervasive forum of modern sociality known as social media, which has insinuated into human relations an enormous amount of clicking on the keyboard, mouse, and screen. For billions of people around the world today, it is now hard to avoid socializing by pressing a virtual "like" button, taking and posting digital photos, typing public and private messages, and defending one's privacy with so many toggled options. But social media mediates not just by way of screens and networks, algorithms and powerful corporations, but also through an unavoidable binary interface that has translated social interactivity into the seemingly individual activity of button pushing.

One wonders where the next stages of switching will lead. Not long ago, virtual buttons on touch screens were unreliable novelties; today my mobile phone cannot function without them. On its small screen, almost everything does double duty as an interactive switch, from photographs to icons to text. Although the screen simulates analog manipulations too, like dragging or stretching, simple clicking still predominates. These recent developments are just the latest stages of a long history in which switching has required less and less human effort. A century ago, a typewriter key required a firm press; soon after, electric typewriters responded to much less; a modern computer key needs only the slightest touch; a virtual key on a screen requires no force at all. I suspect the next steps in switching will advance even further toward an ideal of bodily effortlessness. Microsoft is already experimenting with a technology called Skinput, which maps various interactive regions onto the palm or forearm and detects which points have been touched by analyzing vibrations propagated through the body.[1] In Skinput, the body itself becomes an array of buttons. Other switched interfaces dispense with the fingers entirely. Both Mac and Windows computers already have accessibility features that can track head or eye movements to control the mouse cursor and register mouse clicks. At some stage, who knows, switches may respond to thoughts, which is not hard to fathom in a world where the neural control of prosthetic limbs is already routine. Engineers at MIT even have developed a headset they call AlterEgo that can accurately interpret and respond to "subvocalizations," the subtle contractions of vocal muscles that occur when a person speaks silently and internally.[2] But who can say whether people will welcome an articulation of agency that no longer involves the fingers. If they do, that would amount to a truly radical overhaul of centuries of digitizing techniques, perhaps even compelling a reconsideration of what we mean by the "digital."

Then again, if switching were to dispense with the fingers, it might compel a reconsideration of what we mean by the "human" too. As we have seen, those features that seem so peculiar to the human, such as opposable thumbs, surely evolved as accommodations to early hand tools. People did not start using hand tools because they found themselves fortuitously in possession of opposable thumbs. They evolved opposable thumbs because they were already dependent upon their tools. Flint and flesh chiseled each other into shape. Given the prevalence of switching around the world today, a switching interface that does not use the fingers might not be all that desirable, or all that easy. Indeed, just two centuries of intensive switching might be reshaping human biology already. As some have noted, social media insidiously exploits peculiarities of neural chemistry, the reward–response system that turns social media followers, feedback, and "likes" into something akin to addiction. We drug ourselves with our own dopamine, wrung from the hypothalamus by what Richard Seymour has called a "twittering machine."[3] That machine is usually thought to work through media content algorithmically targeted to users' emotions, but it also might depend on the binary switches that make it all happen. No less than the content of social media, the switches at its interface also might act back powerfully on their users, shaping much more than just opposable thumbs. Who can say how much pleasure people derive from clicking itself? Seymour's most disturbing critiques involve cases in which social media users have goaded others to suicide, but even in cases involving lesser kinds of cruelty, it is not inapt to imagine that, in social media, everyone has normalized the experience of waving around a gun. Such cruelty surely derives from many different sources, but at least some of it might be attributable to the fantasies of individual mastery sourced from the switch.

I suppose defenders of social media will be quick to recall the many admirable things it can do, such as facilitating political organization and protest. They might even imagine more utopian possibilities. I need no persuading that social media sometimes can be an effective instrument of political action or self-determination, at least within existing structures. Similarly, switches also can be empowering, egalitarian, and even liberating, at least within the liberal political system they help legitimize and sustain. After all, individualism might be socially constructed, but that does not mean it is simply an illusion. In the introduction, we saw how switches help many people achieve a degree of independence otherwise impossible, such as blind or autistic writers. Switches also can aid more practical forms of resistance. Because a simple button now triggers a high-quality camera on most cell phones, some people have been able to record abuses of power that too often go unpunished, such as police brutality. Switches also make smaller-scale protests possible in everyday life. We have seen how central the *mute* function was to early television remote controls, and how welcome it was as a defense against mass-media intrusions. For the time being, users still can mute their radios, televisions, and even computers with the touch of a finger. Whether their neural chemistry can endure that deprivation for long is a different story, but most media is still routed through a binary gateway nominally under the user's command. Then again, because it often severs mediated connections to others, switching things off might be the most potently individualizing gesture of all.

For that reason, the quintessential switch might be the kind that is most violent, a switch that simultaneously establishes individual agency in its most extreme form and utterly negates it. I am thinking of certain instruments of suicide, an act that has been automated to an extraordinary degree. In liberal societies, suicide has long registered as the epitome of individual choice. All suicides must be individualized, for it takes very little collaboration to make the death seem like murder. Some even have argued that suicide is the ultimate act of self-determination for some kinds of people— namely, slaves— because it asserts ownership of the self while simultaneously robbing the master.[4] Because nobody has the right destroy my property but me, suicide establishes self-possession through self-destruction. Especially when accomplished with binary switches, such as gun triggers, suicide openly exhibits many of the contradictions that previous pages have pursued through more mundane contexts. In the United States, with its permissive gun culture and millions of available firearms, it is hard to deny that switching and suicide have their own peculiar relationship. Each year almost twenty-four thousand Americans die by suicide from firearms, far more than the fifteen thousand who are shot and killed by others.[5] Among all suicides, more than half are accomplished by a simple squeeze of a trigger, including more than 60 percent of suicides by men. If, as earlier chapters have argued, switching typically connotes both power and powerlessness, then shooting oneself might epitomize the full range of meanings subtly wired into all switching.

Anton Chekov said that, if you hang a rifle on the wall in the first act, you had better fire it by the second or third. The rifle hanging above this story has not sounded yet, but it looms ominously over everyone, for it is big enough to destroy the world: "The Button," or "The Big Red Button," that fanciful synecdoche for a global system of nuclear destruction triggered by a single finger. Never mind that nuclear weapons would be fired by very different procedures; the point is that people have preferred to imagine nuclear destruction in the form of a switch, as if the world had a gun to its head. That telling metaphor compresses many of the qualities of binary switches previously considered and magnifies them to the largest possible proportions. It implies an intuitive grasp of the Anscombe/Davidson theory of action discussed in chapter 14, in which the primitive movement of a finger is already the final result. It magnifies agency to its maximum extent, in a single act that engulfs the world. And it frankly acknowledges that this most extreme model of agency is utterly self-negating, even suicidal. Because every metaphor is an embryonic theory, the metaphor of The Button condenses the deepest contradictions of binary switching and also reveals that most people have an intuitive grasp of them already. It is not just that the button is part of the bomb, but also that the bomb is part of the button.

The Button is all or nothing, but also all *and* nothing. It is the point where dualisms collapse: maximal and minimal, subject and object, master and slave, life and death. Everything becomes nothing. Alternatives fuse into identities. Binary distinctions become analogical equivalents. At the culmination of the logic of binary switching lies this mutually assured destruction of binary options, the destruction of binary opposition *itself,* the point where *on* becomes *off.* Fortunately, we are not quite at that stage yet. But if switching's legacy of rationalization, individualization, and domination is any guide, the germ of destruction might yet lurk in those tiny machines whose promise of human mastery also quietly presages: the end.

Notes

Introduction

1. A number of literary critics have taken an interest in the mechanization of the writing process, from Hugh Kenner's *The Mechanical Muse* (Oxford: Oxford University Press, 1987) to Matthew Kirschenbaum's *Track Changes: A Literary History of Word Processing* (Cambridge, Mass.: Belknap, 2016). See also Thomas S. Mullaney, *The Chinese Typewriter: A History* (Cambridge, Mass.: MIT Press, 2017). For an influential account of the typewriter keyboard as mediating far more than just written language, see Friedrich Kittler, *Gramophone, Film, Typewriter,* trans. Geoffrey Winthrop-Young and Michael Wutz (Stanford, Calif.: Stanford University Press, 1999).

2. The computer buttons are certainly undercounted. The keystroke logging program did not record passwords entered or administrative key combinations. It also did not record the pressing of the backspace or delete key, probably a significant number as well. Other keys not recorded include arrows, function keys to control sound volume, and escape.

3. Ferdinand de Saussure, *Course in General Linguistics,* trans. Wade Baskin (New York: Philosophical Library, 1959), 120.

4. On binary opposition in anthropology, see the introduction to Claude Lévi-Strauss, *The Raw and the Cooked,* trans. John and Doreen Weightman (New York: Octagon, 1979), and Lévi-Strauss, "The Structural Study of Myth," *Journal of American Folklore* 68, no. 270 (1955): 428–44. The influence of Hegelian dialectics is more evident in Lévi-Strauss than in Saussure. For instance, Lévi-Strauss notes in "Structural Study," 440: "Mythical thought always works from the awareness of oppositions towards their progressive mediation."

5. Whitehead's rejection of Cartesian dualism was at the base of his own sweeping attempt to describe reality as an ongoing becoming at the level of experience. Whitehead's system resists summary or condensation, especially by Whitehead himself, but for two sections that convey key ideas with greater than usual clarity and compression see Alfred North Whitehead, *Process and Reality,* ed. David Ray Griffin and Donald W. Sherburne (New York: Free Press, 1978), 79–82 and 208–15. See also Gilles Deleuze and Félix Guattari, *A Thousand Plateaus: Capitalism and Schizophrenia,* vol. 2 of *Capitalism and Schizophrenia,* trans. Brian Massumi (Minneapolis: University of Minnesota Press, 1987), 5, where the authors oppose the rhizomatic to other root metaphors that remain grounded in a "binary logic" of branching. Finally, see Donna J. Haraway, "A Cyborg Manifesto," chapter 8 of *Simians, Cyborgs, and Women: The Reinvention of Nature* (New York: Routledge, 1991), 149–81, where the figure of the cyborg disrupts "certain dualisms" that

"have been persistent in Western traditions," including "self/other, mind/body, culture/nature, male/female," and others.

6. Taylor Dotson, "Technology, Choice and the Good Life: Questioning Technological Liberalism," *Technology in Society* 34, no. 4 (2012): 331.

7. Langdon Winner, "Do Artifacts Have Politics?," *Daedalus* 109, no. 1 (1980): 128.

8. For a useful introduction to cultural techniques, see the special issue of *Theory, Culture & Society* 30, no. 6 (2013), especially Geoffrey Winthrop-Young, "Cultural Techniques: Preliminary Remarks" (3–19), Bernhard Siegert, "Cultural Techniques: Or the End of the Intellectual Postwar Era in German Media Theory" (48–65), Bernard Dionysius Geoghegan, "After Kittler: On the Cultural Techniques of Recent German Media Theory," (66–82), and Jussi Parikka, "Afterword: Cultural Techniques and Media Studies" (147–59).

9. Winthrop-Young, "Cultural Techniques," 5.

10. Bernhard Siegert, *Cultural Techniques: Grids, Filters, Doors, and Other Articulations of the Real,* trans. Geoffrey Winthrop-Young (New York: Fordham University Press, 2015), 9.

11. Siegert, *Cultural Techniques,* xv.

12. Siegert, 6.

13. Siegert, 9.

14. Bernard Stiegler, *Technics and Time: The Fault of Epimetheus,* trans. Richard Beardsworth and George Collins (Stanford, Calif.: Stanford University Press, 1998), 150.

15. Cited from the English translation in Siegert, *Cultural Techniques,* 11; originally in Thomas Macho, "Zeit und Zahl: Kalender- und Zeitrechnung als Kulturtechniken," in *Bild—Schrift—Zahl,* ed. Sybille Krämer and Horst Bredekamp (Munich: Fink, 2003), 179.

16. For one expression of this metaphor, see Marshall McLuhan, *Counterblast* (London: Rapp and Whiting, 1969), 22: "Media effects are environments as imperceptible as water to a fish, subliminal for the most part."

17. Caetlin Benson-Allott, *Remote Control* (New York: Bloomsbury, 2015).

18. Jocelyn Szczepaniak-Gillece, *The Optical Vacuum: Spectatorship and Modernized American Theater Architecture* (Oxford: Oxford University Press, 2018).

19. David Parisi, *Archaeologies of Touch: Interfacing with Haptics from Electricity to Computing* (Minneapolis: University of Minnesota Press, 2018).

20. Roger Moseley, *Keys to Play: Music as a Ludic Medium from Apollo to Nintendo* (Oakland: University of California Press, 2016).

21. Rachel Plotnick, *Power Button: A History of Pleasure, Panic, and the Politics of Pushing* (Cambridge, Mass.: MIT Press, 2018).

22. Plotnick, *Power Button,* xix.

23. See, e.g., Till A. Heilmann, "'Tap, tap, flap, flap': Ludic Seriality, Digitality, and the Finger," *Eludamos* 8, no. 1 (2014): 33–46; Søren Pold, "Button," in *Software Studies: A Lexicon,* ed. Matthew Fuller (Cambridge, Mass.: MIT Press, 2008), 31–36.

1. Origin Stories

1. Jay David Bolter and Richard Grusin, *Remediations: Understanding New Media,* rev. ed. (Cambridge, Mass.: MIT Press, 2000).

2. Ralph Waldo Emerson, *Essays and Lectures* (New York: Library of America, 1983), 457.

3. The origins of the word are not entirely clear. The *Oxford English Dictionary* traces it to Old

French *boton* (Modern French *bouton*), for *bud, knob, button,* with related cognates in other Romance languages. But it also associates it with late Latin **bottare, buttare,* "to thrust, put forth" (* = no record of actual use). By this lineage, Modern English *button* may be recognized as a relative both of the noun *bud* in Modern English, a physical form, and the verb *butt,* a pushing or thrusting motion away from the body.

4. Friedrich Diez suggested that *button* in Modern English and *bouton* in Modern French are derived through the French from Middle High German *bozen,* to strike (*An Etymological Dictionary of the Romance Languages: Chiefly from the German* [London: Williams and Norgate, 1864]).

5. Rachel Plotnick, *Power Button: A History of Pleasure, Panic, and the Politics of Pushing* (Cambridge, Mass.: MIT Press, 2018), xviii.

6. William Fothergill Cooke and Charles Wheatstone, Electric Telegraphs, GB Patent 7,390, issued June 12, 1837, p. 4; Wheatstone and Cooke, Electro Magnetic Telegraph, U.S. Patent 1,622, issued June 10, 1842, p. 2.

7. Wheatstone and Cooke, Electro Magnetic Telegraph, 9. On the Cooke and Wheatstone telegraph, including the equipment depicted in Plate 1, see John Liffen, "The Introduction of the Electric Telegraph in Britain, a Reappraisal of the Work of Cooke and Wheatstone," *Journal for the History of Engineering and Technology* 80, no. 2 (2010): 268–99.

8. Joseph Henry, *The Scientific Writings of Joseph Henry,* vol. 2 (Washington, D.C.: Smithsonian Institution, 1886), 434.

9. Alfred Vail, *The American Electro Magnetic Telegraph: With the Reports of Congress, and a Description of All Telegraphs Known, Employing Electricity or Galvanism* (Philadelphia: Lea and Blanchard, 1847), 22.

10. Samuel Morse, Improvement in Electro-Magnetic Telegraphs, U.S. Patent 4,453, issued April 11, 1846, p. 2; Morse, Improvements in Electric Telegraphs, U.S. Patent 6,420, issued May 1, 1849, p. 2.

11. "Morse's Magnetic Telegraph," *Niles's Weekly Register,* June 22, 1844, 262.

12. "The American Electro Magnetic Telegraph," *Scientific American,* March 22, 1851, 211.

13. Alfred Vail, *The American Electro-Magnetic Telegraph* (Philadelphia: Lea and Blanchard, 1847), 35.

14. See J. Cummings Vail, *Early History of the Electro-Magnetic Telegraph: From the Letters and Journals of Alfred Vail* (New York: Hine, 1914), 24. A. Vail notes in an 1848 letter (recorded here by J. C. Vail) to his father that he had "made an entirely new key by which means there is no need for a button for connecting and closing the circuit independent of the key." He seems to mean that, instead of a metal lever with a raised button on the end, he had designed an interface consisting of the metal level alone.

15. J. J. Baranowski, *Nouveau système de voter au moyen d'un appareil dit: scrutateur mécanique* (Paris: L'Auteur, 1849), 12.

16. Baranowski, 22.

17. Baranowski, 12.

18. Baranowski, *Pour une montre à calcul opérant les quatre règles de l'arithmétique, brevets d'inveintion,* Patent 2,663, issued January 14, 1847, p. 87.

19. J. J. Baranowski, Reckoning Machine, U.S. Patent 5,746, issued September 5, 1848, p. 2.

20. David Hughes's printing telegraph, which we will encounter in ch. 11, reused clothing

fasteners as keys (Ivor Hughes and David Ellis Evans, *Before We Went Wireless: David Edward Hughes FRS, His Life, Inventions, and Discoveries* [Bennington, Vt.: Images from the Past, 2011], 34).

21. For instance, see Taliaferro P. Shaffner, *The Telegraph Manual* (New York: D. Van Nostrand, 1867), 439 and elsewhere, which consistently refers to "circuit closers" or "circuit changers."

22. M. G. Farmer, Electric Telegraph, U.S. Patent 9,634, issued March 29, 1853, p. 3.

23. *Eclectic Medical Journal* 2, no. 1 (1856): 13.

24. George B. Prescott, *History, Theory, and Practice of the Electric Telegraph* (Boston: Ticknor and Fields, 1860), 322.

25. "Permutating Telegraph Switch," *Scientific American,* March 19, 1864, 177. The term *switch* had appeared in a patent for the same device just a year before: John Lewis, Improvement in Switches for Telegraphs, U.S. Patent 40,346, issued October 20, 1863, p. 1. Interestingly, although Shaffner describes a wide range of battery switches as "circuit closers" or "circuit changers," he refers to one apparatus as a "switch board," in quotation marks, and notes that, unlike the others, it was meant to be mounted on a wall so that any operator in the room can see it (*Telegraph Manual,* 441).

26. Anton Huurdeman, *The Worldwide History of Telecommunications* (Hoboken, N.J.: John Wiley and Sons, 2003), 67.

27. Huurdeman, 70.

28. Edward Topsell, *The Historie of Foure-Footed Beastes* (London: William Iaggard, 1607), 509 (available through Early English Books Online, searchable at quod.lib.umich.edu/e/eebogroup/).

29. See Nicholas J. Conrad, Maria Malina, and Susanne C. Münzel, "New Flutes Document the Earliest Musical Tradition in Southwestern Germany," *Nature* 460 (August 6, 2009): 737–40.

2. Designing the Button

1. As Wiebe Bijker suggests in his influential study of the social construction of the bicycle, the "interpretive flexibility" of technical forms expresses the priorities of specific social groups. "There is no artifact not constituted by a relevant social group" (*Of Bicycles, Bakelites, and Bulbs: Toward a Theory of Sociotechnical Change* [Cambridge, Mass.: MIT Press, 1995], 76–77).

2. Beatriz Colomina and Mark Wigley, *Are We Human? Notes on an Archaeology of Design* (Zurich: Lars Müller, 2016), 69.

3. Larry D. Busbea, *The Responsive Environment: Design, Aesthetics, and the Human in the 1970s* (Minneapolis: University of Minnesota Press, 2020), 239.

4. John Harwood, *The Interface: IBM and the Transformation of Corporate Design, 1945–1976* (Minneapolis: University of Minnesota Press, 2011).

5. The number of poles refers to the number of circuits a single switch controls. A single-pole switch turns a single circuit *on* or *off,* while a double-pole switch controls two circuits at the same time. The number of throws indicates the number of output connections. A single-throw switch has only one output, and thus one operable position, such as the kind used to turn on a simple light. A double-throw switch has two outputs, and thus two operable positions, such as the kind used to operate a two-way light bulb, with different *on* positions for two different levels of illumination.

6. See the company's website for detailed descriptions of the various options at cherrymx.de/en.

7. See, however, Rachel Plotnick, *Power Button: A History of Pleasure, Panic, and the Politics of Pushing* (Cambridge, Mass.: MIT Press, 2018); also Till A. Heilmann, "'Tap, tap, flap, flap': Ludic Seriality, Digitality, and the Finger," *Eludamos* 8, no. 1 (2014): 33–46; also Søren Pold, "Button," in *Software Studies: A Lexicon,* ed. Matthew Fuller (Cambridge, Mass: MIT Press, 2008), 31–36.

8. For a useful summary of twentieth-century work on the topic, see James R. Lewis, Kathleen M. Potosnak, and Regis L. Magyar, "Keys and Keyboards," in *Handbook of Human-Computer Interaction,* ed. M. G. Helander, Thomas K. Landauer, and Prasad V. Prabhu (Amsterdam: Elsevier Science, 1997), 1285–315. An influential early study of the ergonomics of QWERTY keyboards was D. G. Alden, R. W. Daniels, and A. F. Kanarick, "Principles for Keyboard Design and Operation—A Summary Review," *Human Factors* 14, no. 4 (1972): 275–93. See also R. D. Kinkead and B. K. Gonzalez, *Human Factors: Design Recommendations for Touch-Operated Keyboards,* Document 12091-FR (Minneapolis, Minn.: Honeywell, 1969); Robert W. Monty, Harry L. Snyder, and Gerald G. Birdwell, "Keyboard Design: An Investigation of User Preference and Performance," in *Proceedings of the Human Factors Society 27th Annual Meeting* (Santa Monica, Calif.: Human Factors Society, 1983), 201–5; Texas Instruments, Inc., *VPI Keyboard Study* (Dallas: Design Development Center, 1983); R. D. Huchingson, L. J. Lampen, and R. J. Koppa, *Keyboard Design and Operation: A Literature Review,* vol. 1 of *Final Report on Project "QWERTY Literature Search and Research Recommendations"* (College Station: Texas Transportation Institute, 1985).

9. See, for instance, Richard C. Bodner, "A Comparison of Identification Rates of Static and Animated Buttons," in *Proceedings of the 1994 Conference of the Centre for Advanced Studies on Collaborative Research (CASCON '94)* (Toronto: IBM, 1994), 5.

10. Jason Alexander, John Hardy, and Stephen Wattam, "Characterising the Physicality of Everyday Buttons," *Proceedings of the Ninth ACM International Conference on Interactive Tabletops and Surfaces,* November 2014, 205–8.

11. John Durham Peters, Florian Sprenger, and Christina Vagt, *Action at a Distance* (Minneapolis: University of Minnesota Press, 2020), ix–xv.

12. See especially the chapter "Daemonic Interfaces, Empowering Obfuscations" in Wendy Hui Kyong Chun, *Programmed Visions: Software and Memory* (Cambridge, Mass.: MIT Press, 2011), 59–95.

3. Analogs and Analogies

1. The history of dualism, dialectics, and structuralist binary opposition is far too large to address adequately here, and these different philosophical matters should not be too quickly conflated. For instance, Saussure's binary opposition betrays little of the Hegelian instinct for synthesis, while Lévi-Strauss's involves a great deal. Similarly, Saussure regards the binary oppositions of language as fundamentally arbitrary, and thus historical and changeable, but other structuralists sometimes imagined them as universal structures of human consciousness.

2. On the innovations of binary computing in EDVAC by Presper Eckert, John Mauchly, and John von Neumann, see Thomas Haigh and Paul Ceruzzi, *New History of Modern Computing* (Cambridge, Mass.: MIT Press, 2021), 15–17. On ENIAC, see Thomas Haigh, Mark Priestley, and Crispin Rope, *ENIAC in Action: Making and Remaking the Modern Computer* (Cambridge, Mass.: MIT Press, 2016).

3. Gilles Deleuze and Félix Guattari, *A Thousand Plateaus: Capitalism and Schizophrenia,* vol. 2 of *Capitalism and Schizophrenia,* trans. Brian Massumi (Minneapolis: University of Minnesota Press, 1987), 5.

4. Jacques Derrida, *Positions,* trans. Alan Bass (Chicago: University of Chicago Press, 1981), 41.

5. John von Neumann quoted in *Cybernetics: The Macy Conferences 1946–53: The Complete Transactions,* ed. Claus Pias (Zurich and Berlin: Diaphanes, 2016), 181.

6. Lev Manovich, *The Language of New Media* (Cambridge, Mass.: MIT University Press, 2000).

7. Benjamin Peters, "Digital," in *Digital Keywords: A Vocabulary of Information Society and Culture,* ed. Benjamin Peters (Princeton, N.J.: Princeton University Press, 2016), 95.

8. Alexander R. Galloway, *Laruelle: Against the Digital* (Minneapolis: University of Minnesota Press, 2014), 52.

9. McKenzie Wark, *Gamer Theory* (Cambridge, Mass.: Harvard University Press, 2007), 92.

10. For an excellent summary of the Macy Conference debates about the analog and the digital, and a clear account of the stakes for the various feuding participants, see Claus Pias, "Analog, Digital, and the Cybernetic Illusion," *Kybernetes* 34, no. 3/4 (2005): 543–50. Many of those trying to expand the analog were partisans on its behalf, Pias argues. In contrast, partisans of the digital like Warren McCulloch and von Neumann tried to turn the analog into "forbidden ground" by treating transitional states as if they did not exist (546).

11. Ralph W. Gerard, "Some of the Problems concerning Digital Notions in the Central Nervous System," in Pias, *Cybernetics,* 181–82.

12. See Jonathan Sterne, "Analog," in Peters, *Digital Keywords,* 31–44. Sterne's decade-by-decade history of the changing meaning of the term *analog* is the most detailed and reliable to date.

13. Jonathan Sterne, "The Death and Life of Digital Audio," *Interdisciplinary Science Reviews* 31, no. 4 (2006): 345–46.

14. Sterne, "Analog," 41. For a strong and technically detailed defense of analog recording in the digital age, see Aden Evans, *Music, Machines, and Experience* (Minneapolis: University of Minnesota Press, 2005).

15. Florian Sprenger, "Temporalities of Instantaneity: Electric Wires and the Media of Immediacy," in *Action at a Distance,* ed. John Durham Peters, Florian Sprenger, and Christina Vagt (Minneapolis: University of Minnesota Press, 2020), 22.

16. Note, however, that the wave metaphor is itself part of the symbolic rehearsal of the analog, which equates the analog with continuous movement of a fluid medium; see Sterne, "Analog," 41.

17. See Keith J. Holyoak and Paul Thagard, *Mental Leaps: Analogy in Creative Thought* (Cambridge, Mass.: MIT Press, 1995), 7: "Analogy must be recognized as a source of plausible conjectures, not irrefutable conclusions." See also Paul Bartha, "Analogy and Analogical Reasoning," in *The Stanford Encyclopedia of Philosophy,* Winter 2016, ed. Edward N. Zalta, plato.stanford.edu/archives/win2016/entries/reasoning-analogy/.

18. Holyoak and Thagard, *Mental Leaps,* 186–88.

19. See, e.g., the essays collected in *The Analogical Mind: Perspectives from Cognitive Science,* ed. Dedre Gentner, Keith J. Holyoak, and Boicho N. Kokinov (Cambridge, Mass.: MIT Press,

2001), especially those by Usha Goswami on children and by Thagard and Cameron Shelley on affect.

20. Nelson Goodman, "Seven Strictures on Similiarity," in *Problems and Projects* (Indianapolis, Ind.: Bobbs-Merrill, 1972), 437.

21. Wendy Hui Kyong Chun, *Programmed Visions: Software and Memory* (Cambridge, Mass.: MIT Press, 2011), 129.

4. The Point of Touch

1. Benjamin Peters, "Digital," in *Digital Keywords: A Vocabulary of Information Society and Culture,* ed. Benjamin Peters (Princeton, N.J.: Princeton University Press, 2016), 93–108.

2. Peters, 99.

3. Those debates are far too extensive to review in any detail here. Suffice it to say, there is considerable disagreement about exactly what counts as an indexical. For recent debates, see the foundational article by David Kaplan, "Demonstratives: An Essay on the Semantics, Logic, Metaphysics, and Epistemology of Demonstratives and Other Indexicals," in *Themes from Kaplan,* ed. Joseph Almog, John Perry, and Howard Wettstein (Oxford: Oxford University Press, 1989), 481–563. For an overview of recent debates, see David Braun, "Indexicals," *The Stanford Encyclopedia of Philosophy,* Summer 2017, ed. Edward N. Zalta, plato.stanford.edu/archives/sum2017/entries/indexicals/.

4. Peirce's semiotics changed over his long career, as did his terminology. For an early account, see "On a New List of Categories," in *The Writings of Charles S. Peirce: A Chronological Edition,* vol. 2, *1867–1871,* ed. Edward C. Moore et al. (Bloomington: Indiana University Press, 1984), 49–59. From his middle career, see "On the Algebra of Logic: A Contribution to the Philosophy of Notation," in *Writings,* vol. 5, *1884–1886,* ed. Edward C. Moore et al. (Bloomington: Indiana University Press, 1993), 163–203. For a later statement (but by no means his last word), see "What is a Sign?," in *The Essential Peirce: Selected Philosophical Writings,* vol. 2, *1893–1913,* ed. Nathan Houser et al. (Bloomington: Indiana University Press, 1998), 4–10.

5. Peirce, "On the Algebra of Logic," 163.

6. Vilém Flusser, *Gestures,* trans. Nancy Ann Roth (Minneapolis: University of Minnesota Press, 2014), 3.

7. There is reason to doubt that the canonical indexical gesture is universal, as some have claimed, but among anthropologists and behavioral psychologists that debate continues. For instance, David Wilkins has objected that, in reference to societies that practice lip pointing rather than finger pointing, two unwarranted assumptions tend to be made. The first is that lip pointing predominates only because taboos restrict finger pointing; the second is that lip pointing is less precise, so less useful ("Why Pointing is Not Universal," in *Pointing: Where Language, Culture, and Cognition Meet,* ed. Sataro Kita [Mahwah, N.J.: Erlbaum, 2003], 174–75). He notes that there simply is no research about whether children in cultures that employ lip pointing exclusively go through a stage of pointing with the index finger first (179).

8. Adam Kendon, *Gesture: Visible Action as Utterance* (Cambridge: Cambridge University Press, 2004), 208.

9. Still, researchers cannot even agree on whether indication is distinctively human, given that some animals seem to understand it and perform it (such as those dogs called *pointers*) and

that some primates can be trained to perform it and may even develop the ability to perform it in captivity. However, primates in their native habitats have not been observed to indicate with a finger thus far. See David A. Leavens, William D. Hopkins, and Kim A. Bard, "Indexical and Referential Pointing in Chimpanzees *(Pan troglodytes),*" *Journal of Comparative Psychology* 110, no. 4 (1996): 346–53.

10. Peirce, "On the Algebra of Logic," 163.

11. The United States adapted Leete's poster for what became an equally famous one featuring the figure of Uncle Sam. In both cases, the clear violation of good manners in the poster signals the exceptional state of affairs.

12. Lev S. Vygotsky, "Genesis of Higher Mental Functions," in *Cognitive Development to Adolescence: A Reader,* ed. Ken Richardson and Sue Sheldon (East Sussex, UK: Open University Press, 1988), 71.

13. See also E. H. Leung and H. L. Rheingold, "Development of Pointing as a Social Gesture," *Developmental Psychology* 17 (1981): 215–20.

14. Elizabeth Bates, Laura Benigni, Inge Bretherton, Luigi Camaioni, and Virginia Volterra, "Cognition and Communication from Nine to Thirteen Months: Correlational Findings," in *The Emergence of Symbols: Cognition and Communication in Infancy* (New York: Academic Press, 1979), 69–140.

15. Nobuo Masataka, "From Index-Finger Extension to Index-Finger Pointing: Ontogensis of Pointing in Preverbal Infants," in Kita, *Pointing,* 69–84.

16. Charles Darwin, *The Expression of the Emotions in Man and Animals* (New York: Appleton, 1873), 272 (see also ch. 2 [50–65], on antithesis).

17. George Butterworth, "Pointing is the Royal Road to Language for Babies," in Kita, *Pointing,* 9–33.

18. Bruno Latour, "Where are the Missing Masses? The Sociology of a Few Mundane Artifacts," in *Shaping Technology / Building Society: Studies in Sociotechnical Change,* ed. Wiebe E. Bijker and John Law (Cambridge, Mass.: MIT Press, 1992), 225–58.

19. otis.com/en/us/products-services/products/ecall.

5. Counting on the Body

1. See, for instance, "Information Technology—Keyboard Layouts for Text and Office systems," in ISO 9995-4:1994 (Geneva: International Organization for Standardization, 1994); ITU-T E161 (Washington, D.C.: International Telecommunications Union, 2001). Arrangement of digits, letters, and symbols on telephones and other devices that can be used for gaining access to a telephone network.

2. See also recent and relevant work on haptic media, especially David Parisi, *Archaeologies of Touch: Interfacing with Haptics from Electricity to Computing* (Minneapolis: University of Minnesota Press, 2018).

3. Bernhard Siegert, *Cultural Techniques: Grids, Filters, Doors, and Other Articulations of the Real,* trans. Geoffrey Winthrop-Young (New York: Fordham University Press, 2015), 9.

4. For a similar account in a different theoretical idiom, see Jonathan Sterne, "Bourdieu, Technique and Technology," *Cultural Studies* 17, no. 3–4 (2003): 367–89. Sterne argues that "technologies are little crystallized parts of habitus," and that "they are structured by human practices so that they may in turn structure human practices" (376–77).

5. See Roger Moseley, *Keys to Play: Music as a Ludic Medium from Apollo to Nintendo* (Oakland: University of California Press, 2016).

6. Alexander R. Galloway, *Laruelle: Against the Digital* (Minneapolis: University of Minnesota Press, 2014), 52.

7. For an example, see Eric W. Rothenbuhler and John Durham Peters, "Defining Phonography: An Experiment in Theory," *Musical Quarterly* 81, no. 2 (1997): 242–64; D. A. Norman, *The Invisible Computer: Why Good Products Can Fail, the Personal Computer Is So Complex, and Information Appliances Are the Solution* (Cambridge, Mass.: MIT Press, 1998). For consideration of analog nostalgia in the context of sound studies, where vinyl records have struck some as more natural or even more real than digital music, see Aden Evans, *Sound Ideas: Music Machines, and Experience,* Theory Out of Bounds 27 (Minneapolis: University of Minnesota Press, 2005), 62–125; Jonathan Sterne, "The Death and Life of Digital Audio," *Interdisciplinary Science Reviews* 31, no. 4 (2006): 338–48.

8. Brian Massumi, *Parables for the Virtual: Movement, Affect, Sensation* (Durham, N.C.: Duke University Press, 2002), 133; Bernard Stiegler, *For a New Critique of Political Economy,* trans. D. Ross (Cambridge: Polity, 2010).

9. Ralph W. Gerard, "Some of the Problems Concerning Digital Notions in the Central Nervous System," in *Cybernetics: The Macy Conferences 1946–53: The Complete Transactions,* ed. Claus Pias (Zurich and Berlin: Diaphanes, 2016), 181–82.

10. For instance, see David Hume, *A Treatise of Human Nature* (Amherst, NY: Prometheus, 1992), where Hume mounts a skeptical critique of the continuousness of the self, which he regards as an absurd fiction designed to knit together the relational parts of experience into "some new and unintelligible principle, that connects the objects together, and prevents their interruption or variation," such as *soul,* and *self,* and *substance* (254). See also Immanuel Kant's challenge to Hume's argument in *Critique of Pure Reason,* ed. and trans. Paul Guyer and Allen W. Wood (Cambridge: Cambridge University Press, 1998), 245–66.

11. See Michael Lambek's recent work on "continuous and discontinuous persons," which distinguishes in more politically nuanced ways what I have been calling analog and digital persons ("The Continuous and Discontinuous Person: Two Dimensions of Ethical Life," *Journal of the Royal Anthropological Institute* 19, no. 4 [2013]: 837–58).

12. See Zoltán Kövecses, *Metaphor and Emotion: Language, Culture, and Body in Human Feeling* (Cambridge: Cambridge University Press, 2000).

13. William James, *The Principles of Psychology* (Cambridge, Mass.: Harvard University Press, 1983), 238.

14. James, 233.

15. Benjamin Peters, "Digital," in *Digital Keywords: A Vocabulary of Information Society and Culture,* ed. Benjamin Peters (Princeton, N.J.: Princeton University Press, 2016), 94.

16. Peters, 104.

17. See Brian Butterworth, *The Mathematical Brain* (London: Macmillan Education, 1999); Virginie Crollen, Xavier Seron, and Marie-Pascale Noël, "Is Finger-Counting Necessary for the Development of Arithmetic Abilities?," *Frontiers in Psychology* 2 (September 2011): 1–3; Andrea Bender and Sieghard Beller, "Nature and Culture of Finger Counting: Diversity and Representational Effects of an Embodied Cognitive Tool," *Cognition* 124, no. 2 (2012): 156–82.

18. See Bender and Beller, "Finger Counting"; Brian Butterworth, Robert Reeve, and Fiona

Reynolds, "Using Mental Representations of Space When Words are Unavailable: Studies of Enumeration and Arithmetic in Indigenous Australia," *Journal of Cross-Cultural Psychology* 42, no. 4 (2011): 630–38.

19. Bender and Beller, "Finger Counting," 158.

20. Bender and Beller, 158.

21. Karl Menninger, *Number Words and Number Symbols: A Cultural History of Numbers,* trans. Paul Broneer (Cambridge, Mass.: MIT Press, 1969), 201–17.

22. On its possible nonnecessity, see Crollen, Seron, and Noël, "Is Finger-Counting Necessary"; on its possible inutility see Elena Nicoladis, Simone Susanne Pika, and Paula Marentette, "Are Number Gestures Easier than Number Words for Pre-Schoolers?," *Cognitive Development* 25, no. 3 (2010): 247–61.

23. Aristotle, *De anima* 3.8, in *De Anima: Books II and III,* trans. D. W. Hamlyn (Oxford: Clarendon, 1968), 65.

24. Mark Paterson, "The Forgetting of Touch: Re-Membering Geometry with Eyes and Hands," *Angelaki* 10, no. 3 (2005): 121.

25. The debate amounts to whether the haptic is more or less rational, and hence more or less modern, than the optic. For an account of the rationalizing of touch in the nineteenth century, see David Parisi, "Tactile Modernity: On the Rationalization of Touch in the Nineteenth Century," in *Media, Technology, and Literature in the Nineteenth Century: Image, Sound, Touch,* ed. Colette Colligan and Margaret Linley (Farnham, UK: Ashgate, 2016), 189–213.

26. Menninger, *Number Words,* 375–77.

27. Paterson, "Forgetting of Touch," 115.

28. For an earlier Roman representation of the extended finger to connote calculation, see the images of the *Testamentum Relief* in Eve D'Ambra, "Mourning and the Making of Ancestors in the *Testamentum Relief,*" *American Journal of Archaeology* 99, no. 4 (1995): 667–81.

29. Stephen Johnston, "Making the Arithmometer Count," *Bulletin of the Scientific Instrument Society* 52 (1997): 12–21.

30. Johnston, 16; M. Sebert, *"Instruments de précision," Bulletin de la Sociéte d'encouragement pour l'industrie Nationale* 6 (1879): 406.

31. Ernst Martin, *The Calculating Machines (Die Rechenmaschinen): Their History and Development,* ed. and trans. Peggy Kidwell and Michael R. Williams (1925; repr. Cambridge, Mass.: MIT Press, 1992).

32. Johnston, "Arithmometer," 18.

33. James White, *A New Century of Inventions, Being Designs and Descriptions of One Hundred Machines Relating to Arts, Manufacture, and Domestic Life* (London: Leech and Cheetham, 1822); Silvio Henin, "Early Italian Computing Machines and Their Inventors," in *Reflections on the History of Computing: Preserving Memories and Sharing Stories,* ed. Arthur Tatnall (Heidelberg: Springer, 2012), 204–30; Dennis Roegel, "An Early (1844) Key-Driven Adding Machine," *IEEE Annals of the History of Computing* 30, no. 1 (2008): 59–65.

34. White, *New Century of Inventions,* 343.

35. White, 348.

36. J. A. V. Turck, *Origin of Modern Calculating Machines* (Chicago: Western Society of Engineers, 1921), 26.

37. Turck, 151.

38. James W. Cortada, *Before the Computer: IBM, NCR, Burroughs, and Remington Rand and the Industry They Created, 1865–1965* (Princeton, N.J.: Princeton University Press, 1993), 33–34.

6. Darth Vader's Nipples

1. See for instance Karel Capek's 1920 play *R.U.R.,* which introduced the word *robot,* and in which robots are artificial laborers. On the mechanical slave in literature and film, see Despina Kakoudaki, *Anatomy of a Robot: Literature, Cinema, and the Cultural Work of Artificial People* (New Brunswick, N.J.: Rutgers University Press, 2014). Sometimes mechanical slaves are part of a critique of the increasingly mechanical nature of modern labor, as in Fritz Lang's 1927 film *Metropolis,* in which the people of Metropolis must escape the deranging powers of industrial and technological modernity associated with the *Maschinenmensch.*

2. Kakoudaki, *Anatomy of a Robot,* 117.

3. Kakoudaki, 83.

4. Susan M. Squier, *Babies in Bottles* (New Brunswick, N.J.: Rutgers University Press, 1994), 95–6.

5. See Anne Balsamo, *Technologies of the Gendered Body: Reading Cyborg Women* (Durham, N.C.: Duke University Press, 1996); Allison Muri, *The Enlightenment Cyborg: A History of Communications and Control in the Human Machine 1660–1830* (Toronto: University of Toronto Press, 2007); Monica Casper, "Fetal Cyborgs and Technomoms on the Reproductive Frontier: Which Way to the Carnival?," in *The Cyborg Handbook,* ed. Chris Hables Gray, Heidi J. Figuerao-Sarrierra, and Steve Mentor (New York: Routledge, 1995), 183–202; Mary Ann Doane, "Technophilia: Technology, Representation, and the Feminine," in *Body/Politics: Women and the Discourses of Science,* ed. Mary Jacobus, Evelyn Fox Keller, and Sally Shuttleworth (New York: Routledge, 1990), 163–76. See also Robyn Rowland, *Living Laboratories: Women and Reproductive Technologies* (Bloomington: Indiana University Press, 1992); Jocelynne Scutt, *The Baby Machine: Reproductive Technology and the Commercialisation of Motherhood* (London: Green Print, 1990); Gena Corea, *The Mother Machine: Reproductive Technologies from Artificial Insemination to Artificial Wombs* (New York: Harper and Row, 1985).

6. Muri, *Enlightenment Cyborg,* 177.

7. For Muri, the mechanical woman shows up less in philosophy than in literature, where there is a long tradition of satirizing the artificiality of fashionable women, whose allegedly natural bodies are, as one poem she quotes puts it, "False in legs, and false in Thighs / False in Breast, Teeth, Hair and Eyes." Muri, *Enlightenment Cyborg,* 185.

8. See Muri, *Enlightenment Cyborg,* 171, where she points out that these ideas have been prevalent in recent science fiction films like Ridley Scott's *Alien,* where the ship's computer is named Mother, or Lana and Lilly Wachowski's *The Matrix,* named with the Latin term for *womb.*

9. On Vaucanson's and Jaquet-Droz's automatons, see Alfred Chapuis and Edmond Droz, *Automata: A Historical and Technological Study,* trans. Alec Reid (Neuchatel: Griffon, 1958), 273–77; Jessica Riskin, *The Restless Clock: A History of the Centuries-Long Argument over What Makes Living Things Tick* (Chicago: University of Chicago Press, 2016), 119–20.

10. Although Vaucanson later did make a human male piper and drummer, he is best remembered for his defecating duck, which consumed corn and simulated its elimination as excreta (Riskin, *Restless Clock,* 132–37).

11. Chapuis and Droz, *Automata,* 284.

12. Chapuis and Droz, 285.

13. Chapuis and Droz, 278.

14. On "philosophical toys," see Daniel Tiffany, *Toy Medium: Materialism and the Modern Lyric* (Berkeley: University of California Press, 2000), 34–62.

15. Peter Webb and Robert Short, *Death, Desire and the Doll: The Life and Art of Hans Bellmer* (Los Angeles: Solar, 2008), 23.

16. Hans Bellmer, *The Doll,* trans. Malcolm Green (London: Atlas, 2005), 40.

17. Bellmer, 37.

18. Bellmer, 40.

19. Bellmer, 42.

20. Freud published his longest statement on sexual fetishism just a few years before Bellmer published *Die Puppe,* and the doll suggests that Bellmer may have had more than a passing familiarity with it. In Freud's account, the sexual fetish results from the horror a young boy feels when he discovers his mother has no penis and fears he might lose his as well. To overcome this castration fear, he restores it to her symbolically, and later to all women, by eroticizing some other part of the body, recasting an alleged absence as an especially charged presence.

21. Sigmund Freud, "Fetishism," in *The Complete Psychological Works of Sigmund Freud,* trans. J. Strachey, vol. 21 (London: Hogarth, 1927), 149–57.

22. Ali Na, "The Fetish of the Click: A Small History of the Computer Mouse as Vulva," *Feminist Media Studies* 18, no. 2 (2018): 221–34.

23. Maryam Kamvar and Shumeet Baluja, "A Large Scale Study of Wireless Search Behavior: Google Mobile Search," *CHI Proceedings 2006,* April 22–27, 2006, 701–9.

24. There were a comparatively small number that continued to aspire to anthropomorphic fidelity, usually as clothed figures with simulated human flesh, such as the Westinghouse's African American 1930 Rastus, Andrew Bober's 1933 Willie, or the Franklin Institute's 1934 Egbert.

25. The heart is not entirely absent then or now, and seems to be more prominent in Japanese than in Western versions. See the example of the famous manga robot boy Atom, who has a heart in a compartment in his chest, and whose moral capacities exceed those of humans. See also the 1960s cyborg superhero Ultraman, whose glowing light on his chest indicates the state of his heart. One notable Western version is Marvel Comics' Iron Man, whose glowing mechanical sternum sustains his damaged heart.

26. Kakoudaki, *Anatomy of a Robot,* 118.

27. For a history of Televox, see Noel Sharkey and Amanda Sharkey, "Artificial Intelligence and Natural Magic," *Artificial Intelligence Review* 25 (2006): 9–19.

28. Few of these machines have been well documented, and photographic evidence of them is not well preserved. The best archive of images, historical documents, and information about twentieth-century androids and robots is Reuben Hoggett's Cybernetic Zoo website (cyberneticzoo.com).

29. "Mechanical Wonder Man is Operated by Radio Control," *Modern Mechanix,* December 1936; "Iron Man Comes to Life with Burgess Portable Power," *QST* 20, no. 8 (1936): 57.

30. See Timothy N. Hornyak, *Loving the Machine: The Art and Science of Japanese Robots* (Tokyo: Kodansha International, 2006), 38–39.

31. The ad ran in several small Texas newspapers as part of a 1930 promotional tour. See, for example, p. 16 of the *Corsicana Daily Sun* on May 21, 1930.

32. See also "'Radio Man' Walks, Talks, and Yodels," *Popular Science,* April 1939, 122.

33. Eric Lanz, *"Mit Sabor Unterwegs,"* in *Die Roboter Kommen! Mensch, Maschine, Kommunikation,* ed. Bodo-Michael Baumunk (Heidelberg: Braus, 2007), 66–67.

34. See Baumunk, *Die Roboter Kommen!,* 58–59, 96, 99, for examples, including Sabor lifting the skirt of a woman and posed with a woman in lingerie and a similar tin-man robot posed with a nude woman in a 1934 Paris striptease act.

35. There is a surprising lack of critical attention to the social meaning of male nipples in western societies. If men's bodies underwent the same scrutiny that women's typically do, much more would have been ventured.

36. In the racist and sexist pseudoscience of the earlier century, the conflation of race and gender categories was standard. Civilized people, identified with the descendants of certain white Europeans, demonstrated their alleged superiority through strict gender differentiation, while nonwhites were deemed to conflate male and female features and behaviors, such as the charge that the women in these societies were too active and violent and the men too emotional (Gail Bederman, *Manliness and Civilization: A Cultural History of Gender and Race in the United States 1880–1917* [Chicago: University of Chicago Press, 1995], 25).

37. *Forbidden Planet,* directed by Fred M. Wilcox (1956; Metro-Goldwyn-Mayer).

38. The cover headline is "Clink. Clank. Think." Although the story itself takes no interest in the potential dangers of autonomous or self-conscious computers, the cover illustration deploys what were, by this time, the rudiments of widely recognizable science fictional themes ("The Brain Builders," *Time,* March 28, 1955, 89–97). On Artzybasheff, see Domenic J. Iacono, "The Art of Boris Artzybasheff," *Scientific American,* November 11, 1993, 72–77.

39. Boris Artzybasheff, "Executive of the Future," *Esquire,* April 1952, 64. For another cover illustration for *Time* (January 23, 1950), Artzybasheff drew an anthropomorphic version of the Navy's Mark III computer with a bank of buttons at its center and two human arms extending from either side. One arm reaches forward to press a button on a teletype terminal, presumably used to control the computer itself.

40. The English translators of *The Book of the Thousand and One Nights* rendered the singular *Jinni* as *genie,* a term borrowed from French attestations dating to the early seventeenth century, and derived from Latin *genius,* or guardian spirit. The hiddenness of the otherwise corporeal *Jinni* thus transferred quite naturally to the invisible spirits of Christianity.

41. Sometimes derided by fans for their clunky appearance, the switches on Vader's chest resemble a less fanciful version that many people would have seen a decade before, on the Remote Control Unit for the Portable Life Support System (RCU-PLSS) used by the last three Apollo missions to the moon. Mounted high on the chest, the RCU is the control interface for a survival suit, with several large switches to control a fan, radio, water pump, and oxygen-purge system; see diagrams at /www.hq.nasa.gov/alsj/alsj-RCU.html.

7. The Keyboard's Checkered Past

1. See Roger Moseley, *Keys to Play: Music as a Ludic Medium from Apollo to Nintendo* (Oakland: University of California Press, 2016).

2. Rosalind Krauss, "Grids," *October* 9 (Summer 1979), 50–64; Curt Meine, "Inherit the Grid," in *Placing Nature: Culture and Landscape Ecology,* ed. Joan Iverson Nassauer (Washington, D.C.: Island, 1997), 45–62.

3. Brian Massumi, *Parables for the Virtual: Movement, Affect, Sensation* (Durham, N.C.: Duke University Press, 2002) 2–3; see also 133–43 (ch. 5: "On the Superiority of the Analog").

4. Bernhard Siegert, *Cultural Techniques: Grids, Filters, Doors, and Other Articulations of the Real,* trans. Geoffrey Winthrop-Young (New York: Fordham University Press, 2015), 98.

5. Leon Battista Alberti first proposed using the grid in perspectival drawing in *De Pictura* (On painting), published in 1435. Dürer's engraving of a draftsman using the grid appeared in his *Underweysung der Messung, mit dem Zirckel und Richtscheyt,* first published in 1525; however, the illustration was not included until the posthumous 1538 edition.

6. On the sexual politics of Dürer's engraving, see Donna Haraway, *Modest_Witness@Second_Millenium.FemaleMan©_Meets_OncoMouse™: Feminism and Technoscience* (New York: Routledge, 1997), 180–82. Summarizing feminist readings of Dürer's engraving, Haraway says this "visual technology was part of the apparatus for the production of modern gender, with its proliferating series of sexually charged oppositions condensed into the tension at the interface between touch and vision" (181). See also Lynda Nead, *The Female Nude: Art, Obscenity, and Sexuality* (New York: Routledge, 1992), 11.

7. Aristotle, *Metaphysics* 1.5, trans. C. D. C. Reeve (Indianapolis: Hackett, 2016).

8. Boethius, *Fundamentals of Music,* trans. Calvin M. Bower, ed. Claude V. Palisca (New Haven, Conn.: Yale University Press, 1989), 7.

9. Johannes Kepler, *Epitome of Copernican Astronomy and Harmonies of the World,* trans. Charles Glenn Wallis (New York: Prometheus, 1995), 209.

10. For a detailed account and lucid critique of Pythagorean theory, see Daniel Heller-Roazen, *The Fifth Hammer: Pythagoras and the Disharmony of the World* (New York: Zone, 2011).

11. Heller-Roazen, 14.

12. Max Weber, *The Rational and Social Foundations of Music,* ed. and trans. Don Martindale, Johannes Riedel, and Gertrude Neuwirth (Carbondale: Southern Illinois University Press, 1958), 117.

13. See David Suisman, "Sound, Knowledge, and the 'Immanence of Human Failure': Rethinking Musical Mechanization through the Phonograph, the Player-Piano, and the Piano," *Social Text* 28, no. 1 (2010): 13–34.

14. Weber, *Foundations of Music,* 122.

15. Bernard Brauchli, *The Clavichord* (Cambridge: Cambridge University Press, 1998), 21.

16. See Curt Sachs, *The History of Musical Instruments* (New York: Norton, 1940), 332. Sachs notes that the shortness of the keys compelled an awkward fingering, with the thumb rarely engaged.

17. For reproductions of those images, see Brauchli, *Clavichord,* 22–28.

18. Geoffrey Hindley, "Keyboards, Crankshafts and Communication: The Musical Mindset of Western Technology," in *Music and Technology in the Twentieth Century,* ed. Hans-Joachim Braun (Baltimore, Md.: Johns Hopkins University Press, 2002), 33–42. Hindley notes that, in Europe since the Middle Ages, the most elite classical instrument "has been an elaborate keyboard mechanism—organ, harpsichord, or piano" (34).

19. Hans-Joachim Braun, "Introduction: Technology and the Production and Reproduction of Music in the 20th Century," in *Music and Technology,* 9.

20. The history of early keyboards is far too large a topic to engage comprehensively here.

On the medieval organ, see especially Peter Williams, *The Organ in Western Culture: 750–1250* (Cambridge: Cambridge University Press, 1993).

21. On the organ recovered at Aquincum, see Werner Walcker-Mayer, *The Roman Organ of Aquincum,* trans. Joscelyn Godwin (Ludwigsburg: Musikwissenschaftliche, 1972). Only two fragments of keys survive. Originally attributed to the Alexandrian engineer and mathematician Ctesibios, the hydraulis was described in some detail by both Hero of Alexander and the Roman architect Vitruvius around the first century CE. Neither describes the keyboard or individual keys in any detail, but both acknowledge that the keys are spring-loaded and return to their original position by means of flexible pieces of horn or iron. Hero seems to be describing sliders that operate the valves, rather than keys. Archaeological evidence confirms that the keys were scaled to the fingers, not the entire hand, and the Aquincum organ from the third century CE clearly has pressable keys rather than slides. For translations and full commentaries of both Hero and Vitruvius, see Jean Perrot, *The Organ from Its Invention in the Hellenistic Period to the End of the Thirteenth Century* (Oxford: Oxford University Press, 1971), 3–22, 112–13.

22. On this history of the early medieval organ, see Williams, *Organ,* 137–53, 187–97.

23. See Williams, 304. On the representation of medieval keyboards in manuscripts, see the excellent compendium of sources in Kimberly Marshall, *Iconographic Evidence for the Late-Medieval Organ in French, Flemish and English Manuscripts,* 2 vols. (New York: Garland, 1989), esp. 1:83–89. Note that, in the Rutland Psalter illustration, another man in the foreground plays the hurdy-gurdy, a small stringed keyboard instrument operated by a crank. Its keys certainly are scaled to the fingers.

24. On the Norrlanda organ, however, B flat is included in the lower row with the naturals.

25. On the keyboard of the Norrlanda organ and the music in the Robertsbridge Codex, see Kimberly Marshall, "Organ," in *A Performer's Guide to Medieval Music,* ed. Ross W. Duffin (Bloomington: Indiana University Press, 2000), 421.

26. Although the original organ does not survive, the church of St. Burchardi in Halberstadt began performing the work, titled *As Slow As Possible,* on Cage's birthday in 2001. It will conclude in the year 2640.

27. David Kinsela, "The Capture of the Chekker," *The Galpin Society Journal* 51 (July 1998): 64. On the gift of the chekker from Edward III, see Brauchli, *Clavichord,* 14.

28. Edwin M. Ripin, "Towards an Identification of the Chekker," *The Galpin Society Journal* 28 (April 1975): 11–25.

29. Edmund A. Bowles, "On the Origin of the Keyboard Mechanism in the Late Middle Ages," *Technology and Culture* 7, no. 2 (1966): 153. See the illustration of the small harpsichord labelled an *"echiquier"* in Bowles's Fig. 1.

30. Kinsela, "Capture of the Chekker," 79–89. For similar arguments see Tess Knighton, "Another Chekker Reference," *Early Music* 8 (1980): 375–76; Christopher Page, "In the Direction of the Beginning," in *The Historical Harpsichord,* ed. Howard Schott, vol. 1 (Stuyvesant, UK: Pendragon, 1984), 111–25.

31. For thirty-one of those references, including the Gerson illustrations, see Ripin, "Towards an Identification." For references to the rest see Kinsela, "Capture of the Chekker," 64–85.

32. Ripin, "Towards an Identification," 14; see also 19–20 for the original French passages about the chekker.

33. Quoted in Kinsela, "Capture of the Chekker," 79.

34. Quoted in Kinsela, 83. The translation is from Nicolas Meeùs, "The Chekker," *The Organ Yearbook* 16 (1985): 5–25. Kinsela clarifies that the Latin name for the chekker is the *scacordum,* a compounding of *scacarium*—or checkerboard—with *chorda,* or strings (83).

35. Kinsela, "Capture of the Chekker," 66.

36. Kinsela, 84.

37. For an edition of the manuscript with a French translation and facsimiles of all illustrations, see *Les traités d'Henri-Arnaut de Zwolle et de divers anonymes (Paris: Bibliothèque nationale, MS Latin 7295),* ed. G. le Cerf (Kassel: Bärenreiter, 1972). Although some have suggested that *clavis* refers to the metal tangent that strikes the string, and not to the keys themselves, it should be noted that Arnaut's Latin description of the clavichord consistently uses *clavis* to refer to the keys.

38. See the examples in Brauchli, *Clavichord,* 21–28. He notes that a depiction from 1425 shows a clavichord with "twelve or thirteen keys, with no accidentals" (21). But just a few years later, around 1440 Arnaut's detailed description of a Clavichord clearly specifies all five accidentals, and surviving instruments from the later fifteenth century distinguish them by color.

39. The appearance of accidentals on keyboards is, in general, poorly documented, but Williams, *Organ,* suggests that large organs had fewer accidentals than smaller organs, perhaps because of their role in playing melody and counterpoint in traditional chants. He also notes that, as late as 1502, a large organ in Freiberg Cathedral still had no keys for C sharp and G sharp (308n22), well after all five accidentals were standard on stringed keyboard instruments. It may be that smaller, more nimble instruments like the chekker pioneered the use of all five accidentals, an arrangement that subsequently migrated back to organs. Williams makes a point that current theorists of cultural techniques would admire: he suggests that keyboards do not have twelve keys per octave because music theory requires it, but rather that "twelve notes produced a theory of music because that is what keyboards were coming to have" (308).

40. See Ripin, "Towards an Identification," 13–14; Meeùs, "Chekker," 5–25; for a dissent see Christopher Page, "The Myth of the Chekker," *Early Music* 7, no. 4 (1979): 482–89.

41. On the use of *abacus* in Mersenne and Kircher, see Ripin, "Identification of the Chekker," 24n15, and Meeùs, "Chekker," 8–10.

42. Karl Menninger, *Number Words and Number Symbols: A Cultural History of Numbers,* trans. Paul Broneer (Cambridge, Mass.: MIT Press, 1969), 340.

43. Menninger, 341.

44. Menninger, 341. The staff seems not to have been entirely unprecedented, however, and historians have debated whether Guido invented the musical staff or simply popularized it and systematized it. For a full account of these qualifications see John Haines, "The Origins of the Musical Staff," *Musical Quarterly* 91, no. 3/4 (2008): 327–78.

45. See Williams, *Organ,* 216. William of Malmesbury also claimed that Gerbert made a deal with the devil to steal mathematical knowledge from the Saracens (*Chronicles of the Kings of England from the Earliest Period to the Reign of King Stephen,* ed. and trans. J. A. Giles [London, 1866], 174).

46. Ripin, "Towards an Identification," 13–14.

47. See also the fourteenth-century illustration from London, British Library Sloane MS 213 depicting a gridded counting table, reproduced in Francis Pierrepont Barnard, *The Casting-*

Counter and the Counting-Board: A Chapter in the History of Numismatics and Early Arithmetic (Oxford: Clarendon, 1916), 237 and Plate XLIV.

48. Richard fitzNigel, *"Dialogus de scaccario": The Dialogue of the Exchequer, and "Constitutio domus regis": The Disposition of the King's Household,* ed. and trans. Emilie Amt and S. D. Church (Oxford: Oxford University Press, 2007), 9.

49. fitzNigel, 11.

50. Quotations from Marin Mersenne, *Harmonie Universelle: The Books on Instruments,* trans. Roger E. Chapman (The Hague: Martinus Nijhoff, 1957), 437.

51. Mersenne, 437.

8. Human Types

1. Emily I. Dolan, "Toward a Musicology of Interfaces," *Keyboard Perspectives* 5 (2012): 11.

2. Margery W. Davies, *Woman's Place Is at the Typewriter: Office Work and Office Workers 1870–1930* (Philadelphia: Temple University Press, 1982), appendix table 1. Data on women office workers does not always distinguish typists from other kinds of workers. In England, for example, census data shows a massive increase in the number of women clerical workers, but it does not subdivide them by the kinds of clerical work performed. See Meta Zimmeck, "Jobs for the Girls: The Expansion of Clerical Work for Women, 1850–1914," in *Unequal Opportunities: Women's Employment in England, 1800–1918,* ed. Angela V. John (Oxford: Blackwell, 1986), 171n4.

3. Women made up just 6 percent of bookkeepers, cashiers, and accountants in 1880, 29 percent by 1900, and about half by 1930 (Davies, *Woman's Place,* 52 and appendix table 1). However, census data does not distinguish between bookkeepers who did routine work and accountants who exercised more judgment and who tended to be men. The user manuals for keyboard calculating machines such as the Comptometer, crucial for basic bookkeeping by the early century, suggest that women gained more rapidly in this area than in others: manuals from the 1890s depict men using the machines; after 1920 they typically depict women.

4. Friedrich Kittler, *Gramophone, Film, Typewriter,* trans. Geoffrey Winthrop-Young and Michael Wutz (Stanford, Calif.: Stanford University Press, 1999), 195.

5. Davies, *Woman's Place,* 55.

6. The machine in the patent drawing and the even smaller patent model did not contain enough keys for every letter of the alphabet. The patent drawing depicts a machine with only twenty-one keys; the working model has even fewer.

7. Michael H. Adler, *The Writing Machine* (London: George Allen and Unwin, 1973), 123–30.

8. Adler, 99.

9. See Leon Plantinga, "The Piano and the Nineteenth Century," in *Nineteenth-Century Piano Music,* ed. R. Larry Todd (New York: Routledge, 2004), 1–15; Laura Vorachek, "'The Instrument of the Century': The Piano as an Icon of Female Sexuality in the Nineteenth Century," *George Eliot–George Henry Lewes Studies* 38/39 (September 2000): 26–43.

10. On the gendering of piano playing and performance, see E. Douglas Bomberger, Martha Dennis Burns, James Parakilas, Judith Tick, Marina Tsvetaeva and Mark Tucker, "The Piano Lesson," in James Parakilas with others, *Piano Roles: Three Hundred Years of Life with the Piano* (New Haven, Conn.: Yale University Press, 1999), 144: "Boys were trespassing into the female realm when they learned to play the piano at all, . . . whereas girls were trespassing into a male realm when they got too good at it." See also Arthur Loesser, *Men, Women, and Pianos: A Social*

History (New York: Simon and Schuster, 1954); Nicholas Temperley, "Ballroom and Drawing-Room Music," in *Music in Britain: The Romantic Age 1800–1914,* ed. Nicholas Temperley (London: Athlone, 1981), 109–34.

11. Although men might have accrued social status from the piano, they had to do so indirectly, because, according to Richard Sennett, even into the twentieth century the piano was considered "not quite manly" and men were encouraged to sing or play other instruments instead ("Pianists in Their Time: A Memoir," in *The Lives of the Piano,* ed. James R. Gaines [New York: Holt, 1981], 197).

12. Cyril Ehrlich, *The Piano: A History* (Oxford: Oxford University Press, 2000), 91. That rate was four times higher than in Germany and more than six times higher than in France, but even those countries had democratized and domesticated the piano to a considerable degree.

13. For the fullest account of the history of the image, see Peter Weil, "Oh, Lillian, We Hardly Knew Ye!: The First Photographic Portrait of a Typist," *ETCetera: Journal of the Early Typewriter Collector's Association,* no. 117 (2017): 9–13.

14. On the early history of Sholes's machine, see Adler, *Writing Machine*; Wilfred Beeching, *Century of the Typewriter* (New York: St. Martins, 1974); Arthur Toye Foulke, *Mr. Typewriter: A Biography of Christopher Latham Sholes* (Boston: Christopher, 1961); Darren Wershler-Henry, *The Iron Whim: A Fragmented History of Typewriting* (Ithaca, N.Y.: Cornell University Press, 2005).

15. The piano is a very decorous instrument, after all. The player touches it only at the extremities of the body, rather than cradling it in the lap, holding against the neck, bracing it between the legs, or touching it with the mouth. The body meets the piano only at the fingertips and feet, a modest technical interaction if ever there was one.

16. Sigmund Freud, "Civilization and Its Discontents," in *The Complete Psychological Works of Sigmund Freud,* trans. J. Strachey, vol. 21 (London: Hogarth, 1927), 91–92.

17. See Vivian Sobchack, "A Leg to Stand on: Prosthetics, Metaphor, and Materiality," in *The Prosthetic Impulse: From a Posthuman Present to a Biocultural Future,* ed. Marquard Smith and Joanne Morra (Cambridge, Mass.: MIT Press, 2006), 17–41; Sarah Jain, "The Prosthetic Imagination: Enabling and Disabling the Prosthesis Trope," *Science, Technology, and Human Values* 24, no. 1 (1999): 31–54; Steven L. Kurzman, "Presence and Prosthesis: A Response to Nelson and Wright," *Cultural Anthropology* 16, no. 3 (2001): 374–87.

18. Sobchack, "Leg to Stand on," 22.

19. For two useful critiques of the prosthetic theory, see Jain, "Prosthetic Imagination," and Mark Coté, "Technics and the Human Sensorium: Rethinking Media Theory through the Body," *Theory & Event* 13, no. 4 (2010), muse.jhu.edu/article/407142.

20. Marshall McLuhan, *Understanding Media: The Extension of Man* (1964; repr., Cambridge, Mass.: MIT University Press, 1994), 42. On prosthetics, see especially McLuhan's ch. 4, "The Gadget Lover" (41–47), and ch. 6, "Media as Translators" (56–61). Electrical technology, he argues, is "a live model of the central nervous system" housed outside the body but designed to protect against "violent and superstimulated" social experience (43). As such, the prosthesis of the electrical system is a "desperate and suicidal amputation" of the body's own organs, an attempt to distance and control what the body can no longer endure on its own.

21. Henry Ford, *My Life and Work* (New York: Doubleday, 1923), 108. For other considerations of this passage, see Mark Seltzer, *Bodies and Machines* (New York: Routledge, 1992), 157; Jain, "Prosthetic Imagination," 34.

22. Many of these positions are indebted to Jacques Derrida's theory of the *supplement* as expressed in his deconstruction of Jean-Jacques Rousseau's romantic humanism (". . . That Dangerous Supplement . . . ," in *Of Grammatology,* trans. Gayatri Chakrovorty Spivak [Baltimore, Md.: Johns Hopkins University Press, 1974], 141–64).

23. For other influential posthumanist developments related to prosthetic theory, see Mark B. N. Hansen, *Bodies in Code: Interfaces with Digital Media* (New York: Routledge, 2006), for whom the very term *prosthesis* signals a traditionalist ontology of the human; Don Ihde, *Bodies in Technology: Electronic Mediations* (Minneapolis: University of Minnesota Press, 2001), who insists that human embodiment presupposes technical mediation; Donna J. Haraway, "Cyborg Manifesto: Science, Technology, and Socialist-Feminism in the Late Twentieth Century," in *Simians, Cyborgs, and Women: The Reinvention of Nature* (New York: Routledge, 1991), 149–82, whose cyborgs unravel myths of the naturalness of racial and gender superiority; and N. Katherine Hayles, *How We Became Posthuman: Virtual Bodies in Cybernetics, Literature, and Informatics* (Chicago: University of Chicago Press, 1999), who sees the human body as the first prosthesis and the model for all that follow. Of these critics, only Hayles retains the term *prosthesis* for serious critical work; the rest are characteristic of more theorists recently, who, like Hansen, see the term as enforcing traditional distinctions between people and technology.

24. Bernard Stiegler, *Technics and Time: The Fault of Epimetheus,* trans. Richard Beardsworth and George Collins (Stanford, Calif.: Stanford University Press, 1998), 150.

25. Stiegler, 136. Note that, for Stiegler, the prosthetic extends to all technical objects. "There is nothing but prostheses: my glasses, my shoes, the pen, the diary, or the money in my pocket" (99).

26. As Raiford Guins and Omayra Zaragoza Cruz have noted in their study of turntables in African American culture, prosthetic theory should focus much more on "media as extensions of specific rather than supposedly universal bodies in time and place" (Raiford Guins and Omayra Zaragoza Cruz, "Prosthetists at 33 1/3," in *The Prosthetic Impulse: From a Post-Human Present to a Biocultural Future,* ed. Marquard Smith and Joanne Morra [Cambridge, Mass.: MIT Press, 2006], 233). The turntable, they imply, mediates whiteness and blackness differently, and may even help differentiate them from each other.

27. The point has been made many times in the history of machine writing, including by Kittler, *Gramophone,* 189: "Knie, Beach, Thurber, Malling Hansen, Ravizza: they all constructed their early typewriters for the blind and/or deaf."

28. Adler, *Writing Machine,* 56–7, and 162–70. Turri's machine does not survive, but its level of sophistication and even some details of its operation can be inferred from several surviving letters from his blind lover Countess Carolina Fantoni, which she wrote on the machine.

29. Adler, 57.

30. Adler, 66, 68, 83.

31. Adler, 85, 88.

32. Adler, 88, 104, 129.

33. Adler, 152. On Nietzsche's use of the writing ball, see Kittler, *Gramophone,* 202–6.

34. F. G. Hubert, Jr. "The Typewriter; Its Growth and Uses," *The Chautauquan* 8, no. 9 (888): 584.

35. "The Type-Writer: A Valuable Machine for Professional Men," *Harvard Register,* October 1880, vi.

36. "The Institute for the Blind: Managers for Next Year Chosen—Sightless Typewriters," *New York Times,* December 25, 1890, 3. See also Charlotte W. Howe, "How Blind Children See," *The Phrenological Journal and Science of Health* 103, no. 5 (1897): 233.

37. Helen Keller, *Selected Writings,* ed. Kim E. Nielsen (New York: New York University Press, 2005), 12 and 126.

38. Jain, "Prosthetic Imagination," 32.

39. Rosemarie Garland-Thomson, "Integrating Disability, Transforming Feminist Theory," in *Gendering Disability,* ed. Bonnie G. Smith and Beth Hutchison (Piscataway, N.J.: Rutgers University Press, 2004), 78.

40. Iris Marion Young, "Throwing Like a Girl: A Phenomenology of Feminine Body Comportment Motility and Spatiality," *Human Studies* 3, no. 2 (1980): 152.

41. Kittler, *Gramophone,* 184.

42. Tamara Plakins Thornton, *Handwriting in America: A Cultural History* (New Haven, Conn.: Yale University Press, 1996), 38.

43. Thornton, 58–59.

44. Davies, *Woman's Place,* appendix table 1.

45. Thomas C. Jepsen, *My Sisters Telegraphic: Women in the Telegraph Office, 1846–1950* (Athens: Ohio University Press, 2000), 9.

46. Jepsen, 53.

47. Margery W. Davies, "Women Clerical Workers and the Typewriter: The Writing Machine," in *Technology and Women's Voices: Keeping in Touch,* ed. Cheris Kramarae (New York: Routledge, 1988), 31.

48. Davies, *Woman's Place,* 56.

49. Davies, "Women Clerical Workers," 32.

50. See a reproduction of the image in Wershler-Henry, *Iron Whim,* 96.

51. For a characteristic summary of arguments that women lacked these qualities, see William T. Sedgwick, "Feminist Revolutionary Principle is Biological Bosh," *New York Times Magazine,* January 18, 1914, 2.

52. Anne Balsamo, *Technologies of the Gendered Body: Reading Cyborg Women* (Durham, N.C.: Duke University Press, 1996), 9.

9. Chording and Coding

1. Douglas C. Engelbart, *Augmenting Human Intellect: A Conceptual Framework* (Menlo Park, Calif.: Stanford Research Institute, 1962), 1.

2. Thierry Bardini, *Bootstrapping: Douglas Engelbart, Coevolution, and the Origins of Personal Computing* (Stanford, Calif.: Stanford University Press, 2000).

3. Bardini, 134, 137.

4. Thomas Haigh, Mark Priestley, and Crispin Rope, *ENIAC in Action: Making and Remaking the Modern Computer* (Cambridge, Mass.: MIT Press, 2016), 74–5.

5. See chapter two of Bardini, *Bootstrapping,* 59–80. Bardini's illuminating account includes a wealth of historical detail on the sources of Engelbart's thinking, the opposition it encountered, and his contemporaries' attempts to understand it.

6. Thomas S. Mullaney, *The Chinese Typewriter: A History* (Cambridge, Mass.: MIT Press, 2017), 240.

7. Mullaney, 239.

8. For the Twiddler3, see twiddler.tekgear.com. The most successful chorded keyset to date was probably the Microwriter from 1978, created by blacklisted Hollywood director Cy Enfield. Not just a keyboard but a portable word processor, the Microwriter enjoyed moderate success but never seriously threatened QWERTY's dominance. Its more modern descendent, the CyKey—named after Enfield—remains on the market through a bare-bones Google site. Many other contenders have come and gone, including the Writehander from 1978, the Ergoplic Octima Chord Keyboard from 1982, and the Infogrip Bat from 1991. For a fuller description of these and other chorded keysets, see Microsoft's online museum of Bill Buxton's collection of input and interactive technologies at microsoft.com/buxtoncollection/.

9. Anna Maria Feit and Antti Oulasvirta, "PianoText: Redesigning the Piano Keyboard for Text Entry," *Proceedings of the 2014 Conference on Designing Interactive Systems* (New York: Association for Computing Machinery, 2014), 1045–54 (for supplemental materials to the article, including a demonstration video, see dl.acm.org/doi/10.1145/2598784.2602800).

10. Feit and Oulasvirta, 1045.

11. Michael H. Adler, *The Writing Machine* (London: George Allen and Unwin, 1973), 105–6. See also the American version of Zucco's patent: A. Michela, Type-Writing Machine, U.S. Patent 212,724, filed November 1, 1878, and issued February 25, 1879. A useful summary of the machine and its system is *Il Sistema Michela* (Rome: Senato della Repubblica, 2018). For other stenographic machines with musical keyboards, see Martin Ernst, *Die Schreibmaschine und Ihre Entwicklungsgeschichte* (Pappenheim: Meyer, 1949), 400–12; these include the 1871 Clafigrafo, the 1878 Stentiposillabica, the 1904 Tacheografo, and the 1905 Shortwriter.

12. Both Berry's and Dujardin's devices also output code consisting of punched or printed dots on rolls of paper; see Adler, *Writing Machine,* 65 and 67.

13. Adler, 123–30.

14. Brian Bowers, *Sir Charles Wheatstone FRS: 1802–1875* (London: Science Museum, 2001), 179–80.

15. See Samuel W. Francis, Writing Machine, U.S. Patent 18,504, issued October 27, 1857. Like Sholes, Pratt also eventually abandoned the musical keyboard for an interface using rows of round buttons. Sholes was aware of Pratt's efforts, which he read about in *Scientific American,* and may have followed his lead (Adler, *Writing Machine,* 99–102; 106–7).

16. This includes Benton Halstead's device, which he reportedly used for six years in his office before patenting it in 1872, and Copenhagen inventor I. A. Peters's device from the 1860s, roughly the size of a table (Adler, *Writing Machine,* 112–13). This excludes the many printing telegraphs that used musical keyboards, some of which will be considered in the next chapter, starting with Jacob Brett's in 1828, Royal House's in 1846, and David Hughes's in 1855 (81).

17. There are twenty-six letter keys, any one of which can be combined with one of the remaining twenty-five keys, for a total of 650 combinations. However, if these can be pressed in any order, $T + H$ registers as the same as $H + T$, which means that half of the combinations (325) would be identical to the other half.

18. The gains are small but add up quickly. For someone who works eight hours each day at the computer, saving just one minute each hour by mastering keyboard shortcuts would free up almost an entire work week over the course of a year.

19. The principle of Livermore's machine loosely resembles that of the Raphigraphe devised

by Pierre Foucauld in France around 1839, which similarly used chorded inputs to print parts of letters as groups of dots (Adler, *Writing Machine,* 68). See also Henri Gensoul's 1866 Presse Stenographique, which purportedly did print both syllables and words with chorded inputs using an interface resembling a piano keyboard (116–17). See also the immense 1864 writing machine by a French inventor named Bryois that employed the chording principle on a ranked interface consisting of almost two hundred round button-shaped keys (114–15).

20. Benjamin Livermore, Hand Printing Device, U.S. Patent 39,296, issued July 21, 1863, p. 1.

21. This modern reproduction of Livermore's typeface is courtesy of Johannes Lang, its designer. The letter forms approximate Livermore's originals as they appear in his patent.

22. There has been very little written about Livermore. For the fullest account, including photographs of Livermore with his machine, see Jos Legrand, "Among the Jostling Crowd: Livermore's Permutation Typograph," *ETCetera* 81 (March 2008): 3–5.

23. N. Katherine Hayles, *How We Became Posthuman: Virtual Bodies in Cybernetics, Literature, and Informatics* (Chicago: University of Chicago Press, 1999), 199.

24. As Aristotle said, "Those things therefore which are in the voice are symbols of the passions of the soul, and when written, are symbols of the (passions) in the voice" (from *On Interpretation,* in *The Organon, Or Logical Treatises of Aristotle,* vol. 1, trans. Octavius Freire Owen [London: Bohn, 1908], 46–47). In other words, speech symbolizes directly while writing is a secondary symbol referring back to speech.

25. Amédée Guillemin, *Electricity and Magnetism,* ed. Silvanus P. Thompson (London: Macmillan, 1891), 653.

26. On the Baudot system, see Robin Boast, *The Machine in the Ghost: Digitality and its Consequences* (London: Reaktion, 2017), 28–35; Ken Beauchamp, *History of Telegraphy* (London: Institute of Engineering and Technology, 2001), 390–96; and various early historical accounts collected in *The Electric Telegraph: An Historical Anthology,* ed. George Shiers (New York: Arno, 1977).

27. Shiers, *Electric Telegraph,* 653.

28. Vilém Flusser, "Crisis of Linearity," trans. Adelheid Mers, *Boot Print* 1, no. 1 (2007): 19–21.

29. Flusser, 21.

30. Friedrich Kittler, "Number and Numeral," *Theory, Culture & Society* 23, no. 7–8 (2006): 56. On related topics, see also his interview in John Armitage, "From Discourse Networks to Cultural Mathematics: An Interview with Friedrich A. Kittler," *Theory, Culture & Society* 23, no. 7–8 (2006): 17–38.

31. Kittler, "Number and Numeral," 57.

32. Murray's machine created paper punch tapes on both ends, so transmitting consisted first of typing a message and creating a paper tape that recorded the message in binary code, then feeding that paper tape through a transmitter to send the code to the receiver. Murray revised Baudot's 5-bit code as well. Baudot designed his code to make fewer demands on operators by associating the most frequently used letters with the simplest finger inputs. The commonly used letter *E* was 01000, requiring the operator to press one key. Less frequently used letters employed more complex multifinger patterns such as 11101 for *V.* Because Murray's typewriter input did not require the operator to incorporate binary patterns, he revised the 5-bit code to reduce wear on the machinery rather than on the operators.

33. Unlike 5-bit Baudot and Murray codes, ASCII is a 7-bit code, which permits 128 individual

combinations, enough for separate upper and lowercase letters, numbers, punctuation marks, and other special characters, but hardly enough for the characters required by all natural languages. The current standard of 16-bit Unicode permits more than 1.1 million combinations.

34. Other typewriting techniques persist onscreen as well, such as the ruler, margin scale, and margin stops, which are little changed from the earliest Remington typewriters.

35. Matthew Kirschenbaum's *Track Changes: A Literary History of Word Processing* (Cambridge, Mass.: Belknap, 2016).

10. The Archeology of QWERTY

1. Paul A. David, "Clio and the Economics of QWERTY," *American Economic Review* 75, no. 2 (1985): 332. A number of other economists, historians, and theorists have eagerly taken up the issue of path dependency, only sometimes using the example of QWERTY, including Paul Krugman, *Peddling Prosperity* (New York: Norton, 1995); Robert H. Frank and Philip J. Cook, *The Winner-Take-All Society: Why the Few at the Top Get So Much More Than the Rest of Us* (New York: Penguin, 1995).

2. For a skeptical history of neoclassical orthodoxy, see Philip Mirowski, *More Heat than Light: Economics as Social Physics, Physics as Nature's Economics* (Cambridge: Cambridge University Press, 1991). The claim that economy is a version of physics is, for Mirowski, an ideologically laden denial that economic relations are social and historical.

3. Bruce Bliven Jr., *The Wonderful Writing Machine* (New York: Random House, 1954).

4. For the most influential version of this counter argument, see Stan Liebowitz and Stephen E. Margolis, "The Fable of the Keys," *Journal of Law and Economics* 33, no. 1 (1990): 1–25. Liebowitz and Margolis's argument is based heavily on the lack of advantages by the Dvorak keyboard in particular, but it is less persuasive in regard to the potential advantages of other possible alternatives.

5. For an instructive contrast, see Thomas S. Mullaney, *The Chinese Typewriter: A History* (Cambridge, Mass.: MIT Press, 2017), 237–43. Mullaney shows how the QWERTY keyboards used for Chinese computers operate on very different principles, involving more complexly mediated "input method editors," by which QWERTY input triggers the appearance of a menu of possible Chinese characters from which the user can select.

6. Other staggered arrangements are possible. Compact computer keyboards often use a regular 1/4 key stagger for all rows. The Commodore 128, an early personal computer introduced in 1985, used a 1/2 key stagger across all rows. Other variations are also possible, but even slight changes from the traditional form can compel painful adjustments for touch typists.

7. The claim reappears repeatedly in popular sources, and in some professional typewriter histories both early and late. See William E. Cooper, "Introduction," in *Cognitive Aspects of Skilled Typewriting*, ed. William E. Cooper (New York: Springer, 1983), 6. See also Wilfred Beeching, *Century of the Typewriter* (New York: St. Martins, 1974), 40. Beeching implausibly claims that Sholes designed QWERTY so that "a maximum distance has to be covered by the hands," slowing down entry.

8. Henry W. Roby, *Henry W. Roby's Story of the Invention of the Typewriter*, ed. Milo M. Quaife (Menasha, Wis.: George Banta, 1925), 44. It is not clear what Roby means by associating the QWERTY layout with a printer's font, other than the implication that the layout has something to do with letter frequency in English, which determined how printers arranged their fonts.

9. See Beeching, *Century of the Typewriter,* 40.

10. This count uses printers' traditional, though debatable, ranking of letter frequencies in English. This ranking was later canonized in type-casting machines that arranged the first two columns of the most common letters in descending order of frequency, which spelled the nonsense phrase "etaoin shrdlu."

11. Michael H. Adler, *The Writing Machine* (London: George Allen and Unwin, 1973), 205.

12. James Densmore, "Typewriting and Telegraphy," *The Phonographic World* 2, no. 1 (1886): 6.

13. "The Type Writer," *Scientific American,* August 10, 1872, 1. Note that the positions of *C* and *X* are reversed from their later arrangement, and *P* is located at the right side of the bottom row. In Sholes's QWE.TY and QWERTY layouts, *M* is located at the right side of the middle row, not the right side of the bottom row, as on later QWERTY arrangements.

14. Adler, *Writing Machine,* 205.

15. See Koichi Yasuoka and Motoko Yasuoka, "On the Prehistory of Qwerty," *ZINBUN,* no. 42 (March 2011): 161–74.

16. The masthead initially featured an illustration of a man operating a Morse terminal at its center, flanked on one side by an office containing several printing telegraphs with musical keyboards. By 1869, the masthead depicted a Morse terminal at the center operated only by a single human hand, but flanked on one side by a printing telegraph and on the other by assorted telegraphic receivers. By 1873, the Morse transmitter had disappeared from the masthead entirely.

17. Alvin F. Harlow, *Old Wires and New Waves: The History of the Telegraph, Telephone, and Wireless* (New York: Appleton, 1936), 246.

18. The machines were much improved by a third engineer hired by the American Telegraph Company, George Phelps. On Phelps's refinements, see James D. Reid, *The Telegraph in America: Morse Memorial* (New York: John Polhemus, circa 1884), 406.

19. In France, Paul-Gustave Froment manufactured Hughes printers, in what turned out to be a productive relationship. Froment had created his own telegraph in the 1850s that also employed a piano keyboard with only white keys, but which displayed the message at the receiving end with a dial that pointed to the transmitted letter. For illustrations and a detailed description of Froment's device, see Taliaferro Preston Shaffner, *Telegraph Manual: A Complete History and Description of the Semaphoric, Electric and Magnetic Telegraphs of Europe, Asia, Africa, and America, Ancient and Modern* (New York: Van Nostrand, 1876), 373–79.

20. Hughes and Evans, *Before We Went Wireless,* 111.

21. Hughes and Evans, 126. See also Erich Hausmann, *Telegraph Engineering: A Manual for Practicing Telegraph Engineers and Engineering Students* (New York: Van Nostrand, 1915), 133. Hausmann claims about 3,000 Hughes instruments were in use in Europe at the time of publication, well into the twentieth century.

22. The mechanism was ingenious but complex, consisting of a rotating cylinder beneath the keyboard precisely synchronized to a rotating type wheel on the other end. If the wheels were perfectly synchronized, pressing the key with a certain letter would trigger the machine on the other end to print at precisely the moment when the desired letter had rotated into position. For one of the most detailed accounts of the actual functioning of the Hughes printer, see appendices I-1 to I-3 of Hughes and Evans, *Before We Went Wireless,* 279–324.

23. See Hughes and Evans, 284, 302. Because Hughes and Phelps collaborated on the early

improvements to Hughes's model, it is not possible to know with full certainty which man introduced which alterations.

24. See Hughes and Evans, 34. Hughes's notebooks inventory the many objects he repurposed for his first model, including vest buttons.

25. Hughes was hardly the only telegrapher attempting to replace the musical keyboard with an array of round buttons. See ch. 2 on the buttons of the Cooke and Wheatstone five-wire telegraph. Similarly, dial telegraphs by Polydoor Lippens of Belgium and Michel Gloesener of Luxembourg all feature circular rings of small, round buttons to input individual Roman letters.

26. Not all American printing telegraphs followed this rule. George Anders's Magneto Printing Telegraph produced in Boston from the 1870s arranged the alphabet in a single row from left to right, with alternating letters on white and black keys. European printing telegraphs also differed. Froment's French printing telegraph had two rows of small keys that look only a little like piano keys, with *A–M* from left to right on the top row, and *N–Z* from left to right on the bottom.

27. Note that, on Hughes telegraphs, the black keys ended with *N*, so that the first white key at the far right was *O*. However, on the Sholes and Glidden Type-Writer, the middle row is mostly faithful to the order of the black keys but ends with *M*.

28. Densmore describes Porter's efforts in detail in "Typewriting and Telegraphy."

29. "Porter's Telegraph College," *Dodgeville Chronicle* (Dodgeville, Wis.), November 20, 1868, 4, chroniclingamerica.loc.gov/lccn/sn85033019/1868-11-20/ed-1/seq-4/.

30. "Porter's Telegraph College," 4.

31. As early as 1867, Sholes's one-key prototype employed an actual Morse-Vail telegraph key, as he experimented with alternatives to musical keyboards. The only record of that prototype appears in a remembrance written by Charles Weller, who was, tellingly, the chief operator for the Western Union Telegraph Company in Milwaukee, Wisconsin, where Sholes lived and worked (Weller, *The Early History of the Typewriter* [La Porte, Ind.: Chase & Shepherd, 1918], 8–9).

11. The Toys of Dionysus

1. For the title of this chapter and for its account of the relationship between Dionysus and toys, I am indebted to Olga Levaniouk, "The Toys of Dionysos," *Harvard Studies in Classical Philology* 103 (2007): 173–75.

2. As quoted in Levaniouk, "Toys," 167–68. See also her citations from other classical authorities on the toys of Dionysus, including Arnobius and Apuleius (168–71).

3. See Roger Moseley, *Keys to Play: Music as a Ludic Medium from Apollo to Nintendo* (Oakland: University of California Press, 2016).

4. Plato, *The Laws*, trans. Trevor J. Saunders (New York: Penguin, 1975), 72.

5. John Locke, *Some Thoughts concerning Education and of the Conduct of the Understanding*, ed. Ruth W. Grant and Nathan Tarcov (Indianapolis, Ind.: Hackett, 1996), 99–100.

6. Locke, 114.

7. On the German toy industry, including emphasis on the rise of mechanical toys, see David D. Hamlin, *Work and Play: The Production and Consumption of Toys in Germany, 1870–1914* (Ann Arbor: University of Michigan Press, 2007). See also Anthony Burton, "Design History and the History of Toys: Defining a Discipline for the Bethnal Green Museum of Childhood," *Journal of Design History* 10, no. 1 (1997): 1–21; Karl Ewald Fritzsche and Manfred Bachmann, *An Illustrated History of Toys*, trans. Ruth Michaelis-Jena (London: Abbey Library, 1966).

8. Amy F. Ogata, *Designing the Creative Child: Playthings and Places in Midcentury America* (Minneapolis: University of Minnesota Press, 2013), esp. ch. 2 (35–70).

9. Brian Sutton-Smith, *The Ambiguity of Play* (Cambridge, Mass.: Harvard University Press, 1997), 24–5.

10. Allison J. Pugh recently noted that, while advertisements for more than thirty-five hundred toys depict images of warm companionship between mothers and children, many nonetheless promised mothers that they would still be good mothers without having to be physically present with the child ("Selling Compromise: Toys, Motherhood, and the Cultural Deal," *Gender and Society* 19, no. 6 [2005]: 729–49).

11. The commercial life cycle of a toy is short, so by the time this book reaches print, these toys will likely have been replaced by newer versions. I suspect that browsing any online toy store will reveal an equal number of similar kinds of toys.

12. Roger Caillois, *Man, Play and Games,* trans. Meyer Barash (Urbana: University of Illinois Press, 2001), 7.

13. Plato, *Laws,* 76.

14. "Designs for Leaving," direct by Robert McKimson and Tedd Pierce, *Looney Toons* (Burbank, Calif.: Warner Bros. Pictures, 1954).

15. By the end of the episode, the "red one" has been labeled "In Case of Tidal Wave," and when Elmer finally does press it the house rises into the air on a giant pneumatic lift. Daffy meets him at the front door in a helicopter and says, "For a small price I can install this little blue button to get you down."

16. David Hume, *An Enquiry Concerning Human Understanding,* ed. P. H. Nidditch, 3rd ed. (1777; repr. Oxford: Clarendon, 1975), 39.

17. Sigmund Freud, *Beyond the Pleasure Principle,* trans. James Strachey (New York: Norton, 1961), 9.

18. Jacques Lacan, "The Mirror Stage as Formative of the Function of the I as Revealed in Psychoanalytic Experience," in *Écrits: A Selection,* trans. Alan Sheridan (New York: Norton, 1977), 4.

19. Melanie Klein "The Psychoanalytic Play Technique," *American Journal of Orthopsychiatry,* 25, no. 2 (1955): 223–37; D. W. Winnicott, "Transitional Objects and Transitional Phenomena—A Study of the First Not-Me Possession," *International Journal of Psychoanalysis* 34 (1953): 89–97.

20. Julia Kristeva similarly recognizes that some objects might form before the objectification of the mother, "if not objects at least *pre*-objects, poles of attraction of a demand for air, food, and motion" (*Powers of Horror: An Essay on Abjection,* trans. Leon S. Roudiez [New York: Columbia University Press, 1982], 32). Her point is to connect the object to fear more fundamentally, and thus to discern deeper anxieties about deprivation apart from the withdrawal of the actual mother.

21. See Jean Piaget, *Play, Dreams and Imitation in Childhood,* trans. C. Gattegno and F. M. Hodgson (1951; repr. Milton Park, UK: Routledge, 1999). For a critique of Piaget's theory of play, see Brian Sutton-Smith, "Piaget on Play: A Critique," *Psychology Review* 73, no. 1 (1966): 104–10.

22. Jean Piaget, *The Child's Conception of Physical Causality* (New Brunswick: Transaction, 2001 [1930]), 253–54.

23. Piaget, 196.

24. Piaget, 203.

25. Piaget, 244.

26. See Piaget, 60–63. Note that Winnicott, cited above, similarly sees primitive childhood states persisting within religious and spiritual life and the creative arts.

27. Piaget, 128.

28. Piaget, 131.

29. Piaget, 235.

30. Piaget, 235.

31. Hans-Georg Gadamer, *Truth and Method,* trans. Joel Weinsheimer and Donald G. Marshall (New York: Seabury, 1975), 92. Gadamer similarly notes, "The structure of play absorbs the player into itself, and thus takes from him the burden of initiative" (94).

32. Johan Huizinga, *Homo Ludens: A Study of the Play-Element in Culture* (London: Routledge & Kegan Paul, 1949), 2–3.

33. Martin Heidegger, *Being and Time,* trans. Joan Stambaugh (Albany: State University of New York Press, 2010), 69 (§15). I retain the more familiar English translation of *Zuhandenheit* as *ready-to-hand-ness,* even though Stambaugh prefers "handiness."

34. Kristeva, *Powers of Horror,* 3.

12. Pinball Wizards

1. Chris Morris, "We Spent Over $43 Billion on Video Games Last Year," *Fortune Magazine,* January 23, 2019 (fortune.com/2019/01/23/2018-video-game-sales-totals/). Global gaming revenue for 2019 from Nielsen SuperData's *2019 Year in Review: Digital Games and Interactive Media,* direc.ircg.ir/wp-content/uploads/2020/01/SuperData2019YearinReview.pdf, p. 22. By comparison, box-office revenue for movies in all of North America in the same year amounted to just over $11 billion, and just over $42 billion worldwide, according to Pamela McClintlock of *The Hollywood Reporter,* hollywoodreporter.com/news/2019-global-box-office-hit-record-425b-4-percent-plunge-us-1268600.

2. In addition to those works cited later in this chapter, see also 265–321 (ch. 5) of David Parisi, *Archaeologies of Touch: Interfacing with Haptics from Electricity to Computing* (Minneapolis: University of Minnesota Press, 2018), and Till A. Heilmann, "'Tap, tap, flap, flap': Ludic Seriality, Digitality, and the Finger," *Eludamos* 8, no. 1 (2014): 33–46.

3. Michael Newman, *Atari Age: The Emergence of Video Games in America* (Cambridge, Mass.: MIT Press, 2017), 29–43. See also Carly A. Kocurek, *Coin-Operated Americans: Rebooting Boyhood at the Video Game Arcade* (Minneapolis: University of Minnesota Press, 2019), 27–35.

4. Richard M. Bueschel, *Pinball: Illustrated Historical Guide to Pinball Machines,* vol. 1 (Wheat Ridge, Colo.: Hoflin, 1988), 17–18. For general histories of pinball, see also Roger C. Sharpe and James Hamilton, *Pinball!* (New York: Dutton, 1977), and Adam Ruben, *Pinball Wizards: Jackpots, Drains, and the Cult of the Silver Ball* (Chicago: Chicago Review Press, 2018).

5. Charles Dickens, *The Pickwick Papers* (Oxford: Oxford University Press, 2008), 159.

6. In one late-nineteenth-century advertisement by the Paris billiards manufacturer G. Caro for "Billards Anglais," the table looks rather like a pinball machine, with tilted board and a narrow chute on the right in which the player propels the ball forward toward a curved top, in an attempt to drop it back down into various holes numbered with different points. "Billards Americains" differs only in that it provides even more obstructions and point scoring options; "Billards Chinois" obstructs the holes with the numerous pins that later gave pinball its Anglo-American name (Bueschel, *Pinball,* 31).

7. M. Redgrave, Improvement in Bagatelles, U.S. Patent 115,357, issued May 30, 1871, p. 1.

8. J. S. Nicholas, Game, U.S. Patent 1,730,523, filed March 16, 1928, issued Oct. 8, 1929. On Nicholas's patent, see Bueschel, *Pinball,* 75.

9. Herbert B. Jones, *Coin-Operated Amusements: An Historical and Technological Survey* (Chicago: Bally, circa 1976), 31.

10. Bueschel, *Pinball,* 204.

11. Heribert Eiden and Jurgen Lukas, *Pinball Machines* (West Chester, Pa.: Schiffer, 1992), 20.

12. As late as the 1970s, machines sometimes displayed written affirmations of skill and, hence, moral innocence, such as some versions of Bally's game Space Odyssey, which disclaimed in small type on the back-glass: "Play flipper skill games for fun and recreation"; see Bueschel, *Pinball,* 243.

13. Bueschel, 204.

14. For more detailed descriptions of internal flipper mechanisms, see Marco Rossignoli, *The Complete Pinball Book: Collecting the Game and Its History,* rev. 3rd ed. (Atglen, Pa.: Schiffer, 2011), 47–48. See also Jones, *Coin-Operated Amusements,* 32n1: "The speed of Flipper movements is not variable."

15. Roger Sharpe demonstrated the game to the New York City Council in April 1976 (Sharpe, *Pinball!,* 62–63). See also "Council Approves Pinball Measure: Vote Would End Ban Ordered by La Guardia in 1942," *New York Times,* May 14, 1976, 33.

16. David Hume, *An Enquiry Concerning Human Understanding,* ed. P. H. Nidditch, 3rd ed. (1777; repr. Oxford: Clarendon, 1975), 28–9, 78.

17. Warren Susman, *Culture as History: The Transformation of American Society in the Twentieth Century* (Washington, D.C.: Smithsonian, 2003), 197–98.

18. See Roger Moseley, *Keys to Play: Music as a Ludic Medium from Apollo to Nintendo* (Oakland: University of California Press, 2016).

19. The Who, "Pinball Wizard," side 3, track 3, on *Tommy,* Decca, 1969, LP.

20. The Who, "Pinball Wizard."

21. The Who, "We're Not Gonna Take It," side 4, track 6, on *Tommy.*

22. However, see an excellent synopsis of existing work and a brief history of computer-game controllers in Sheila C. Murphy, "Controllers," in *The Routledge Companion to Video Game Studies,* ed. Mark J. P. Wolf and Bernard Perron (New York: Routledge, 2014), 19–24. See also the essays in *Ctrl-Alt-Play: Essays on Control in Video Gaming,* ed. Matthew Wysocki (Jefferson, N.C.: McFarland, 2013), some of which will be cited below. For a detailed account of historical examples, including photographs and detailed analysis of functionality, see Nicolas Nova and Laurent Bolli, *Joypads! The Design of Game Controllers* (San Bernadino, Calif.: CreateSpace, 2014).

23. Ian Bogost, "Persuasive Games: Gestures as Meaning," *Gamasutra: The Art and Business of Making Games,* June 30, 2009, gamedeveloper.com/design/persuasive-games-gestures-as-meaning.

24. For instance, Rory McGloin and Kirstie Farrar have argued that "more naturalistic controllers help to narrow the gap between the user's existing real-world mental models and the game's models," which generates both immersion and pleasure ("The Impact of Controller Naturalness on Spatial Presence, Gamer Enjoyment, and Perceived Realism in a Tennis Simulation Video Game," *Presence* 20, no. 4 [2011]: 320).

25. J. R. Parker, "Buttons, Simplicity, and Natural Interfaces," *Loading . . .* 2, no. 2 (2008), journals.sfu.ca/loading/index.php/loading/article/view/33.

26. See, e.g., Suzanne de Castell, Jennifer Jenson, and Kurt Thumlert, "From Simulation to Imitation: Controllers, Corporality, and Mimetic Play," *Simulation & Gaming* 45, no. 3 (2014): 332–55; Bart Simon, "Wii Are Out of Control: Bodies, Game Screens and the Production of Gestural Excess," *Loading . . .* 3, no. 4 (2009), journals.sfu.ca/loading/index.php/loading/article/view/65; Stephen N. Griffin, "Push. Play: An Examination of the Gameplay Button," in *Words in Play: International Perspectives on Digital Game Research,* ed. Suzanne de Castell and Jennifer Jenson (New York: Peter Lang, 2007), 153–57.

27. Not everyone has been so enthusiastic. One insightful dissent from Andreas Gregersen and Torben Grodal argues that the Wiimote might actually accentuate, in an unpleasurable way, the difference between the simulated and the real tennis swing. Precisely because the player swings the Wiimote only in the crudest approximation of a tennis swing—often amounting only to a flick of the wrist—the experience may evoke what they call "incongruent motor realism," a disturbing sense of the gap between the bodily motion and what it simulates ("Embodiment and Interface," in *The Computer Game Theory Reader 2* [New York: Routledge, 2009], 76). See also Peter McDonald, "On Couches and Controllers: Identification in the Video Game Apparatus," in Wysocki, *Ctrl-Alt-Play,* 108–20. McDonald argues that binary switches persist in gaming controls because of their semiotic arbitrariness, and hence flexibility.

28. For instance, see the joint treatment of pinball with slots in a special issue on gambling: Anonymous, "Slot Machines and Pinball Games," *Annals of American Academy of Political and Social Science* 269 (May, 1950): 62–70.

29. J. M. Graetz, "The Origin of Spacewar," *Creative Computing,* August, 1981, 56–67. On the Western Electric office buzzers see, Computer History Museum, "Oral History of Steve Russell," interview by Al Kossow, August 9, 2008, 16, archive.computerhistory.org/resources/access/text/2012/08/102746453-05-01-acc.pdf.

30. See the game pads for less successful consoles such as VTech Socrates from 1988 and the Bit Corporation's Telegames Personal Arcade from 1986, a ColecoVision clone also known as the Dina.

31. These include the Pioneer LaserActive (1993), the Atari Jaguar (1993), the Neo-Geo CD (1994), the Sony Play Station (1994), and even the short-lived Apple Bandai Pippin (1995).

32. Statista, "Current Generation Video Game Console Unit Sales Worldwide from 2017 to 2021 (in Million Units)," statista.com/statistics/276768/global-unit-sales-of-video-game-consoles/.

13. The Control Panel of Democracy

1. Mary Shelley, *Frankenstein: The 1818 Text* (New York: Penguin, 2018), 45.

2. See Phillip Garner and Enrico Spolaore, "Why Chads? Determinants of Voting Equipment Use in the United States," *Public Choice* 123, no. 3/4 (2005): 363–92. On voting technology in the 2000 election, see also Roy G. Saltman, *The History and Politics of Voting Technology: In Quest of Integrity and Public Confidence* (New York: Palgrave, 2006), 1–37; Michael Lynch, Stephen Hilgartner, and Carin Berkowitz, "Voting Machinery, Counting and Public Proofs in the 2000 US Presidential Election," in *Making Things Public: Atmospheres of Democracy,* ed. Bruno Latour and Peter Weibel (Cambridge, Mass.: MIT Press, 2005), 814–25.

3. William Thomas, *The Historie of Italie* (1549), 79 (available through Early English Books Online, searchable at quod.lib.umich.edu/e/eebogroup/). Thomas also notes here: "This maner of geuying thyr voices by ballotte, is one of the ladablest thynges used amongest them. For there is no man can knowe what an other doeth."

4. See Malcolm Crook and Tom Crook, "The Advent of the Secret Ballot in Britain and France, 1789–1914: From Public Assembly to Private Compartment," *History* 92, no. 4 (2007): 449–71; Crook and Crook, "Reforming Voting Practices in a Global Age: The Making and Remaking of The Modern Secret Ballot in Britain, France, and the United States, c. 1600–c. 1950," *Past & Present* 212 (August 2011): 199–237.

5. Crook and Crook, "Advent of the Secret Ballot," 457.

6. Benjamin Jolly proposed another early voting machine, a mechanization of the *ballotta* in which voters dropped a ball through a hole by the proper candidate's name, which a mechanical counter then tallied. A description and illustration of the device appear in an 1848 reprint of the *People's Charter,* first published a decade earlier, but it is not clear when Jolly devised his machine. It seems unlikely to have been used in any actual elections. See *The People's Charter; with the Address to the Radical Reformers of Great Britain and Ireland, and A Brief Sketch of Its Origin* (London: Elt and Fox, 1848), 2.

7. "Ballot-Voting," *The Spectator,* February 25, 1837, 184. The article also contains illustrations of Grote's voting machine.

8. "Ballot-Voting," *Spectator,* 184.

9. Harriet Grote, *The Personal Life of George Grote,* 2nd ed. (London: John Murray, 1873), 109.

10. Crook and Crook, "Advent of the Secret Ballot," 457–60.

11. Crook and Crook, 466–67.

12. Crook and Crook, "Reforming Voting Practices," 203–6.

13. Male tenant farmers, shopkeepers, and some others could vote in England after the 1832 Reform Act, which also officially excluded all women from the electorate for the first time.

14. On Britain's public elections and the changes wrought by the secret ballot, see Frank O'Gorman, "The Secret Ballot in Nineteenth-Century Britain," in *Cultures of Voting: The Hidden History of the Secret Ballot,* ed. Romain Bertrand, Jean-Louis Briquet, and Peter Pels (London: Hurst, 2007), 16–42.

15. O'Gorman, 38.

16. Baranowski was not the only designer of early calculating machines to also design voting machines. The Swedish inventor Willgodt Theophil Odhner, who invented the Odhner Arithmometer in 1873 as discussed briefly in ch. 5, also produced a voting machine, the Ballotierapparat, in 1902.

17. Jan J. Baranowski, *Nouveau Système de Voter au Moyen d'un Appareil Dit: Scrutateur Mécanique* (Paris: L'Auteur, 1849), 5–6.

18. Baranowski's reset mechanism may be the first *reset button.* On the reset button, see Donato Ricci, "Don't Push That Button," in *Reset Modernity!,* ed. Bruno Latour and Christophe Leclercq (Cambridge, Mass.: MIT Press, 2016), 24–41.

19. Thomas Alva Edison, Electric Vote Recorder, U.S. Patent 90,646, issued June 1, 1869.

20. The topic is too vast to engage in detail here. For an excellent and succinct history of English and American tort law, see John Fabian Witt, "Toward a New History of American Accident

Law: Classical Tort Law and the Cooperative First Party Insurance Movement," *Harvard Law Review* 114, no. 3 (2001): 690–841, esp. 699–713.

21. For an influential argument along these lines, see Oliver Wendell Holmes Jr., *The Common Law* (Boston: Little and Brown, 1923): 94–97. In arguing against a consistent standard of strict liability, Holmes concludes, "As action cannot be avoided, and tends to the public good, there is obviously no policy in throwing the hazard of what is at once desirable and inevitable upon the actor" (95).

22. Saltman, *History and Politics of Voting Technology,* 117.

23. Crook and Crook, "Reforming Voting Practice," 226–7.

24. H. W. Spratt, Voting Apparatus, U.S. Patent 158,652, filed August 10, 1874, issued January 12, 1875. For brief historical accounts of many other nineteenth-century American voting machines and their creators, see Saltman, *History and Politics of Voting Technology,* 105–18.

25. A. B. Roney, Registering Ballot Box, U.S. Patent 211,056, filed September 14, 1878, issued December 17, 1878, p. 1.

26. A. C. Beranek, Voting Apparatus, U.S. Patent 248,130, filed June 20, 1881, issued October 11, 1881, p. 2.

27. For a helpful review of early voting machines, including attention to the influence of Beranek's design on Myers's later more successful machines, see Douglas W. Jones, "Early Requirements for Mechanical Voting Systems," presented at the First International Workshop on Requirements Engineering for e-Voting Systems, Atlanta, Ga., August 2009, 1–8, ieeexplore.ieee .org/document/5460390/.

28. See, for instance, J. A. Gray, Voting Machine, U.S. Patent 620,767, filed November 9, 1898, issued March 7, 1899; J. W. Rhines, Vote Recording Machine, U.S. Patent 422,891, filed December 7, 1888, issued March 4, 1890; S. E. Davis, Voting Machine, U.S. Patent 549,901, filed March 4, 1895, issued November 19, 1895.

29. Joseph P. Harris, *Election Administration in the United States* (Washington, D.C.: Brookings, 1934), 247.

30. Some early voting machines employed paper records to comply with prevalent state laws after 1872 that required that voting be conducted by written ballot. A federal law from 1871 required written ballots for all congressional elections. The intent, according to Saltman, was to restrict oral voting, not mechanical voting, but such laws limited mechanical options nonetheless: mechanical voting machines that did not use paper faced court challenges, including the Myers machines in New York, which did not modify state laws requiring written ballots until 1899. See Saltman, (*History and Politics of Voting Technology,* 82, 117–18).

31. The dimensions are reported in "Voting by Machinery," *The Sun* (New York City), November 11, 1894, 1. Saltman estimates a much larger size, however, roughly nine feet square (*History and Politics Voting Technology,* 111), but this seems implausibly large, and does not correspond to the ratio of height to width shown in the patent illustration in Figure 30. If the patent illustration in Figure 30 is to scale, given that the door was reported to be 18 inches wide, the chamber may have been even smaller than the *Sun* reported, roughly 4 or 4.5 feet square and about 6 feet high.

32. Saltman, *History and Politics of Voting Technology,* 111.

33. Jacob H. Myers, "Arise, Americans, Arise!" pamphlet, circa 1890, p. 1.

34. Myers, 2.

35. John Crowley, "The Secret Ballot in the American Age of Reform," in *Cultures of Voting: The Hidden History of the Secret Ballot,* ed. Romain Bertrand, Jean-Louis Briquet, and Peter Pels (London: Hurst, 2007), 43–68.

36. Crowley, 60.

37. Similarly, Alain Garrigou has described the French voting *isoloir* as a disciplinary device for generating secrecy rather than surveillance, a "technology of intimization," and a way of "sanctifying the electoral expression of individual opinion" (*"Le Secret de l'isoloir," Actes de la Recherche en Sciences Sociales* 71–72 [March 1988]: 44).

38. Michel Foucault, *Discipline and Punish: The Birth of the Prison,* trans. Alan Sheridan, 2nd ed. (New York: Vintage, 1995), 200.

39. For the token-operated machine, see Jacob H. Myers, Voting Machine, U.S. Patent 415,548, filed April 29, 1889, issued November 19, 1889. It bears remembering that, although the term *button* was available in English at this time in something like its current usage, it seems to have been only intermittently used, and as with the telegraph, more often by observers describing these new machines than by their inventors, most of whom refer to their interfaces as *knobs, push keys, keys, push pieces, handles,* or *slides.* Only John Rhines refers to a "series of push-buttons or keys on the key-board" on his 1890 voting machine, terminology that drew instead on the analogy of the typewriter (Vote Recording Machine, U.S. Patent 422,891, issued March 4, 1890).

40. The fact that Myers patented a machine that uses *ballotte* and another that uses buttons within about one month of each other suggests that his push keys are in fact analogical elaborations of the ball that one drops into the box; the button is a ball that remains attached.

41. Jacob H. Myers, Voting Machine, U.S. Patent 415,549, issued November 19, 1889.

42. "Voting by Machinery," *The Sun,* 1.

43. Sylvanus Davis, Voting Machine, U.S. Patent 549,901, filed March 4, 1895, issued November 19, 1895, p. 1.

44. See "Myers Machine Won," *Rochester Democrat and Chronicle,* October 28, 1896, 11.

45. "Myers Machines Far from Perfect," *Rochester Democrat and Chronicle,* November 6, 1896, 7.

46. "Didn't Work," *Buffalo Evening News,* November 7, 1896.

47. "The Voting Machine in Use," *New York Times,* November 4, 1896, 2.

48. Saltman, *History and Politics of Voting Technology,* 117.

49. Isaiah Berlin, "Two Concepts of Liberty," in *Four Essays on Liberty* (Oxford: Oxford University Press, 2002), 166–217.

14. Switching Philosophies

1. Bruno Latour, "Where Are the Missing Masses? The Sociology of a Few Mundane Artifacts," in *Shaping Technology / Building Society: Studies in Sociotechnical Change,* ed. Wiebe E. Bijker and John Law (Cambridge, Mass.: MIT Press, 1992), 225–258. See also Latour, *Science in Action* (Cambridge, Mass.: Harvard University Press, 1987); Jane Bennett, *Vibrant Matter: A Political Ecology of Things* (Durham, N.C.: Duke University Press, 2020), 3.

2. Bennett, *Vibrant Matter,* 34.

3. Langdon Winner, "Do Artifacts Have Politics?," *Daedalus* 109, no. 1 (1980): 121–36.

4. Donald Davidson, "Actions, Reasons, and Causes," in *Essays on Actions and Events* (Oxford: Oxford University Press, 2001), 4.

5. Davidson, "Agency," in *Essays on Actions and Events,* 54.

6. Davidson, "Freedom to Act," in *Essays on Actions and Events,* 65.

7. Davidson, 78.

8. Davidson, "Agency," 59.

9. Davidson, 53n10.

10. Philippa Foot, "The Problem of Abortion and the Doctrine of the Double Effect," in *Virtues and Vices and Other Problems in Moral Philosophy* (Oxford: Oxford University Press, 2002), 19–32.

11. Carlos D. Navarrete, Melissa M. McDonald, Michael L. Lott, and Benjamin Asher, "Virtual Morality: Emotion and Action in a Simulated Three-Dimensional 'Trolley Problem,'" *Emotion* 12, no. 2 (2012): 365–70.

12. G. E. M. Anscombe, *Intention* (Cambridge, Mass.: Harvard University Press, 1963).

13. Antonio Rainone, "Thirty-Five Years after 'Actions, Reasons, and Causes': What has Become of Davidson's Causal Theory of Action?," in *Interpretations and Causes: New Perspectives on Davidson's Philosophy,* ed. Mario de Caro (Dordrecht: Springer, 1999), 125–35.

14. Nor should *wanting* be mistaken for something as specific as *desiring.* A person may have personally desired a dark room, but felt obligated *for some reason* to turn on the light. The person simply has what Davidson calls a "pro-attitude" toward illuminating the room, which is all that he means by *wanting* ("Actions, Reasons, and Causes," 4).

15. The Anscombe/Davidson thesis has special relevance to possible mental acts such as trying. If I try to turn on a light, is that a separate act—*trying* rather than *turning*—that causes light to go on? The Anscombe/Davidson thesis forbids that division of one act into two by insisting that trying and turning amount to the same action.

16. It is worth noting that any reason will not do. In "Freedom to Act," Davidson clarifies that reasons must cause action "in the right way" (79). He gives the example of a climber holding another man by a rope. The climber might desire to lighten his burden and his risk by releasing the rope, and if he acts on those reasons, the act will be intentional. But if the very thought of killing his partner for those reasons so unnerves him that he loses his grip on the rope, and yet thereby achieves the same result, the reasons did not cause the result "in the right way."

17. Ludwig Wittgenstein, *Philosophical Investigations,* trans. G. E. M. Anscombe, P. M. S. Hacker, and Joachim Schulte (London: Blackwell, 2009), 169 (§621).

18. See Michael Scott, "Wittgenstein's Philosophy of Action," *The Philosophical Quarterly* 46, no. 184 (1996): 347–63. Scott quotes from Wittgenstein's early manuscript versions of this claim, which are rather more direct than the later *Philosophical Investigations.* For instance, Scott quotes Wittgenstein's claim that the difference "is not one lying in the action or the accompaniment of it, but in the surrounding circumstances, the environment of the Action" (363).

19. See Anscombe, *Intention,* 21–25.

20. Davidson, "Actions, Reasons, and Causes," 19.

21. Davidson, 19.

22. Davidson, 4.

23. See Latour, "Where are the Missing Masses?"

24. Davidson, "Agency," 55, 60.
25. Davidson, "Intending," in *Essays on Actions and Events,* 86.
26. Davidson, "Agency," 55. Davidson draws his descriptions of puffing out action through the "accordion effect" from Joel Feinberg, "Action and Responsibility," in *Philosophy in America,* ed. Max Black (1964; repr. New York: Routledge, 2013), 134–60.
27. Davidson, "Actions, Reasons, and Causes," 13.
28. Davidson, "Agency," 50.
29. Davidson, 59.
30. Davidson, 49–50.
31. Davidson, 49, 50, 53, 55, 60.
32. Davidson, 55.
33. Davidson, 58.
34. Davidson, "The Individuation of Events," in *Essays on Actions and Events,* 177.
35. Davidson, 177.
36. Davidson, "Agency," 43.
37. Davidson, 43.
38. Davidson, "The Logical Form of Action Sentences," in *Essays on Actions and Events,* 120.
39. Mario Blaser, "Ontology and Indigeneity: On the Political Ontology of Heterogeneous Assemblages," *Cultural Geographies* 21, no. 1 (2014): 49–58. For a related claim, see Harvey A. Feit, "James Bay Crees' Life Projects and Politics: Histories of Place, Animal Partners and Enduring Relationships," in *In the Way of Development: Indigenous Peoples, Life Projects and Globalization,* ed. Mario Blaser, Harvey A. Feit, and Glenn McRae (London: Zed, 2004), 92–110.

15. Pistolgraphs

1. Susan Sontag, *On Photography* (New York: Delta, 1977), 14–15.
2. Christian Metz, "Photography and Fetish," *October* 34 (Autumn 1985): 84.
3. Friedrich Kittler, *Gramophone, Film, Typewriter,* trans. Geoffrey Winthrop-Young and Michael Wutz (Stanford, Calif.: Stanford University Press, 1999), 124.
4. André Bazin, "The Ontology of the Photographic Image," in *What Is Cinema?* trans. Hugh Gray, vol. 1 (Berkeley: University of California Press, 2004), 12n, 14.
5. Siegfried Kracauer, "Photography," in *The Past's Threshold: Essays on Photography,* ed. Philippe Despoix and Maria Zinfert (Zurich: Diaphanes, 2014), 37, 40.
6. For instance, Finis Dunaway writes, "Nature photography would appear to be a form of taxidermy, an art that predicts and remembers death" ("Hunting with the Camera: Nature Photography, Manliness, and Modern Memory, 1890–1930," *Journal of American Studies* 34 [2000]: 228).
7. See Gerald F. Linderman, *Embattled Courage: The Experience of Combat in the American Civil War* (New York: Free Press, 1987).
8. Russell Gilmore, "'The New Courage': Rifles and Soldier Individualism, 1876–1918," *Military Affairs* 40, no. 3 (1976): 97–102. For a fuller consideration of Gilmore and the history of military tactics, see my previous work on the politics of rifles in Jason Puskar, *Accident Society: Fiction, Collectivity and the Production of Chance* (Stanford, Calif.: Stanford University Press, 2012), 84–93.
9. The figures include purchases by law enforcement. Peter Henderson and Daniel Trotta,

"What's Missing in U.S. Gun Control Scramble? Bullets," *Reuters,* Jan. 20, 2013, reuters.com/article/us-usa-guns-ammunition/whats-missing-in-u-s-gun-control-scramble-bullets-idUSBRE90J02K20130120.

10. Bruno Latour, "On Technical Mediation—Philosophy, Sociology, Genealogy," *Common Knowledge* 3, no. 2 (1994): 32–33.

11. Richard Wright, "The Man Who Was Almost a Man," in *Eight Men: Short Stories* (New York: Harper Perennial, 1996), 10.

12. Donna J. Haraway, "A Cyborg Manifesto," chapter 8 of *Simians, Cyborgs, and Women: The Reinvention of Nature* (New York: Routledge, 1991), 177.

13. Judith Butler, *Giving an Account of Oneself* (New York: Fordham University Press, 2005), 26.

14. Butler, 64.

15. Very little is known about this camera except for the names of the inventor and manufacturer etched on a few surviving models, along with the date. See for instance the Eastman Museum's example, accession number 1974.0037.0090.

16. Helmut Gernsheim and Alison Gernsheim, *The History of Photography: From the Camera Obscura to the Beginning of the Modern Era* (New York: McGraw-Hill, 1969), 260.

17. Thomas Skaife, *Instantaneous Photography, Mathematical and Popular Including Practical Instructions on the Manipulation of the Pistolgraph* (Greenwich, UK: Henry S. Richardson, 1860), 9.

18. Skaife, 4 (appendix).

19. Skaife, 8 (appendix).

20. The anecdote is often repeated, but not usually well substantiated. However, it is documented in the early 1860s in "Talk in the Studio," *Photographic News* 7, no. 247 (1863): 263, in which the author reports that Skaife "related an anecdote of his photographing with his little instrument, Her Majesty as she was at full speed on her route to Wimbledon, and the risk incurred of being apprehended for an attempt to shoot the queen."

21. On Skaife's photograph, see John Darius, *Beyond Vision: One Hundred Historic Scientific Photographs* (Oxford: Oxford University Press, 1984); Peter Geimer, "Picturing the Black Box: On Blanks in Nineteenth-Century Paintings and Photographs," *Science in Context* 17, no. 4 (2004): 467–501. The *carte de visite* was usually an albumen print, and often remarkably crisp and clear, at least in slightly later examples from the early 1860s. In contrast, Skaife's is so poorly rendered and so painterly in its effects that we may be justified in suspecting significant manual alterations. There must even be some question about whether he truly was able to capture such a brief moment photographically, given that we can make a rough estimate of the necessary shutter speed based on the length of the blur of the moving ball. A mid-nineteenth-century thirteen-inch mortar would have been about four feet long. The blur of the ball leaving the barrel appears to be about the length of the mortar itself, which means that it travelled about four feet while the shutter was open. Mortars have a low muzzle velocity compared with other artillery, which some estimate to have been about six hundred feet per second. So each foot of travel visible in the blurred ball would have taken 1/600th of a second. If it travelled about four feet, the shutter must have been open for about 1/150th of a second. During the 1/10th of a second that the Pistolgraph's shutter was open, the shell would have travelled a full sixty feet, not four. Conversely, if the shell did travel four feet in 1/10th of a second, the muzzle velocity must have been about forty feet per second, or less than thirty miles per hour, which is about the speed that most people can throw a ball. Something is clearly wrong here. Even if cameras and emulsions made

this exposure technically feasible, which seems unlikely, Skaife would have had to trip the shutter with extraordinary precision to record the ball at just the right fraction of a second.

22. Thomas Skaife, letter, *London Times,* July 14, 1858, 12.

23. John Herschel, "Instantaneous Photography," *Photographic News* 4, no. 88 (1860): 13.

24. N. H. Harrington and Edmund Yates, *Your Likeness—One Shilling: A Comic Sketch in One Act* (London: Lacy, circa1858), 10.

25. Dion Boucicault, *The Octoroon* (Upper Saddle River, N.J.: Gregg, 1970). I am grateful to Daniel Novak for bringing the role of the camera in *The Octoroon* to my attention.

26. Boucicault, 13.

27. Boucicault, 19.

28. Boucicault, 18.

29. Boucicault, 13.

30. On early instantaneous photography, see Phillip Prodger, *Time Stands Still: Muybridge and the Instantaneous Photography Movement* (Oxford: Oxford University Press, 2003), and the essays by Corey Keller and Tom Gunning in *Brought to Light: Photography and the Invisible 1840–1900,* ed. Corey Keller (New Haven, Conn.: Yale University Press, 2008).

31. In a *Punch* satire on strange and foreign words in English, the voice of an outraged mother objects, "What *ever* is a pistolgram? Is it some new-fangled fire-arm, like an Armstrong breech propeller, and can it be intended really to *go off*? If so, I'm sure *infanticide* will be alarmingly increased, and it will be a mercy if but one out of dozen of one's babies is not shot" (Arabella Araminta Angelina Smith, "The Slipslop of the Shops," *Punch,* April 6, 1861, 138). For an example of Skaife's London newspaper advertisements, see "Pistolgrams of Babies," *London Daily News,* June 14, 1862: 4.

32. Thomas Skaife, advertisement, *British Journal of Photography,* December 16, 1861, ii.

33. Thomas Skaife, "Address to the London Photographic Society," *The Photographic Journal,* no. 145 (1864): 38.

34. At the same time, there is something accidental about these captures as well, in that the photographer determines when to take the picture but cannot possibly determine everything that will be recorded in the fraction of a second when the picture is taken. For an analysis of pressing the shutter button as an *event* rather than a properly agentive *action,* and also akin to gambling, see Walter Benn Michaels, *The Gold Standard and the Logic of Naturalism* (Berkeley: University of California Press, 1987), 217–44.

35. For descriptions and images of these cameras, and others, see Michel Auer, *The Illustrated History of the Camera,* trans. D. B. Tubbs (Boston: New York Graphic Society, 1975), 103–5; Brian Coe, *Cameras: From Daguerreotypes to Instant Pictures* (Gothenburg, Sweden: AB Nordbok, 1978), 53–55.

36. *Photography* 1, no. 2 (1884): 9.

37. Matthew Brower, *Developing Animals: Wildlife and Early American Photography* (Minneapolis: University of Minnesota Press, 2011).

38. J. Traill Taylor, *The Photographic Amateur,* 2nd ed. (New York: Scovill Manufacturing, 1883), 22 (appendix).

39. By the 1890s, camera hunting flourished even with devices that did not resemble guns. George Bird Grinnell, cofounder with Theodore Roosevelt of the conservationist Boone and

Crockett Club, published two influential articles on the subject: "Hunting with a Camera," *Forest and Stream,* May 5, 1892, 1; "Shooting without a Gun," *Forest and Stream,* October 6, 1892, 1.

40. W. B. Devereux, "Photographing Wild Game," *American Big-Game Hunting: The Book of the Boone and Crockett Club,* ed. Theodore Roosevelt and George Bird Grinnell (New York: Forest and Stream, 1893), 306.

41. See especially the essays in Keller, *Brought to Light*; Prodger, *Time Stands Still*; ch. 5 of Jimena Canales, *A Tenth of a Second: A History* (Chicago: University of Chicago Press, 2009).

42. Thomas Skaife, "Photography," *London Times,* July 14, 1858, 12. Note that the *Times* misspells his name as "Kaife," an error that has, unfortunately, been replicated in some subsequent histories.

43. Jessica Lake, *The Face That Launched a Thousand Lawsuits: The American Women Who Forged a Right to Privacy* (New Haven, Conn.: Yale University Press, 2016), 34–35.

16. First-Person Shooters

1. Caetlin Benson-Allott, *Remote Control* (New York: Bloomsbury, 2015).

2. Lawrence A. Wenner and Maryann O'Reilly Dennehy, "Is the Remote Control Device a Toy or a Tool? Exploring the Need for Activation, Desire for Control, and Technological Affinity in the Dynamic of RCD Use," in *The Remote Control in the New Age of Television,* ed. James R. Walker and Robert V. Bellamy Jr. (Westport, Conn.: Praeger, 1993), 122.

3. Harry G. Frankfurt, "Freedom of the Will and the Concept of a Person," *Journal of Philosophy* 68, no. 1 (1971): 6.

4. Jacques Lacan, "The Mirror Stage as Formative of the Function of the I as Revealed in Psychoanalytic Experience," in *Écrits: A Selection,* trans. Alan Sheridan (New York: Norton, 1977), 4.

5. *Operating Guide and Technical Manual: Flash-Matic Remote Control* (Chicago: Zenith Radio Corporation, n.d.), 2.

6. "You Have to See It to Believe It!" (advertisement), *Saturday Evening Post,* September 10, 1955, 14–15.

7. "Long Finger of Light Reaches Across Room to Tune This TV Set," *Popular Science,* September 1955, 162

8. Elana Levine, *Her Stories: Daytime Soap Opera and U.S. Television History* (Durham, N.C.: Duke University Press, 2020). On television and women's domestic labor, see also Lynn Spigel, *Make Room for TV: Television and the Family Ideal in Postwar America* (Chicago: University of Chicago Press, 1992). For a period assessment of women's television consumption in the house that suggests that many were only looking at the screen about half the time, see F. L. Whan, "Special Report: Daytime Use of TV by Iowa Housewives," *Journal of Broadcasting* 2 (1958): 142–48.

9. Rachel Plotnick, *Power Button: A History of Pleasure, Panic, and the Politics of Pushing* (Cambridge, Mass.: MIT Press, 2018), 54.

10. On Warren Johnson and the first thermostat, see Johnson Controls, *Right for the Times: Johnson Controls 100th Anniversary Brochure, 1885–1985* (Milwaukee, Wis.: Johnson Controls, ca. 1985), 3–4, www.hevac-heritage.org/electronic_books/controls/johnson/1-johnson-R.pdf.

11. Bruno Latour, "Where Are the Missing Masses? The Sociology of a Few Mundane Artifacts," in *Shaping Technology / Building Society: Studies in Sociotechnical Change,* ed. Wiebe E. Bijker and John Law (Cambridge, Mass.: MIT Press, 1992), 225–58.

12. “Stop Annoying TV Commercials,” *Field and Stream,* November 1955, 47.

13. In many advertisements, including “Zenith Expected You to Like Them,” *Chicago Daily Tribune,* September 12, 1955, A3.

14. “Long Finger of Light Reaches Across Room to Tune This TV Set,” *Popular Science,* September 1955, 162.

15. Benson-Allott, *Remote Control,* 41.

16. RCA had devised a wireless radio remote using audibly vibrating reeds, but seems never to have brought it to market: H. F. Olson, Radio Remote Control System, U.S. Patent 2,293,166, filed January 31, 1939, issued August 18, 1942 (Patrick Parsons, “The ‘Most Thrilling Invention Since the Radio Itself!’: The Evolution of the Radio Remote Control in the 1920s and 1930s,” *Journal of Radio and Audio Media* 21, no. 1 [2014]: 74–75).

17. Benson-Allott, *Remote Control,* 53–66. She observes that, around this time, images of women or families in advertising gave way to images of solitary men using the remote to improve their own leisure time.

18. “Star Trek: 45 Years of Designing the Future,” Star Trek Question and Answer, Part 3, John Jefferies, Art Director’s Guild, September 27, 2009, video, 15:10, vimeo.com/253313576. Jefferies says: “Magnavox had just come out with a remote control that was actually high-pitched air sound, and that sort of started as the basic shape except the phaser, the hand phaser was smaller” (10:10).

19. See Janet H. Murray, *Hamlet on the Holodeck: The Future of Narrative in Cyberspace* (New York: Free Press, 2016), 24–26.

20. On SAGE and its precursor, MIT’s Project Whirlwind, see Thomas P. Hughes, *Rescuing Prometheus* (New York: Pantheon, 1998). See also Arthur L. Norberg, Judy E. O’Neill, and Kerry J. Freedman, *Transforming Computer Technology: Information Processing for the Pentagon, 1962–1986* (Baltimore, Md.: Johns Hopkins University Press, 1996); Kent C. Redmond and Thomas M. Smith, *Project Whirlwind: The History of a Pioneer Computer* (Bedford, Mass.: Digital, 1980).

21. Hughes, *Rescuing Prometheus,* 51.

22. Hughes, 54.

23. The first was developed for the precursor to SAGE, MIT’s Whirlwind computer, which began in the mid-1940s as a computerized flight simulator for the Navy, but ended as a computerized air-traffic-control system. For that later function, designers fitted Whirlwind with a cathode ray tube that indicated the location of aircraft and a light pen to interact with it. See Michael R. Williams, *A History of Computing Technology,* 2nd ed. (Los Alamitos, Calif.: IEEE Computer Society, 1997). See also Hughes, *Rescuing Prometheus,* 30–40.

24. Thierry Bardini traces the origins of the mouse back to the SAGE light gun, among other imagined and attempted precedents from the 1940s through the 1960s (*Bootstrapping: Douglas Engelbart, Coevolution, and the Origins of Personal Computing* [Stanford, Calif.: Stanford University Press, 2000], 86–89, 247n4). Lev Manovich similarly argues that “the light pen, designed in 1949, can be considered a precursor of the contemporary mouse,” and that the resulting interactive graphical system amounts to “the birth of the computer screen” (*The Language of New Media* [Cambridge, Mass.: MIT Press, 2001], 102).

25. Ralph G. Mork, Light Gun Assembly, U.S. Patent 2,915,643, filed June 21, 1956, issued December 1, 1959.

26. Quoted in Bardini, *Bootstrapping,* 86.

27. J. C. R Licklider and Welden Clark noted in 1962 that, as early as 1958, at least thirteen different companies were manufacturing interactive computer consoles with light pens, though most of these were probably for interactions with analog oscilloscopes ("On-Line Man-Computer Communication," *Proceedings of the AFIPS Spring Joint Computer Conference, May 1–3, 1962,* May 1962, 113).

28. The arrangement is complex, in that the light gun both emits red light for positioning the gun and detects blue light from the screen. See Mork's patent for the Light Gun Assembly for a fuller description.

29. For instance, Wendy Hui Kyong Chun has insightfully examined those automated software processes known as *daemons,* and what it means that users have grown so comfortable imagining that their technology is inhabited by supernatural beings (*Programmed Visions: Software and Memory* [Cambridge, Mass.: MIT Press, 2011], 87–89).

30. There are also significant differences between them, Bardini notes in *Bootstrapping,* 86–89.

31. Bardini, 90 and 93. Bardini is extending Kittler's claim about typewriting to other aspects of the computer interface (see 71).

32. The arrow also had accumulated centuries of meaning before it was used as a marker of selection or direction. In Europe the arrow was first widely used in the form of the "broad arrow," a mark indicating the property of the British Board of Ordinance, and later of the War Department and Ministry of Defense. See John Monk, "Arrows Can Be Dangerous," *Triple-C: Communication, Capitalism & Critique* 11, no. 1 (2013): 67–92.

33. On the history of the arrow as a graphical marker, see Ernst H. Gombrich, "Pictorial Instructions," in *Images and Understanding,* ed. Horace Barlow, Colin Blakemore, and Miranda Weston-Smith (Cambridge: Cambridge University Press: 1990), 28. Gombrich reports he has not located an earlier example than a French treatise on hydraulic architecture from 1737.

34. See William H. Sherman, *Used Books: Marking Readers in Renaissance England* (Philadelphia: University of Pennsylvania Press, 2009), 25–52. Sherman notes that the mark goes by no fewer than fifteen different names in English alone, including "pointer," "digit," "mutton fist," "bishop's fist," "indicule," and "maniple." There is surprisingly little historical attention to the manicule, but see Charles Hasler, "A Show of Hands," *Typographica* 8 (1953): 4–11; Keith Houston, *Shady Characters: The Secret Life of Punctuation, Symbols, and Other Typographical Marks* (New York: Norton, 2013), 167–86. The manicule in the *Domesday Book* may have been added by a later annotator at some point after the book's creation in 1086. It appears in the entry for West Wratting, Cambridgeshire on fol. 199r.

35. Other Coleco Telstar models also included light guns, including the Ranger, Sportsman, and Arcade models, which included pistols resembling a Colt .45.

36. According to the Entertainment Software Association, 21 percent of all video games purchased in 2018 were shooter games., and an additional 27 percent were action games, which can also include shooting ("Essential Facts about the Computer and Video Game Industry," May 2019, 21, theesa.com/wp-content/uploads/2019/05/ESA_Essential_facts_2019_final.pdf).

37. There were earlier first-person shooters, including Wolfenstein 3D in 1992. The difference between a true first-person shooter and over-the-shoulder or view-from-behind third-person shooters is not a stark one. Strictly speaking, showing the avatar's body means that the player cannot be inhabiting it within a simulated first-person visual perspective, but rather is following

along just above or behind from a location that is otherwise not occupied in the game. Still, what is technically a third-person perspective has first-person effects, because the two vantage points are so tightly yoked together, share movement and perspective, and only sometimes detach. The arrangement might be likened to free indirect discourse in narrative, which is similarly phrased in the third person but which freely occupies the consciousness of a character from that very slight remove.

38. On the history of *Spacewar!,* see Steven Levy, *Hackers: Heroes of the Computer Revolution* (New York: Delta, 1984), 33–50; J. M. Graetz, "Origin of Spacewar!," *Creative Computing,* August, 1981, 56–67; Devin Monnens and Martin Goldberg, "Space Odyssey: The Long Journey of Spacewar! from MIT to Computer Labs around the World," *Kinephanos,* June 2015, kinephanos.ca/2015/space-odyssey-the-long-journey-of-spacewar-from-mit-to-computer-labs-around-the-world/.

39. On the first four-button controller, see Computer History Museum, "Oral History of Steve Russell," interview by Al Kossow, August 9, 2008, 1–38, archive.computerhistory.org/resources/access/text/2012/08/102746453-05-01-acc.pdf.

40. Alexander R. Galloway, "Origin of the First-Person Shooter," in *Gaming: Essays on Algorithmic Culture* (Minneapolis: University of Minnesota Press, 2006), 39–69.

41. Lacan, "Mirror Stage," 1–7.

42. See ch. 10 of Charles Brockden Brown, *Wieland, or The Transformation* (1798; repr. New York: Prometheus, 1997).

Epilogue

1. See microsoft.com/en-us/research/project/skinput-appropriating-the-body-as-an-input-surface/.

2. See media.mit.edu/projects/alterego/overview/.

3. Richard Seymour, *The Twittering Machine* (London: Indigo, 2019). The reference is to the title of Paul Klee's surrealist painting of three birds turning a mechanical crank above a yawning red abyss. On dopamine and the pleasures of social media, see 63–76.

4. See chapter 6 of Richard Bell, "The Problem of Slave Resistance," in *We Shall Be No More: Suicide and Self-Government in the Newly United States* (Cambridge, Mass.: Harvard University Press, 2012).

5. Averages compiled by Brady: United Against Gun Violence from data provided by the Centers for Disease Control from 2015 to 2019, bradyunited.org/key-statistics.

Index

Jason Puskar is professor of English at the University of Wisconsin–Milwaukee. He is author of *Accident Society: Fiction, Collectivity, and the Production of Chance.*